I0096494

M44A2 Series 2.5 Ton Military Truck Transmission, Transfer Case, and PTO Maintenance Manual
TM 9-2520-246-34-1

With

TM 9-2520-246-34P

edited by
Brian Greul

The M44A2 series of military trucks is affectionately known as the deuce and a half, or simply the deuce. These ubiquitous trucks were first produced during WWII and General Eisenhower credited them as being one of the most important US Army vehicles.

This book is 1 of 2 manuals for transmission, transfer, and power take off maintenance at the direct and general support levels.
The Parts manual is appended to the end of this manual for convenience.

It is being printed for enthusiasts, restores, and collectors who may wish to own a quality paperback copy.

The editor has endeavored to minimize changes, but the following changes were made: Feedback forms are omitted, the fonts have been updated for printing purposes on modern equipment.

Should you have suggestions or feedback on ways to improve this book please send email to Books@OcotilloPress.com

---

Edited 2021 Ocotillo Press
ISBN 978-1-954285-42-2

No rights reserved. This content of this book is in the public domain as it is a work of the US Government. It is reproduced by the publisher as a convenience to enthusiasts and others who may wish to own a quality copy of it. It has been adjusted to accomodate the printing and binding process.

Printed in the United States of America

Ocotillo Press
Houston, TX 77017
Books@OcotilloPress.com

Disclaimer: The user of this book is responsible for following safe and lawful practices at all times. The publisher assumes no responsibility for the use of the content of this book. The publisher has made an effort to ensure that the text is complete and properly typeset, however omissions, errors, and other issues may exist that the publisher is unaware of.

TM 9-2520-246-34-1
T.O. 36Y33-39-2-1

TECHNICAL MANUAL

MAINTENANCE

DIRECT SUPPORT AND GENERAL SUPPORT LEVEL

TRANSMISSION

MODEL 3053A, NSN 2520-00-884-4833

TRANSMISSION TRANSFERS

MODEL T-136-21, NSN 2520-00-001-7855

MODEL T-136-27, NSN 2520-00-089-8287

POWER TAKEOFFS

MODEL WN-7-28, NSN 2520-00-706-1137

MODEL WND-7-28, NSN 2520-00-706-1136

MODEL P-136-C, NSN 2520-00-229-5673

Chapter 1
 General
 Maintenance
 Information

Chapter 2
 Equipment
 Group
 Maintenance

Appendix A
 References

DEPARTMENTS OF THE ARMY AND THE AIR FORCE

JANUARY 1981

WARNING

Dry cleaning solvent is flammable.   Do not use near an open flame. Keep a fire extinguisher nearby when solvent is used.   Use only in well-ventilated places. Failure to do this may result in injury to personnel and damage to equipment.

Do not use more than 30 psi of air pressure for drying parts. Eye shields must be worn when using compressed air.   Eye injury can occur if eye shields are not used.

Technical Manual
No. 2520-246-34-1
Technical Order
No. 36Y33-39-2-1

DEPARTMENTS OF THE ARMY
AND
THE AIR FORCE
Washington, DC, 30 January 1981

TECHNICAL MANUAL

MAINTENANCE

DIRECT SUPPORT AND GENERAL SUPPORT LEVEL

TRANSMISSION

MODEL 3053A, NSN 2520-00-884-4833

TRANSMISSION TRANSFERS

MODEL T-136-21, NSN 2520-00-001-7855

MODEL T-136-27, NSN 2520-00-089-8287

POWER TAKEOFFS

MODEL WN-7-28, NSN 2520-00-706-1137

MODEL WND-7-28, NSN 2520-00-706-1136

MODEL P-136-C, NSN 2520-00-229-5673

REPORTING OF ERRORS AND RECOMMENDING IMPROVEMENTS

You can help improve this manual.   If you find any mistakes or if you know
of a way to improve the procedure, please let us know. Mail your letter,
DA Form 2028 (Recommended Changes to Publication and Blank Forms), or
DA Form 2028-2 located in the back of this manual direct to: Commander,
U.S. Army Tank Automotive Materiel Readiness Command, ATTN: DRSTA -
MB, Warren, Michigan 48090.   A reply will be furnished to you.

---

*This manual supersedes so much of TM 9-2520-246-34, 12 May 1978, as pertains to
Transfers, Power Takeoffs,   and Transmission model 3053A.

LIST OF TABLES

LIST OF TABLES-CONT

CHAPTER 1

GENERAL MAINTENANCE INFORMATION

1-1. SCOPE.

a. This manual contains the direct support and general support maintenance instructions for transmission model 3053A, transmission transfer models T– 136-21 and T-136-27, and power takeoff models WN-7-28, WND-7-28, and P-136-C. This manual includes procedures for disassembly, cleaning, inspection, repair, test, adjustment, and overhaul as authorized by the maintenance allocation chart.

b. Appendix A gives a list of current and applicable references to the transmission, transfers, and power takeoffs used on the 2 1/2-ton, 6x6 series trucks, equipped with multifuel engines.

c. Refer to TM 9-2520-246-34P for a listing of parts and special tools for the maintenance of the transmission, transmission transfers, and power takeoffs for the 2 1/2-ton, 6x6 series trucks.

1-2. GENERAL MAINTENANCE. General maintenance tasks relating to inspection care and maintenance of antifriction bearings are given in TM 9-214. Welding procedures that apply to this type of equipment are given in TM 9-237. For the lubrication of the transmission, transfers, and power takeoffs covered in this manual, refer to LO 9-2320-209-12/1.

1-3. TROUBLESHOOTING. Troubleshooting a fault within the transmission, transmission transfers, transmission power takeoff and transmission transfer power take-off is done as part of the repair procedures. Regardless of the symptom, all replaceable parts must be inspected and/or tested for serviceability. Any parts found to be faulty must be replaced during the reassembly process. In addition, test procedures are given to make sure the reassembled equipment is working properly.

1-4. CLEANING. Refer to TM 9-247 for cleaning materials for this type of equipment. Special cleaning procedures for the transmission, transmission transfers, and power takeoffs are given in chapter 2.

1-5. PAINTING. For painting instructions for field use of the equipment covered in this manual, refer to TM 43-0139.

1-6. TORQUE VALUES. Critical torque values for a particular component are given in the maintenance procedures in chapter 2. When torque values are not given, bolts, screws and nuts are to be tightened as given in table 1-1.

1-7. SPECIAL TOOLS AND EQUIPMENT. Special tools and equipment are provided to make it easier to do particular maintenance tasks and to keep the equipment in good repair. Table 1-2 lists the special tools and equipment and gives a reference to the maintenance paragraph where they are used and what they are used for.

1-8. SAFETY INSPECTION AND TESTING OF LIFTING DEVICES. Refer to TB 43-0142 for safety inspection and testing of lifting devices used in this manual.

1-9. FORMS AND RECORDS. Maintenance forms, records, and reports which are to be used by maintenance personnel at all levels are listed in and prescribed by TM 38-750.

1-10. EQUIPMENT IMPROVEMENT REPORT AND MAINTENANCE DIGEST (EIR MD) AND EQUIPMENT IMPROVEMENT REPORT AND MAINTENANCE SUMMARY (EIR MS). The quarterly Equipment Improvement Report and Maintenance Digest, TB 43-0001-39 series, contains valuable field information on the equipment covered in this manual.

Table 1-1.   Standard Torque Specifications

USAGE	MUCH USED	MUCH USED	USED AT TIMES	USED AT TIMES
CAPSCREW DIAMETER AND MINIMUM TENSILE STRENGTH PSI [KG/SQ CM]	To 1/2 – 69,000 [4850.7000] To 3/4 – 64,000 [4499.2000] To 1 – 55,000 [3866.5000]	To 3/4 – 120,000 [8436.0000] To 1 – 115,000 [8084.5000]	To 5/8 – 140,000 [9842.0000] To 3/4 – 133,000 [9349.9000]	150,000 [10545.0000]
QUALITY OF MATERIAL	INDETERMINATE	MINIMUM COMMERCIAL	MEDIUM COMMERCIAL	BEST COMMERCIAL
SAE GRADE NUMBER	1 or 2	5	6 or 7	8

CAPSCREW HEAD MARKINGS

Manufacturer's marks may vary. These are all SAE Grade 5 (3 line).

CAPSCREW BODY SIZE (INCHES)–(THREAD)	TORQUE FT·LB [KG M]		TORQUE FT·LB [KG M]		TORQUE FT·LB [KG M]		TORQUE FT·LB [KG M]	
1/4 – 20	5	[0.6915]	8	[1.1064]	10	[1.3830]	12	[1.6596]
– 28	6	[0.8293]	10	[1.3830]			14	[1.9362]
5/16 – 18	11	[1.5213]	17	[2.3511]	19	[2.6277]	24	[3.3192]
– 24	13	[1.7979]	19	[2.6277]			27	[3.7341]
3/8 – 16	18	[2.4894]	31	[4.2873]	34	[4.7022]	44	[6.0852]
– 24	20	[2.7660]	35	[4.8405]			49	[6.7767]
7/16 – 14	28	[3.8132]	49	[6.7767]	55	[7.6065]	70	[9.6810]
– 20	30	[4.1490]	55	[7.6065]			78	[10.7874]
1/2 – 13	39	[5.3937]	75	[10.3725]	85	[11.7555]	105	[14.5215]
– 20	41	[5.6703]	85	[11.7555]			120	[16.5960]
9/16 – 12	51	[7.0533]	110	[15.2130]	120	[16.5960]	155	[21.4365]
– 18	55	[7.6065]	120	[16.5960]			170	[23.5110]
5/8 – 11	83	[11.4789]	150	[20.7450]	167	[23.0961]	210	[29.0430]
– 18	95	[13.1385]	170	[23.5110]			240	[33.1920]
3/4 – 10	105	[14.5215]	270	[37.3410]	280	[38.7240]	375	[51.8625]
– 16	115	[15.9045]	295	[40.7985]			420	[58.0860]
7/8 – 9	160	[22.1280]	395	[54.6285]	440	[60.8520]	605	[83.6715]
– 14	175	[24.2025]	435	[60.1605]			675	[93.3525]
1 – 8	235	[32.5005]	590	[81.5970]	660	[91.2780]	910	[125.8530]
– 14	250	[34.5750]	660	[91.2780]			990	[136.9170]

1. Always use the torque values listed above when specific specifications are not available.

Note: Do not use above values in place of those specified in the engine groups of this manual. special attention should be observed in case of SAE Grade 6, 7 and 8 capscrews.

2. The above is based on use of clean and dry threads.

3. Reduce torque by 10% when engine oil is used as a lubricant.

4. Reduce torque by 20% if new plated capscrews are used.

Caution: Capscrews threaded into aluminum may require reductions in torque of 30% or more, unless inserts are used.

TA 113439

Table 1-2.　Special Tools and Equipment

Item	Part No.	National Stock No.	Reference Paragraph	Use
ADAPTER: Mechanical Puller	7083254	5120-00-708-3254	2-4	Used with puller 5120-00-313-9496 to remove transmission reverse idler gear shaft.
HANDLE	7083241	5120-00-708-3241	2-15 2-21	Used with remover and replacer 5120-00-708-3247 to remove and replace transmission transfer idler shaft front bearing cup.
PULLER KIT: Mechanical (Companion Flange)	8708724	5120-00-338-6721	2-4 2-14	Used to remove companion flanges from transmission and transmission transfers.
REMOVER AND REPLACER	7083247	5120-00-708-3247	2-15 2-21	Used with handle 5120-00-708-3241 to remove and replace transmission transfer idler shaft front bearing cup l
TOOL KIT: Special Power Train Rebuild	7950356	4910-00-795-0356	2-4 2-8 2-18 2-19 2-22 2-28 2-36 2-38	Used to make the right measurements required during the backlash checks, wear limit inspection and assembly of the transmission, transmission transfer, transmission power takeoff and transmission transfer power takeoff.

The information in the TB 43-0001-39 series is compiled from some of the Equipment Improvement Reports that you prepared on the equipment covered in this manual. Many of these articles result from comments, suggestions, and improvement recommendations that you submitted to the EIR program. The TB 43-0001-39 series contains information on equipment improvements, minor alterations, proposed Modification Work Orders (MWO'S) , warranties (if applicable), actions taken on some of your DA Form 2028's (Recommended Changes to Publications), and advance information on proposed changes that may affect this manual. In addition, the more maintenance significant articles, including minor alterations, field–fixes, etc., that have a more permanent and continuing need in the field are republished in the Equipment Improvement Report and Maintenance Summary (EIR MS) for TARCOM Equipment (TM 43-0143). Refer to both of these publications (TB 43-0001-39 series and TM 43-0143) periodically, especially the TB 43–0001– 39 series, for the most current and authoritative information on your equipment. The information will help you in doing your job better and will help in keeping you advised of the latest changes to this manual. Also refer to DA Pam 310-4, Index of Technical publications, and Appendix A, References, of this manual.

1-11. REPORTING IMPROVEMENT RECOMMENDATIONS . If your transmission, transfer or power takeoff needs improvement, let us know. Send us an EIR. you, the user, are the only one who can tell us what you don't like about your equipment. Let us know why you don't like the design. Tell us why a procedure is hard to perform. Put it on an SF 368 (Quality Deficiency Report). Mail it to us at: Commander, U.S. Army Automotive Material Readiness Command, ATTN: DRSTA-MT, Warren, Michigan 48090. We'll send you a reply.

1-12. METRIC SYSTEM. The equipment/system described herein is nonmetric and does not require metric common or special tools. Therefore, metric units are not supplied. Tactical instructions, for sake of clarity, will also remain nonmetric.

1-13. QUALITY ASSURANCE/QUALITY CONTROL. Repair and replacement standards for transmissions, transmission transfers, and power takeoffs are given in chapter 2. By replacing the parts that do not meet the tolerances given in this manual for wear limits and adjustments, quality control of the equipment will be maintained.

1-14. DESCRIPTION.

a. General. A short description covering each transmission, transmission transfer: and power takeoff is given in the following paragraphs. The descriptions are used to show direct support, general support, and depot maintenance personnel the different types and models of each unit. Differences between models are also given.

b. Transmissions.

(1) Transmission model 3053A (fig. 1-1) is a manual shift, synchromesh, selective gear type with five speeds forward and one reverse. The outer case is made of cast iron and gives support for the different bearing shafts and other parts of the transmission gear train.

(2) Transmission model 3053A is a five-speed transmission with fourth speed being a direct drive and fifth speed an overdrive. The shift pattern for the transmission is shown in figure 1-2.

c. Transmission Transfers.

(1) Transmission transfer model T-136-21 and T-136-27 are two-speed synchromesh units driven from the transmission through a propeller shaft. The transfers send power to front and rear axles through propeller shafts.

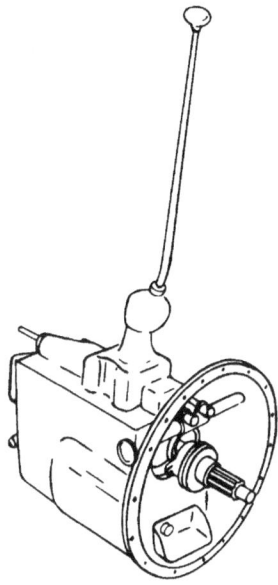

RIGHT FRONT VIEW

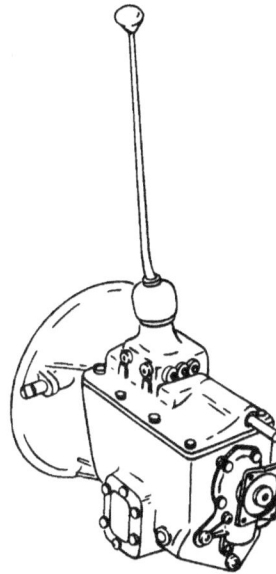

LEFT REAR VIEW

TA 120866

Figure 1–1.   Transmission Model 3053A

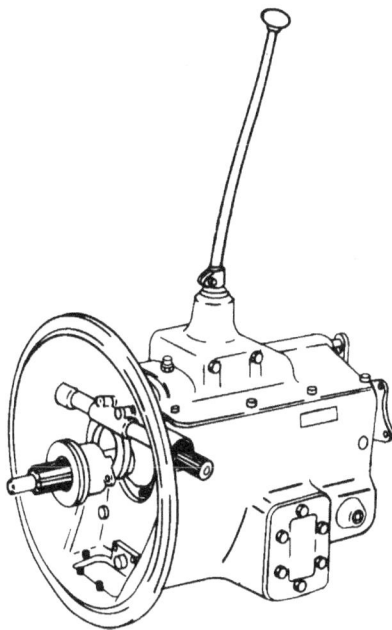

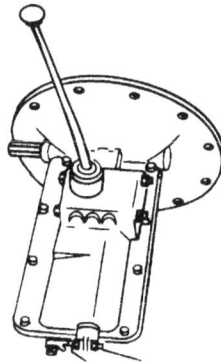

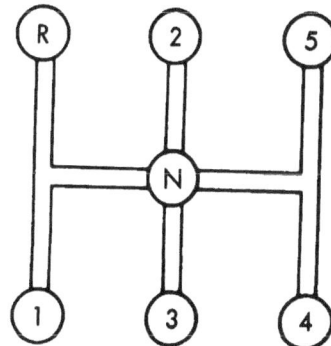

TA 118967

Figure 1-2.   Transmission Model 3053A, Five Speed Shift Pattern

(2) Positive locking transfer, model T-136-27, shown in figure 1-3 is a two-speed synchromesh unit driven by the transmission propeller shaft. This transfer has an air selector valve that lets the air locking assembly engage the front output shaft to the rear output shaft for driving the front axle. Selection of the high or low ranges of the transfer gives the truck operator a total of ten forward speeds and two reverse speeds.

(3) Transmission transfer model T-136-21, figure 1-4 is designed to drive the front axle slightly slower than the rear axles and this difference in speed is taken up by an overrunning sprag unit on the front output shaft during normal operation. When the rear wheels lose traction, the sprag unit automatically engages and the front wheels also drive the truck.

d. Power Takeoffs (Models WN-7-28, WND-7-28, and P-136-C).

(1) Power takeoff models WN-7-28 and WND-7-28, (figure 1-5 and 1-6) are driven by the transmission and give power to drive the truck winches, pumps, and other special or auxiliary equipment. Both power takeoffs are made to be mounted on a Society of Automotive Engineers' (SAE) standard, six-bolt hole power takeoff open-ing. However, when needed, these units may be mounted on four-hole openings if suitable filler blocks and driving gears are used. Both units are of the heavy-duty type, having two forward speeds and one reverse speed. The units can be used for either left or right hand mounting on the drive mechanism. The shifter shaft goes through both ends of the case so that the shift linkage can be hooked up to either end of the shaft.

(2) Power takeoffs WN-7-28 and WND-7-28 are the same except that power takeoff model WND-7-28 uses two output shafts, and model WN-7-28 has only one output shaft.

(3) Power takeoff model P-136-C (fig. 1-7) is a single drive, one-speed unit which is made to mount on transmission transfers only. The outer carriers are circular in shape and support the main drive shaft and driveshaft bearings. This unit is used to drive many different types of auxiliary truck equipment.

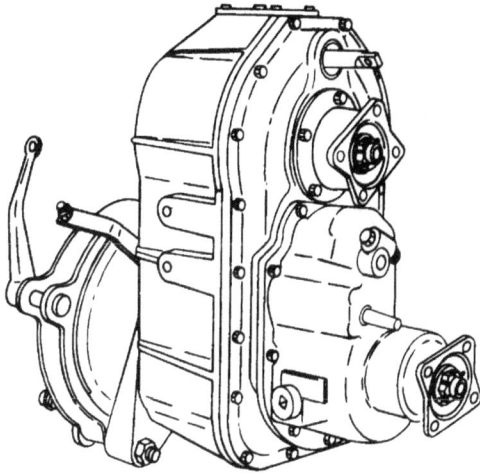 

FRONT VIEW

REAR VIEW

TA 120867

Figure 1-3.   Positive Locking Transfer Model T-136-27

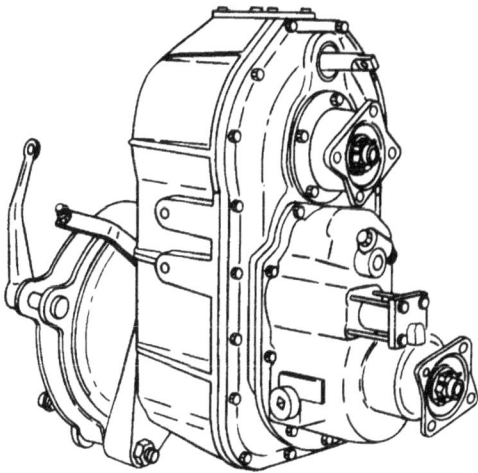 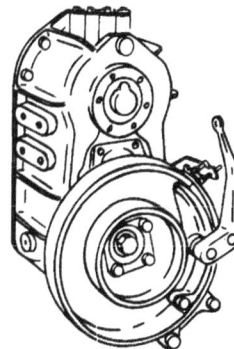

FRONT VIEW

REAR VIEW

TA 120868

Figure 1-4.   Sprag Equipped Transfer Model T-136-21

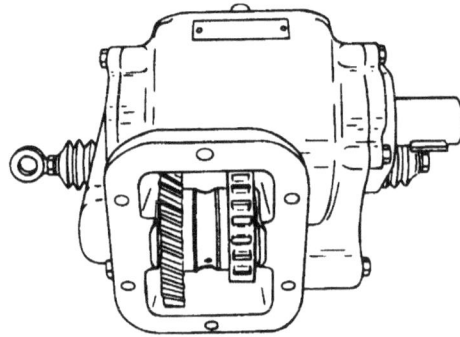

TA 120869

Figure 1-5.   Power Takeoff Model WND-7-28

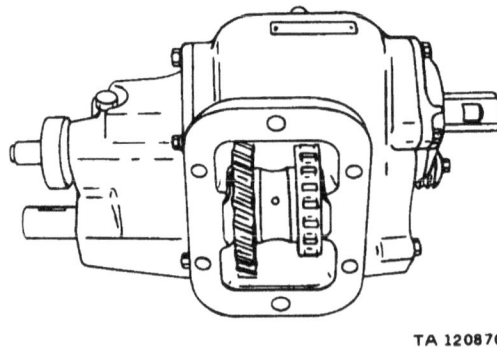

TA 120870

Figure 1-6.   Power Takeoff Model WN-7-28

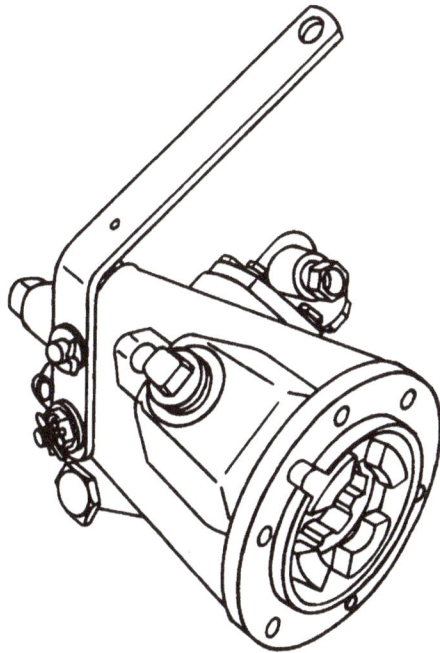

TA 120871

Figure 1-7.   Power Takeoff Model P-136-C

CHAPTER 2

EQUIPMENT GROUP MAINTENANCE

Section I. SCOPE

2-1. EQUIPMENT ITEMS COVERED. Transmission model 3053A, transfer models T-136-21 and T-136-27, and power takeoff models WN-7-28, WND-7-28, and P-136-C used on the 2 1/2-ton, 6x6 series trucks are covered in this manual.

2-2. EQUIPMENT ITEMS NOT COVERED. All equipment items are covered in this manual.

Section II. MAINTENANCE OF TRANSMISSION

TOOLS: Mechanical puller kit, pn 8708724
Mechanical puller adapter, pn 7083254
Power train rebuild tool kit, pn 7950356

SUPPLIES: Dry cleaning solvent, type II (SD-2), Fed. Spec P-D-680
Multipurpose lubricant, GO 85/140, MIL-L-2105
Lubricating oil, ICE, OE/HDO 10, MIL-L-2104
Artillery and automotive grease, type GAA, MIL-G-10924
Safety wire MS20995F47
Transmission assembly gasket set
Clutch housing seal
Rear bearing mainshaft seal
Tags
Wood block 6 inches x 6 inches x 1 inch
Wood block 12 inches x 4 inches x 1 inch (2)
Compressed air source, 30 psi max

PERSONNEL: Two

EQUIPMENT CONDITION: Transmission on workbench.

2-3. CLEANING BEFORE DISASSEMBLY. This paragraph gives instructions for cleaning the transmission assembly before disassembly. Note and scribe transmission case in places that have oil soaked road mud. It is not necessary to mark oil spots around gaskets or seals since new gaskets and seals will be put in. Scrape, brush, and steam clean all dirt and road mud from transmission assembly.

2-4. DISASSEMBLY OF TRANSMISSION INTO SUBASSEMBLIES. The following paragraphs give instructions to disassemble the transmission into subassemblies and also have the backlash checks needed.

    a.   <u>Shifter Shaft Cover.</u>

FRAME 1

1.   Put gear shift lever (1) in neutral position.   Refer to TM 9-2320-209-10.

2.   Take out eight capscrews (2) and eight lockwashers (3).

3.   Pull shifter shaft cover (4) straight up and off.

4.   Take off and throw away cover gasket (5).

END OF TASK

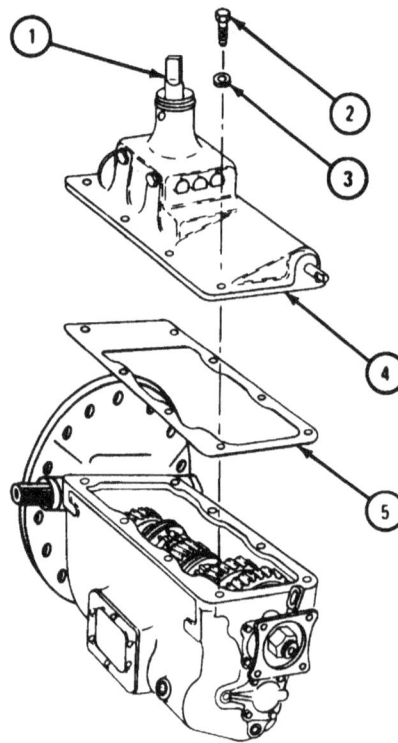

TA 087800

b.   Backlash Check.

FRAME 1

NOTE

This frame tells how to check backlash for all gears.
Do this frame when measuring backlash for each set
of gears in frames 2 and 3.

1.   Mount dial indicator on housing and set stem against side of gear tooth (1)
as shown.

NOTE

When measuring backlash make sure that gear (2) does
not turn.   If gear turns,   backlash readings will be
wrong.

2.   Turn gear (3) away from dial indicator until gear tooth (4) touches gear
tooth (5) as shown in view A.

3.   Set dial indicator to O.

4.   Turn gear (3) towards dial indicator until gear tooth (6) touches other side
of gear tooth (5) as shown in view B.

5.   Check that dial indicator readings are within wear limits given for each set
of gears.

GO TO FRAME 2

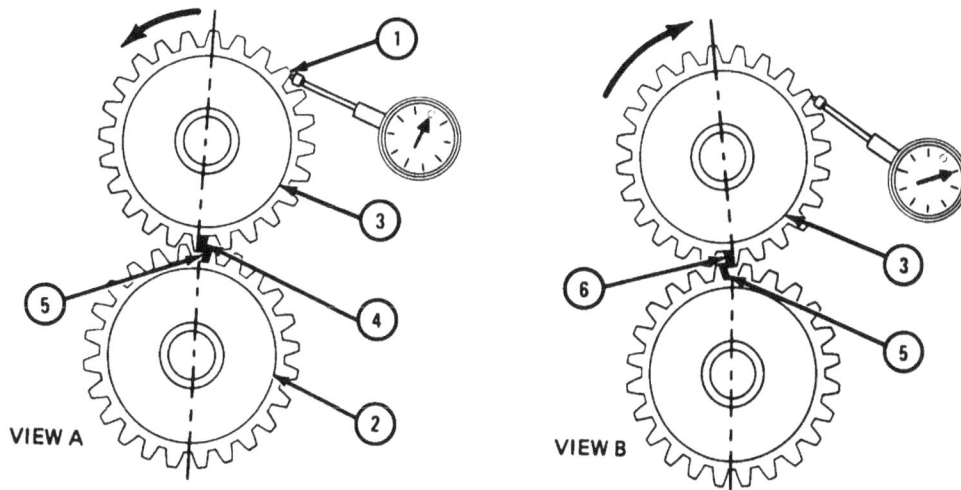

VIEW A   VIEW B

TA 087801

FRAME 2

NOTE

Readings must be within limits given in table 2-1. The
letter L indicates a loose fit.   If readings are not within
given limits,   throw away both gears and get new ones.
Some gears on the countershaft cannot be taken off. If
these gears are damaged, throw away countershaft and
get a new one.

1.   Measure backlash between   input shaft gear (1) and countershaft drive gear (2)

2.   Measure backlash between   fifth gear (3) and countershaft fifth gear (4).

3.   Measure backlash between   third gear (5) and countershaft third gear (6).

4.   Measure backlash between   second gear (7) and countershaft second-reverse
     gear (8) .

GO TO FRAME 3

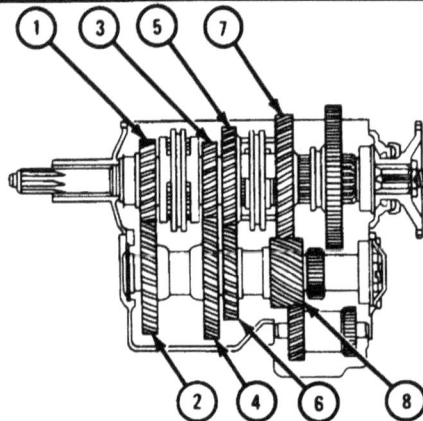

TA 087802

Table 2-1.   Transmission Forward Speed Backlash Wear Limits

Index Number	Item /Point of Measurement	Size and Fit of New Parts (inches)	Wear Limit (inches)
1 and 2	Input shaft gear to counter-shaft gear	0.006 to 0.012	0.016
3 and 4	Fifth gear to countershaft fifth gear	0.006L to 0.012L	0.019L
5 and 6	Third gear to countershaft third gear	0.006L to 0.012L	0.016L
7 and 8	Second gear to countershaft second-reverse idler gear	0.006L to 0.012L	0.019L

FRAME 3

NOTE

Readings must be within limits given in table 2-2. The letter L indicates a loose fit.   If readings are not within given limits,   throw away both gears and get new ones. Some gears on the countershaft cannot be taken off. If these gears are damaged, throw away countershaft and get a new one.

1. Slide first-reverse gear (1) forward to mesh with countershaft first gear (2). Measure backlash between two gears.

2. Slide first-reverse gear (1) back to mesh with reverse idler gear (3). Measure backlash between two gears.   Slide first-reverse gear forward so that it is not meshed with any gears.

3. Measure backlash between second-reverse gear (4) and reverse idler gear (5).

END OF TASK

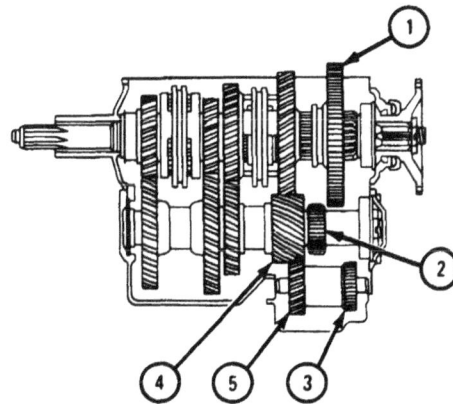

TA 087803

Table 2-2. Transmission Reverse Speed Backlash Wear Limits

Index Number	Item/Point of Measurement	Size and Fit of New Parts (inches)	Wear Limit (inches)
1 and 2	First-reverse gear to counter-shaft first gear	0.008L to 0.014L	0.019L
1 and 3	First-reverse gear to reverse idler gear	0.008L to 0.0146L	0.019L
4 and 5	Second-reverse gear to reverse idler gear	0.006L to 0.012L	0.019L

c. <u>Clutch Release Assembly.</u>

FRAME 1

1. Unhook two support springs (1).
2. Slide off release bearing carrier assembly (2).

GO TO FRAME 2

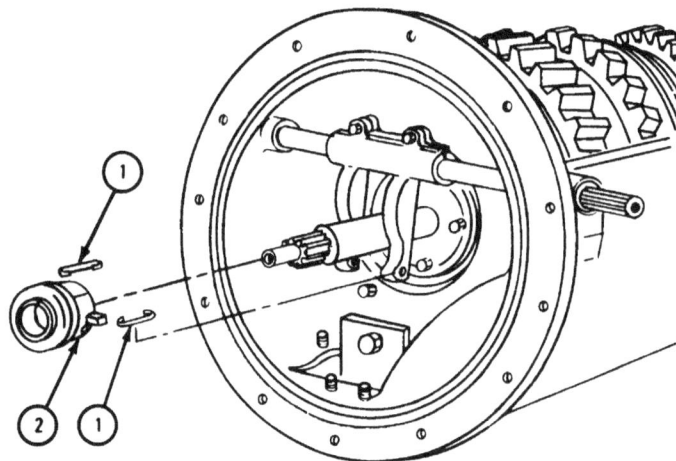

TA 087804

FRAME 2

1. Take out two capscrews (1) with two lockwashers (2).
2. Hold release fork (3) and pull out lever shaft (4) just enough to take out two keys (5). Take out two keys.
3. Hold release fork (3) and pull lever shaft (4) out all the way. Take out release fork.

GO TO FRAME 3

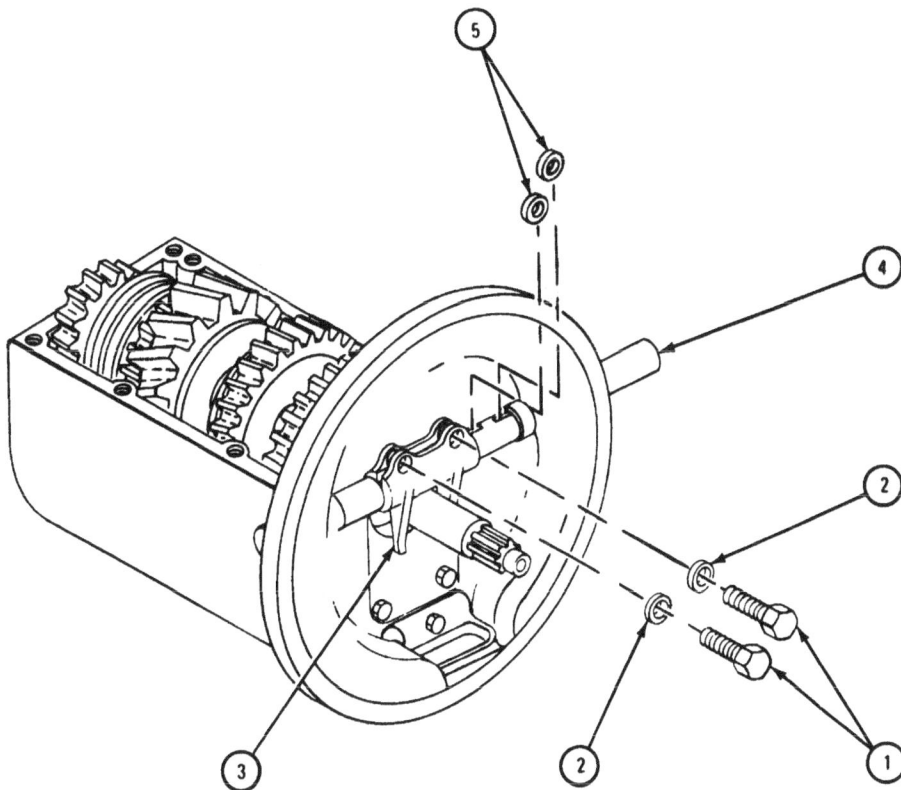

TA 087805

FRAME 3

1. Take out five capscrews (1) with five lockwashers (2).

2. Using soft faced hammer, tap off clutch housing (3) with gasket (4). Throw away gasket.

3. Take out six capscrews (5).

4. Take off cover (6) and gasket (7). Throw away gasket.

END OF TASK

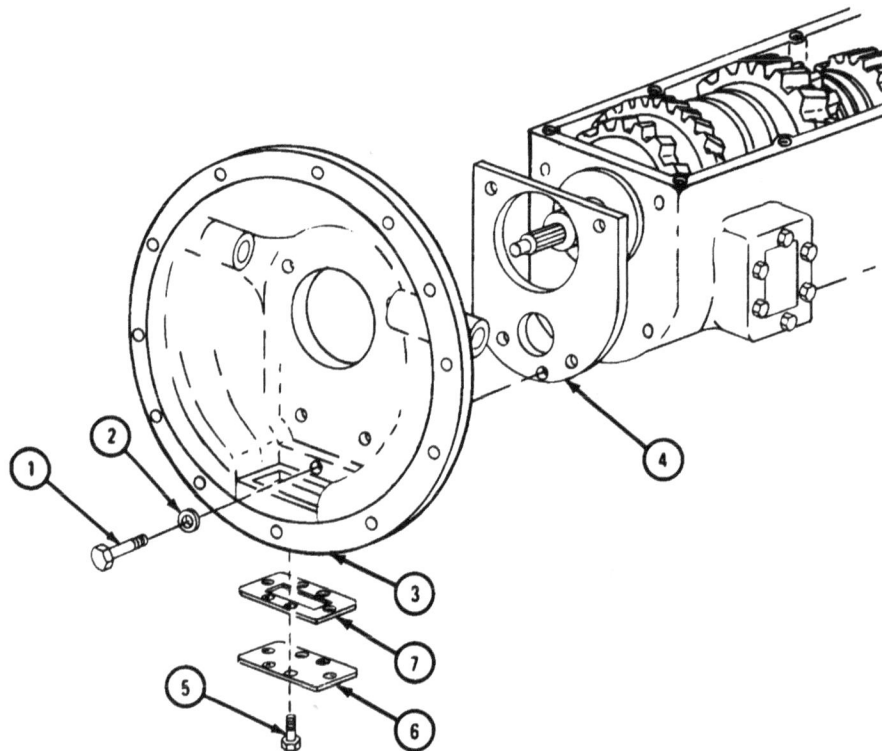

TA 087806

d. <u>Companion Flange.</u>

FRAME 1

NOTE

If synchronizer (1) and first and reverse gear (2) will not mesh,
turn input shaft (3) while meshing synchronizer or gear.

1. Push second and third speed synchronizer (1) forward to mesh.

2. Push first and reverse gear (2) all the way to rear to mesh.

3. Take out cotter pin (4).

4. Take off slotted nut (5) and washer (6).

GO TO FRAME 2

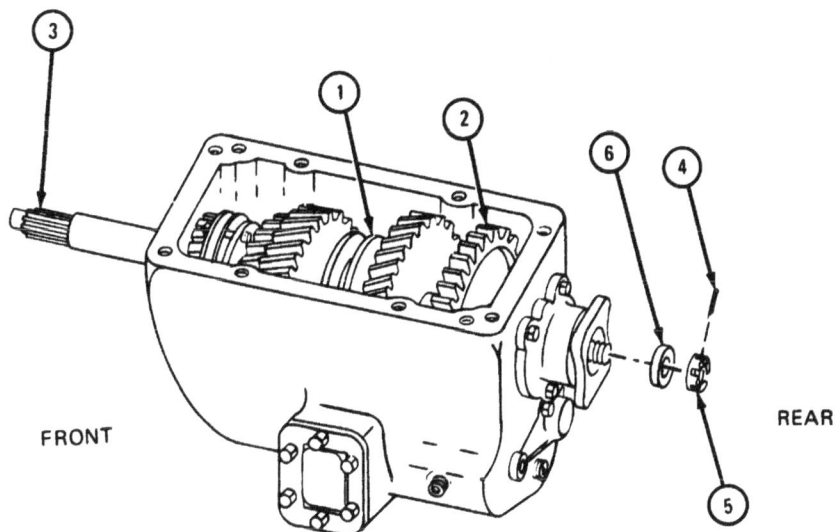

FRONT

REAR

TA 087807

FRAME 2

NOTE

If companion flange (1) cannot be pulled off by hand, setup
mechanical puller as shown and pull off companion flange.

1.   Full off companion flange (1).

END OF TASK

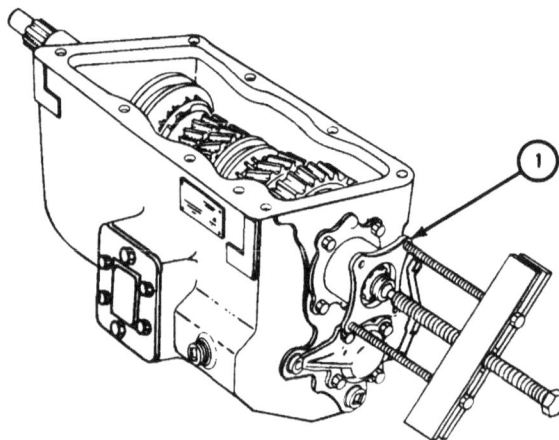

TA 087808

e. <u>Mainshaft Rear Bearing Cap.</u>

FRAME 1

1. Take out four capscrews (1) with four lockwashers (2).

NOTE

If bearing cap (3) is stuck to transmission case (4), tap sides
of bearing cap with soft faced hammer.

2. Take off bearing cap (3), gasket (5), and spacer (6). Throw away gasket.

END OF TASK

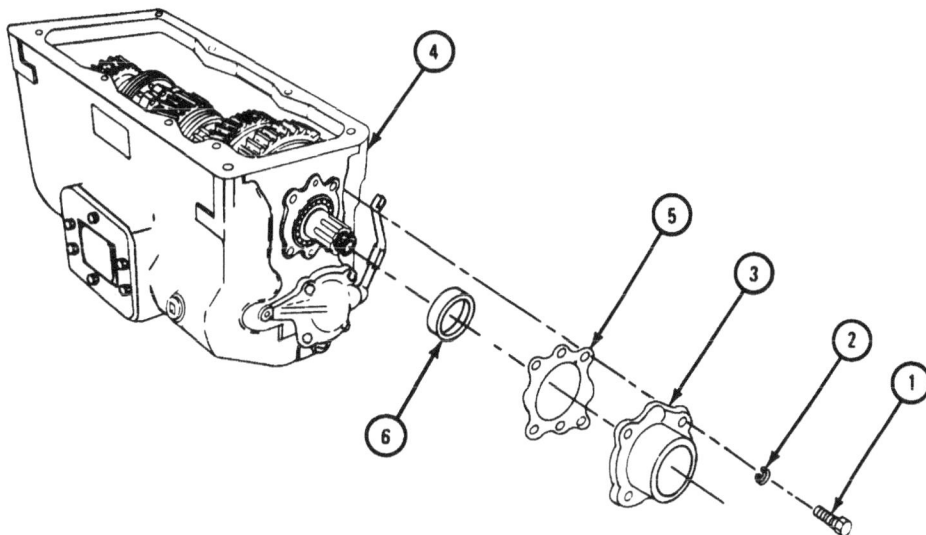

TA 087809

f. Countershaft Rear Bearing and Cover.

FRAME 1

1. Take off locknut (1), washer (2), and lever (3).
2. Take out four capscrews (4) with four lockwashers (5).

NOTE

If bearing cover (6) is stuck to transmission case (7), tap sides
of bearing cover with soft faced hammer.

3. Take off bearing cover (6) and gasket (8). Throw away gasket.

GO TO FRAME 2

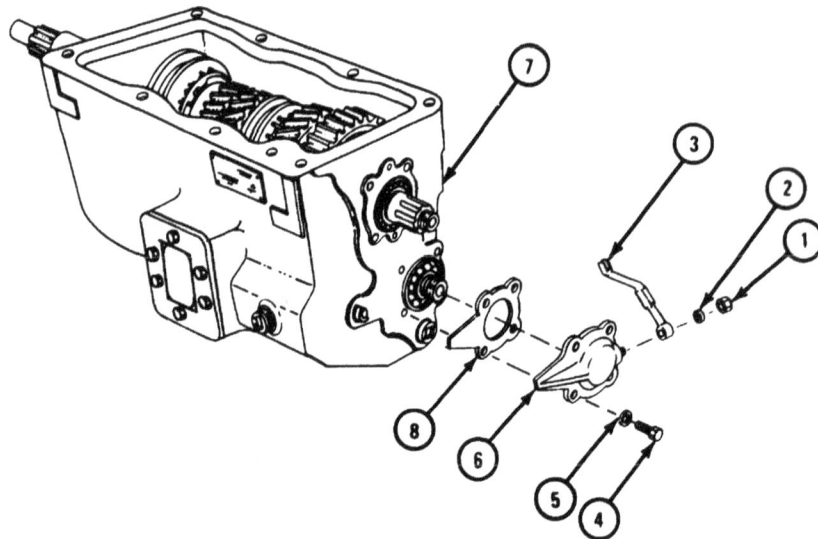

TA 067810

FRAME 2

1. Take out cotter pin (1).
2. Take off slotted nut (2) and washer (3).

END OF TASK

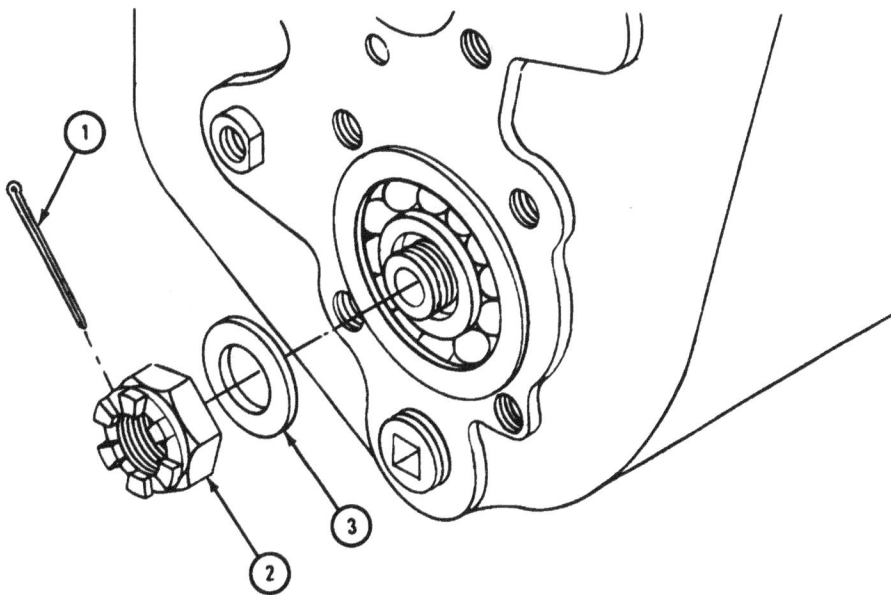

TA 087811

g. <u>Power Takeoff Access Cover (Trucks Without Power Takeoff Installed).</u>

FRAME 1

1. Take off six capscrews (1) with six lockwashers (2).

NOTE

If cover (3) is stuck to transmission case (4),   push blade of
scraper between cover and transmission case   to break gasket
(5) loose.

2. Take off cover (3) and gasket (5). Throw away gasket.

END OF TASK

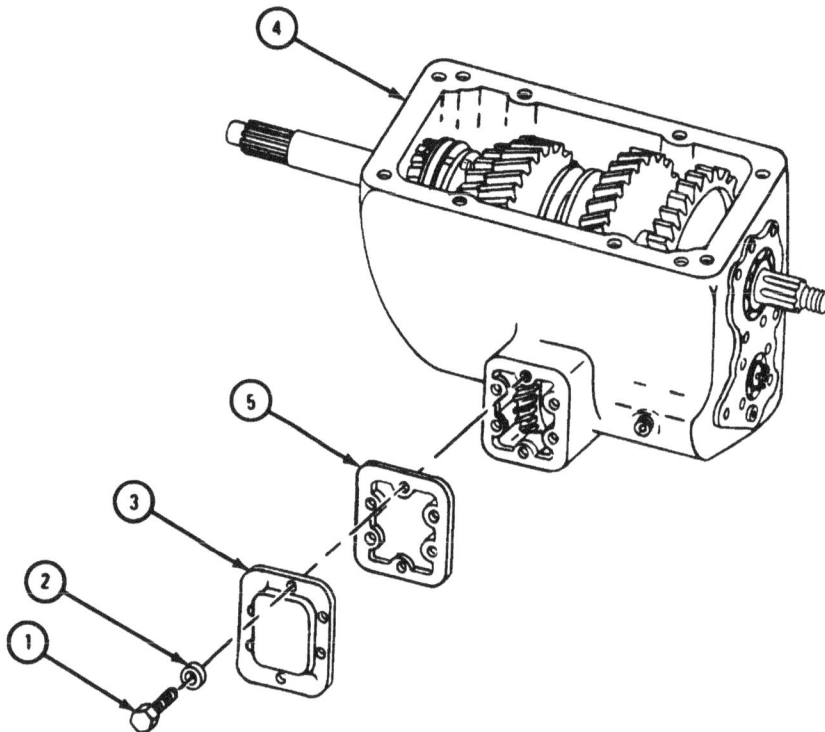

TA 087812

---

h. Input Shaft Bearing Cap.

FRAME 1

---

1. Take out four capscrews (1) with four lockwashers (2).

NOTE

If bearing cap (3) is stuck to transmission case (4), tap sides
of bearing cap with soft faced hammer.

2. Take off bearing cap (3) and gasket (5). Throw away gasket.

END OF TASK

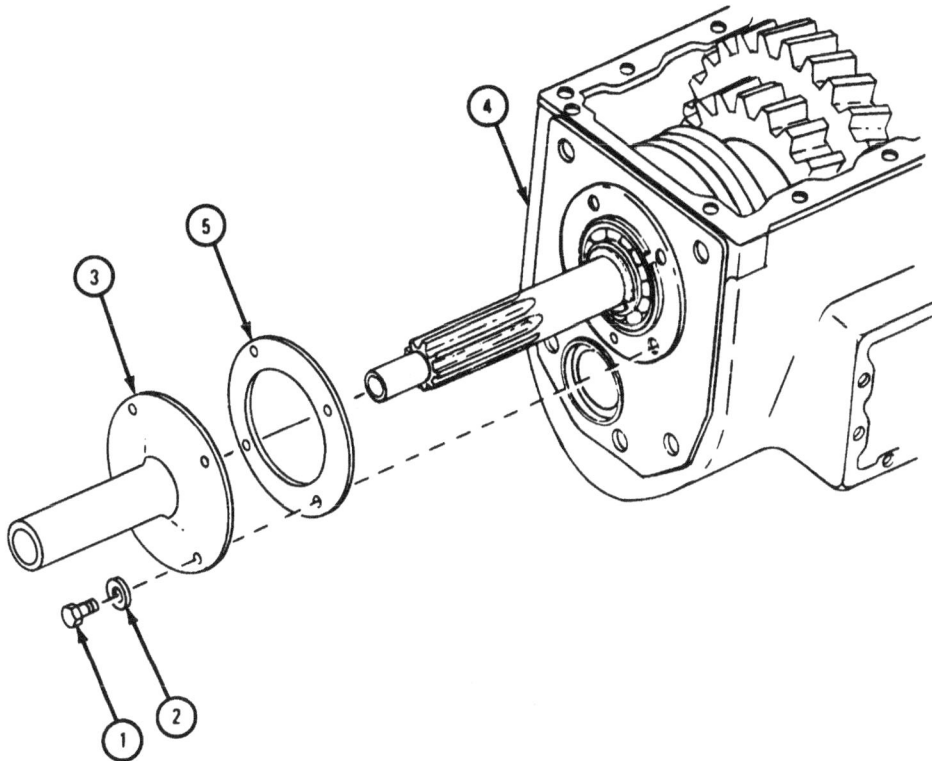

TA 087813

i. Mainshaft Assembly.

FRAME 1

1. Working at front of transmission case (1), take off input shaft bearing snapring (2).

2. Using hammer and wood block, tap input shaft (3) into transmission case (1) until mainshaft rear bearing (4) comes out of rear of transmission case as shown.

GO TO FRAME 2

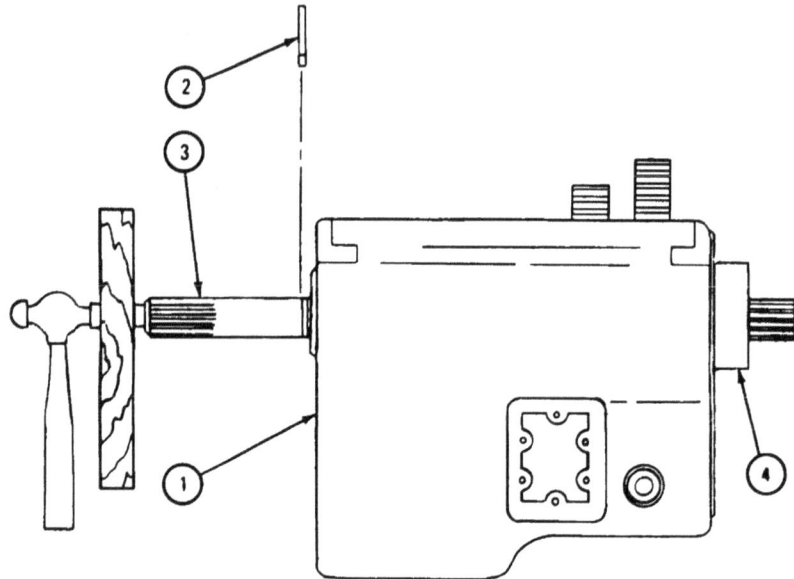

TA 087878

FRAME 2

1. Set up mechanical puller on mainshaft rear bearing (1) as shown.
2. Pull off rear bearing (1).

GO TO FRAME 3

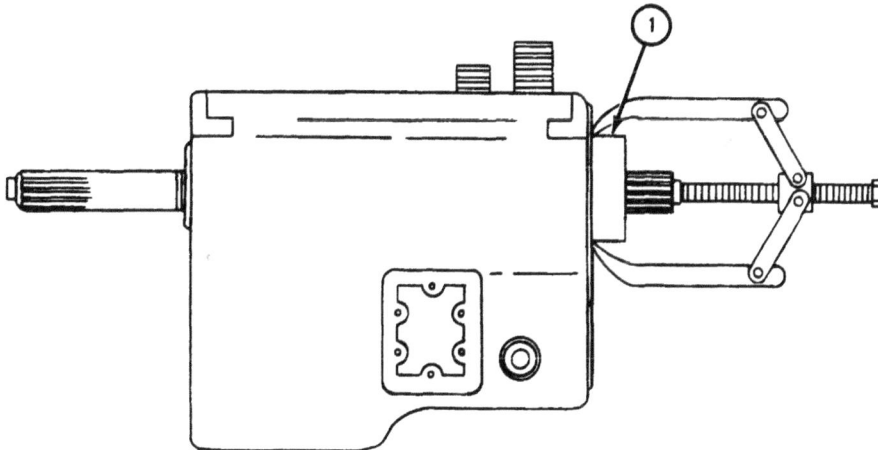

TA 087815

TM 9-2520-246-34-1

FRAME 3

NOTE

When input shaft (1) is moved away from mainshaft assembly
(2) some of the 14 pilot roller bearings inside input shaft
may fall into bottom of transmission case (3). They can be
taken out of transmission case after it is taken apart.

1. Pull input shaft (1) back out of transmission case (3). Using rubber hammer,
tap gear (4) on input shaft forward so that gear is up against front of transmission
case.

GO TO FRAME 4

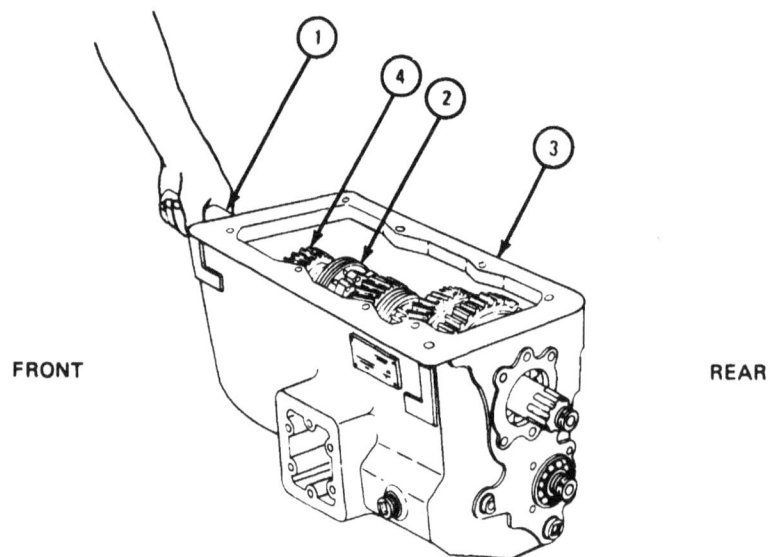

FRONT

REAR

TA 087816

2-18

FRAME 4

1. Holding both ends of mainshaft assembly (1), lift up front of mainshaft assembly as shown.

2. Slide mainshaft assembly (1) out of transmission case (2), leaving first and reverse gear (3) inside transmission case.

3. Take out first and reverse gear (3).

END OF TASK

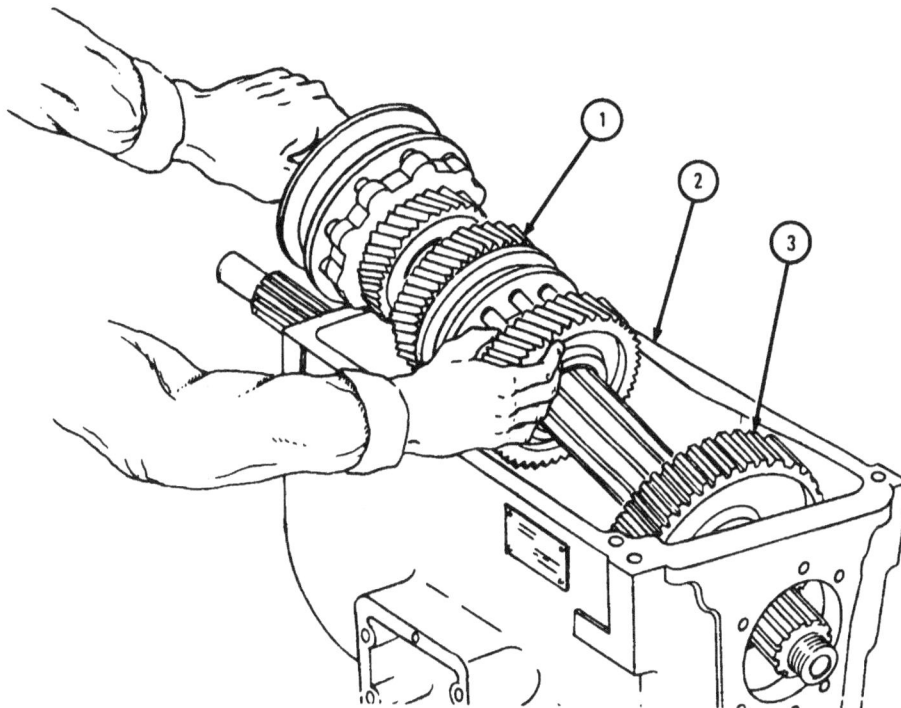

TA 087817

j. <u>Input Shaft Assembly.</u>

FRAME 1

1. Using rubber hammer, tap input shaft (1) into transmission case (2).
2. Take out input shaft (1).

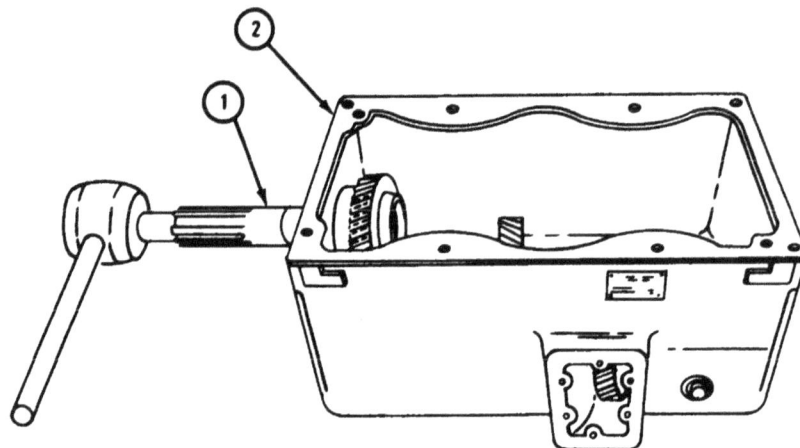

END OF TASK

TA 087818

k. Countershaft Assembly.

FRAME 1

1. Take off snapring (1).

2. Working from inside transmission case (2), push out countershaft (3) until puller can be put on bearing (4).

3. Using mechanical puller, pull off bearing (4).

GO TO FRAME 2

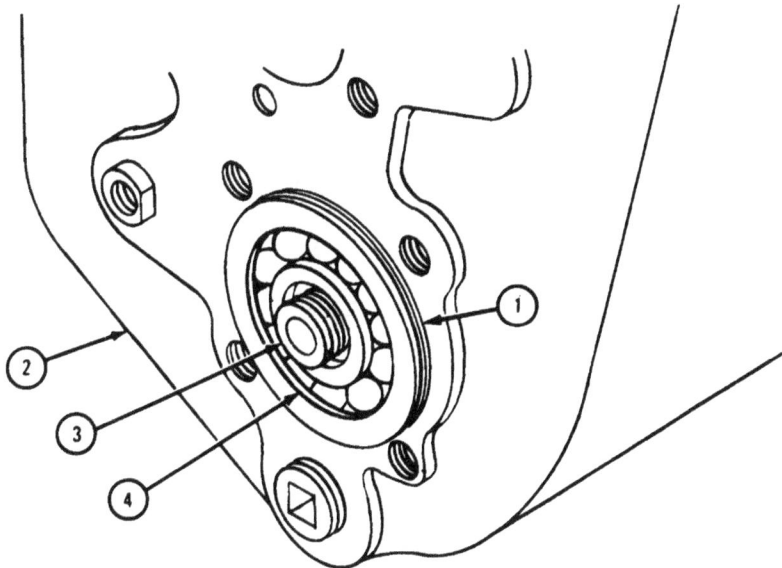

TA 087819

FRAME 2

NOTE

When countershaft assembly (1) is pulled out of front bearing (2), thrust washer (3) will fall into bottom of transmission case (4). It can be taken out after countershaft assembly is taken out.

1. Slide countershaft assembly (1) out of front bearing (2) and lift it out of transmission case (4) as shown.

2. Take out thrust washer (3).

GO TO FRAME 3

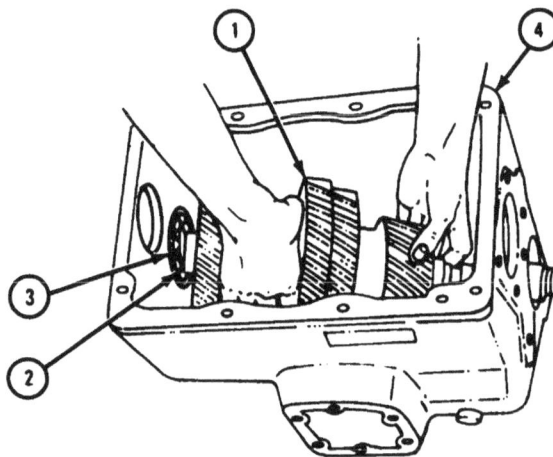

TA 087820

FRAME 3

1. Using hammer and brass drift and working from inside of transmission case (1), drive out bearing retainer (2).

2. Working from outside of transmission case (1) and using hammer and brass drift, tap out countershaft front bearing (3) and take bearing out of transmission case.

END OF TASK

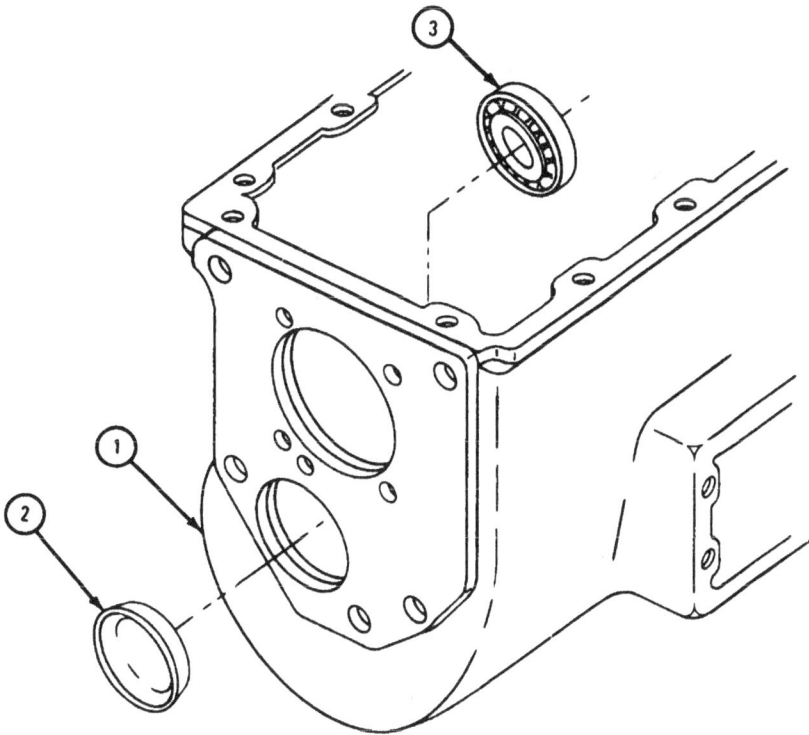

TA 087821

1. Reverse Idler Gear Assembly.

FRAME 1

1. Screw idler gear adapter (1) into reverse idler gear shaft (2).

2. Using slide handle puller, pull out reverse idler gear shaft (2).

3. Take out reverse idler gear assembly (3) and two thrust washers (4).

4. Check bottom of transmission case (5) to see if any of 14 pilot roller bearings fell in when input shaft was taken out. Take out pilot roller bearings.

END OF TASK

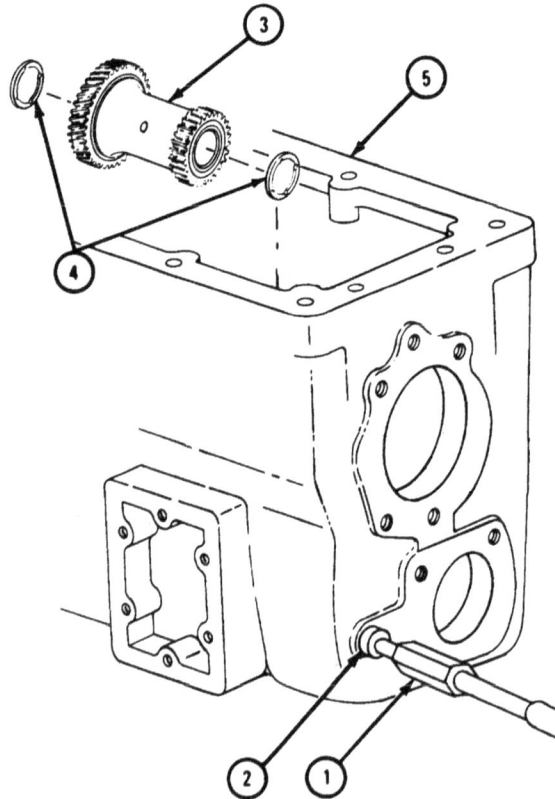

TA 087822

2-5.   DISASSEMBLY OF SUBASSEMBLIES. The following paragraphs give instructions to disassemble the transmission subassemblies.

a.   Input Shaft Assembly.

FRAME 1

Soldier A 1. Take off snapring (1).

2. Set up input shaft (2) in hydraulic press as shown.

Soldier B 3. Working from under press, hold bottom of input shaft (2) to keep it from falling when bearing (3) is pressed off. Tell soldier A when ready.

Soldier A 4. Using hydraulic press, press input shaft (2) out of bearing (3).

END OF TASK

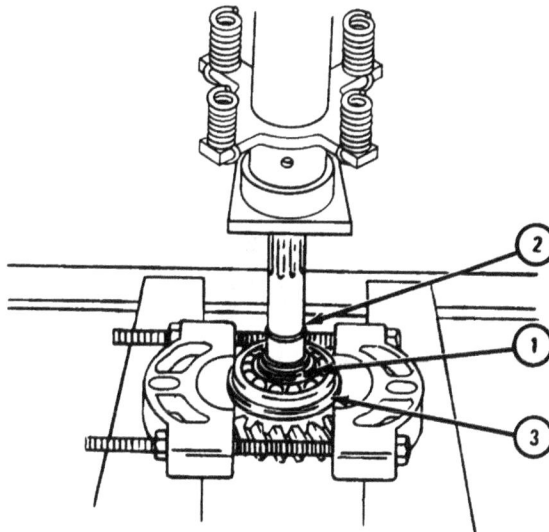

b. Mainshaft.

FRAME 1

1. Set up mainshaft assembly (1) in hydraulic press as shown.
2. Slide off fourth and fifth speed synchronizer (2).
GO TO FRAME 2

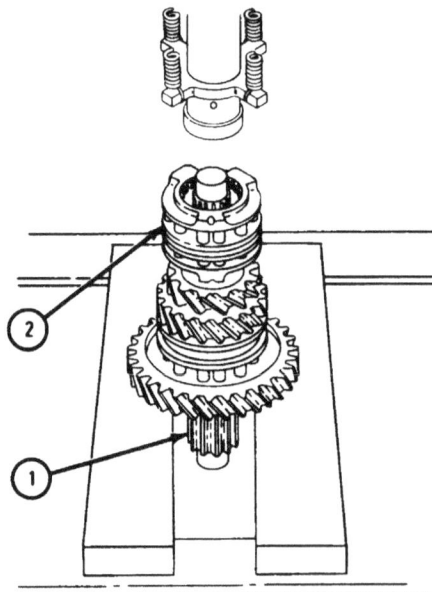

TA 087824

FRAME 2

1. Take off snapring (1).
2. Slide off thrust washer (2) and fifth speed gear (3).
GO TO FRAME 3

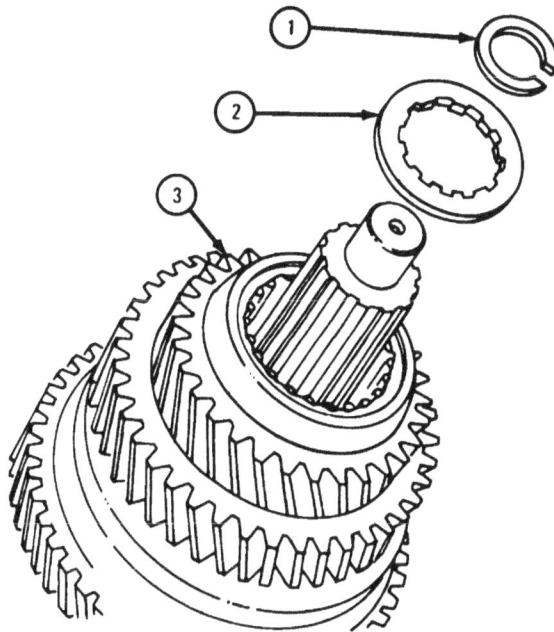

TA 087825

FRAME 3

Soldier A 1.   Setup mainshaft assembly (1) in hydraulic press as shown.

Soldier B 2.   Working from under press, hold bottom of mainshaft (1) to keep it from falling when third speed gear (2) and fifth speed gear sleeve (3) are pressed off.   Tell soldier A when ready.

Soldier A 3.   Press out mainshaft (1).

4.   Slide off second and third speed gear synchronizer (4).

GO TO FRAME 4

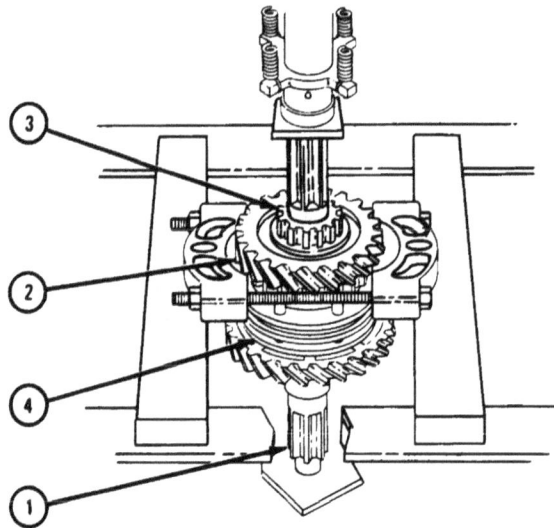

TA 087826

FRAME 4

1. Take off snapring (1).

2. Set up hydraulic press as shown and press second and third speed synchronizer sleeve (2) 1/4 inch. Do not press second speed gear (3) onto two keys in mainshaft (4).

GO TO FRAME 5

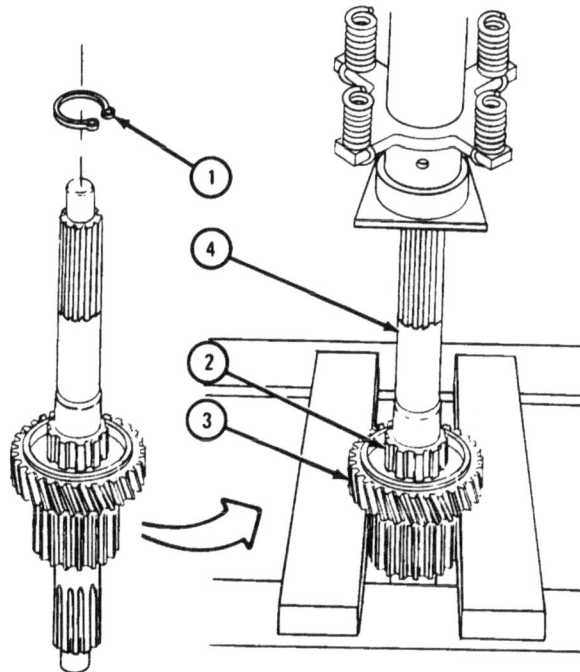

TA 087827

FRAME 5

Soldier A 1.   Set up hydraulic press as shown.

           2.   Put clamp (1) on second and third speed synchronizer sleeve (2) as shown.

Soldier B 3.   Working from under press, hold bottom of mainshaft (3) to keep it from falling when second and third speed synchronizer sleeve (2) is pressed off.   Tell soldier A when ready.

Soldier A 4.   Press out mainshaft (3).

GO TO FRAME 6

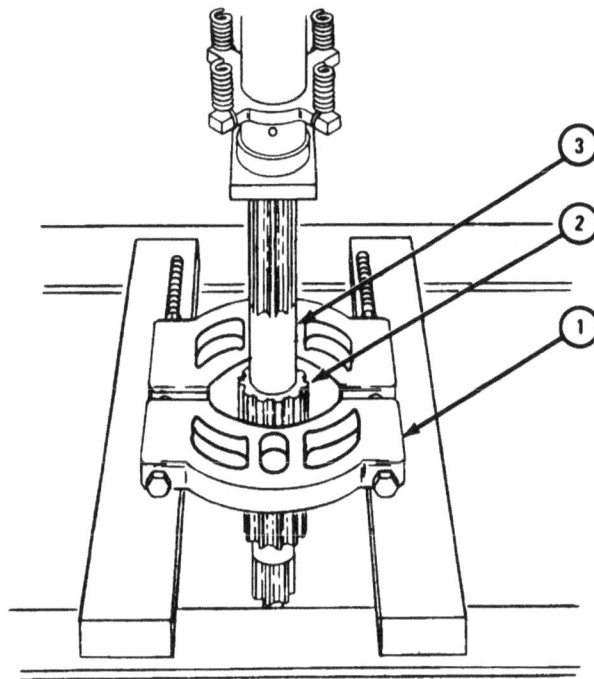

TA 087828

FRAME 6

Soldier A 1.   Set up hydraulic press as shown.

2.   Take out key (1).

Soldier B 3.   Working from under press, hold bottom of mainshaft (2) to keep it from falling when second speed gear (3) is pressed off. Tell soldier A when ready.

Soldier A 4.   Press out mainshaft (2).

5.   Take out key (4).

END OF TASK

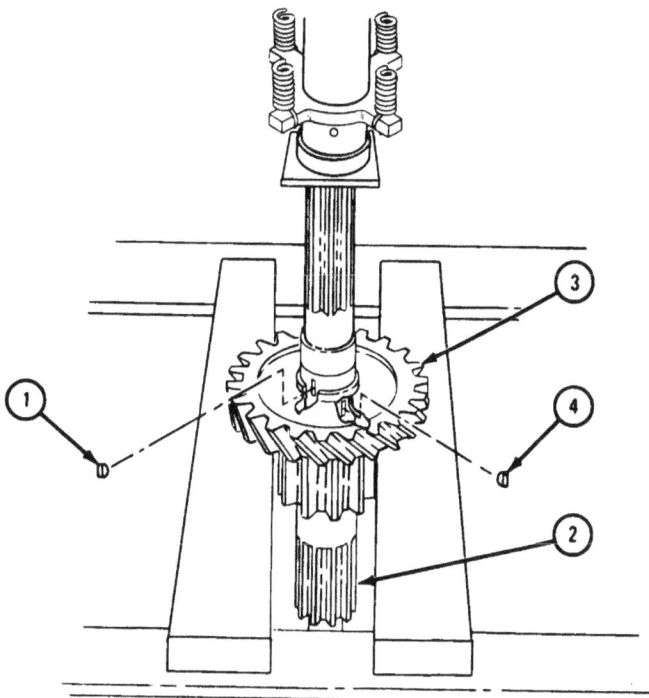

TA 087829

c. Countershaft Assembly.

FRAME 1

Soldier A 1. Take off snapring (1).

2. Set up countershaft assembly (2) in hydraulic press as shown.

Soldier B 3. Working from under press, hold bottom of countershaft (2) to keep it from falling when drive gear (3) is pressed off. Tell soldier A when ready.

Soldier A 4. Press out countershaft (2).

5. Take out key (4).

GO TO FRAME 2

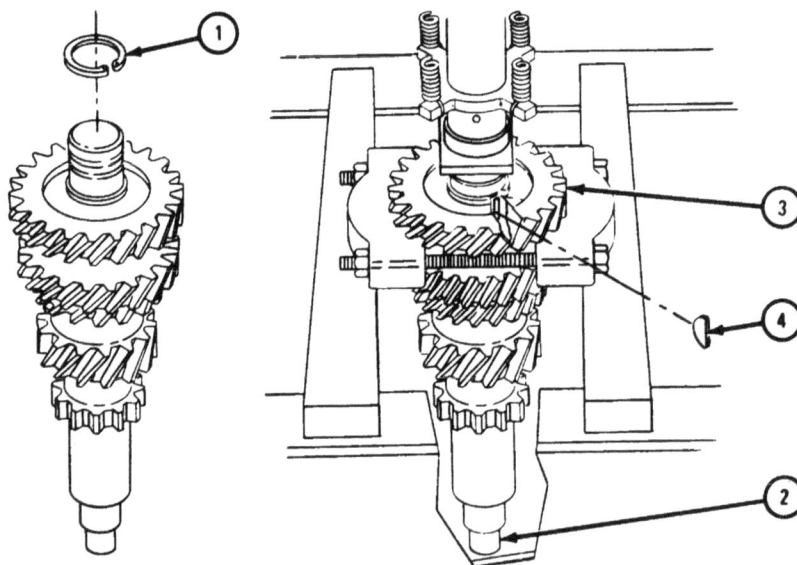

TA 087830

FRAME 2

Soldier A 1.   Take off snapring (1).

        2.   Set up hydraulic press as shown.

Soldier B 3.   Working from under press, hold bottom of countershaft (2) to keep it from falling when fourth speed gear (3) is pressed off. Tell soldier A when ready.

Soldier A 4.   Press out countershaft (2).

        5.   Take out key (4).

END OF TASK

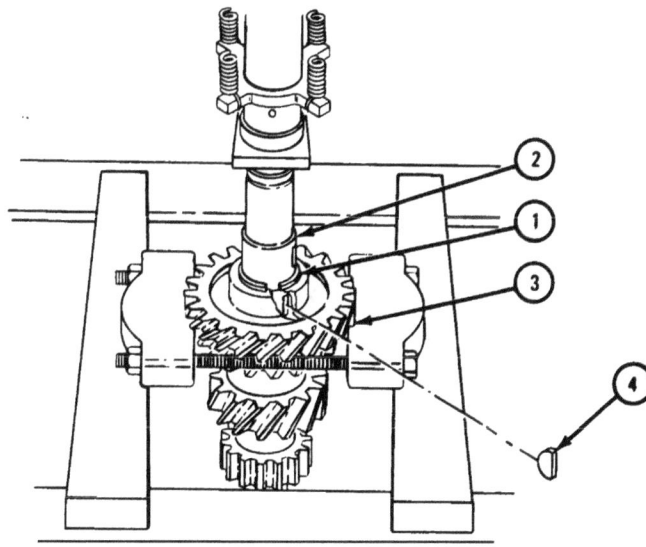

TA 087831

d. Reverse Idler Gear Assembly.

FRAME 1

1. Take out two bearing assemblies (1), one from each end of reverse
    idler gear assembly (2).

END OF TASK

TA 087832

e.   <u>Mainshaft Rear Bearing Cap.</u>

FRAME 1

1.   Put bearing cap (1) between two wood blocks as shown.

2.   Using hammer and brass drift as shown, drive out oil seal (2).

END OF TASK

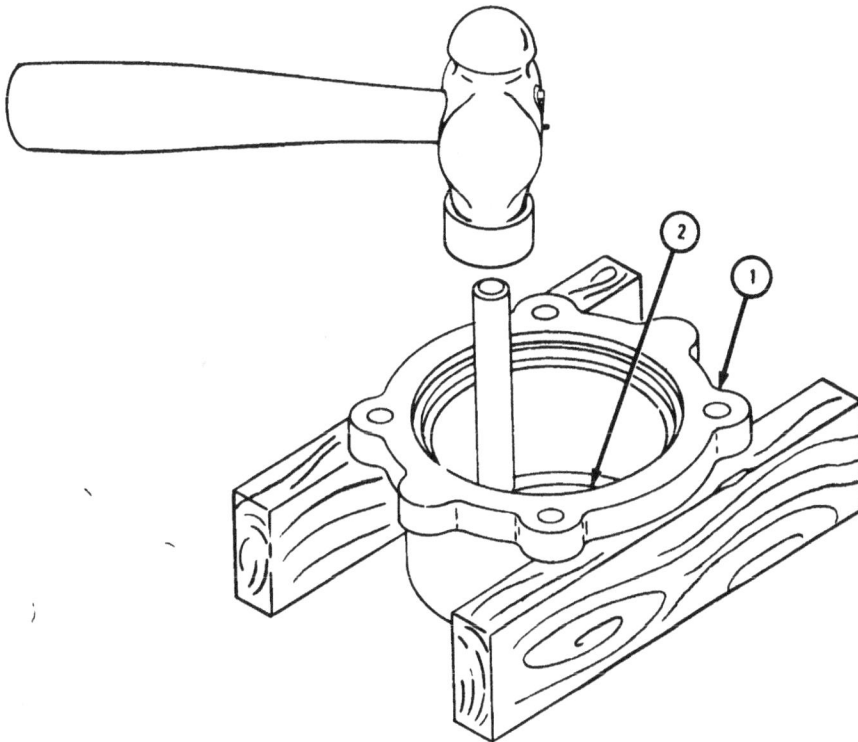

TA 087833

f.   Shifter Shaft Cover Assembly.

FRAME 1

---

1.   Working on inside of cover, cut and take out four safety wires (1). Throw away safety wires.

2.   Take out four setscrews (2).

GO TO FRAME 2

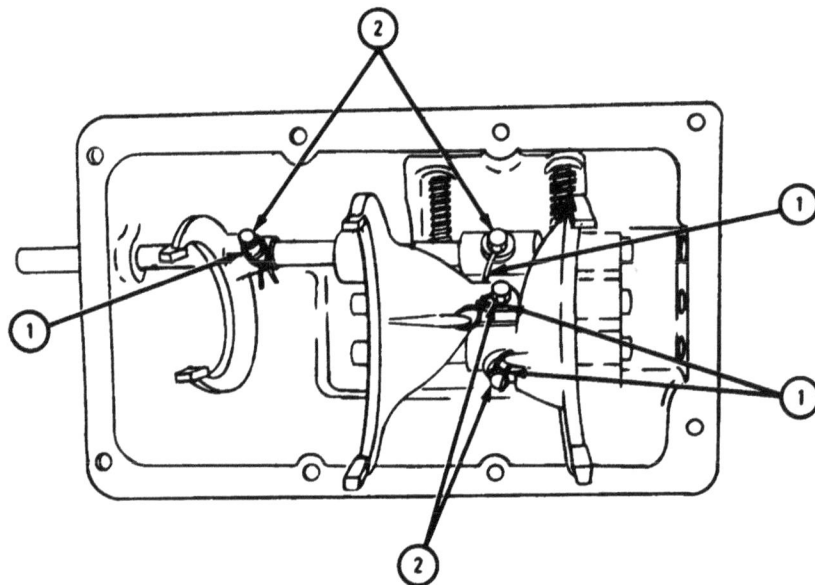

TA 087834

FRAME 2

1. Using hammer and brass drift, tap reverse shifter shaft (1) until expansion plug (2) comes out. Set plug aside.

2. Tap reverse shifter shaft (1) until ball (3) pops up. Take out ball with spring.

3. Tap shaft (1) all the way out and tag it.

4. Take out fork (4) and bracket (5). Tag fork.

GO TO FRAME 3

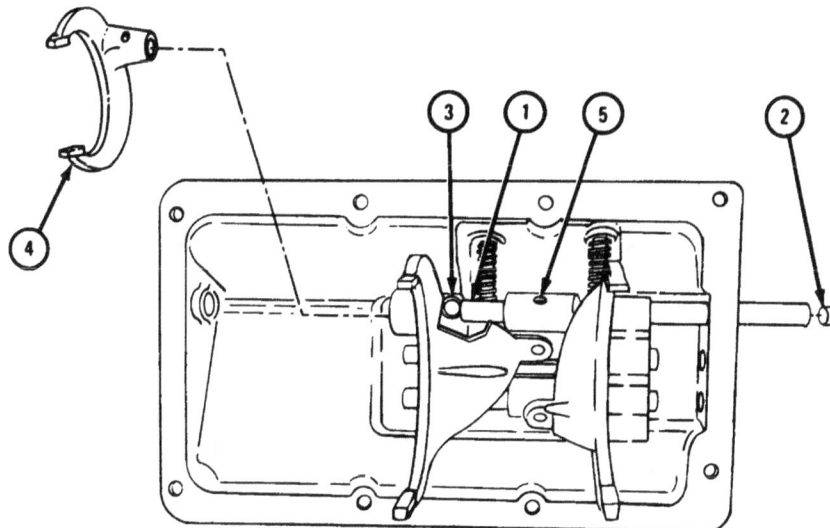

TA 087835

FRAME 3

1. Using hammer and brass drift, tap out second and third speed shifter shaft (1) until ball (2) pops up. Take out ball with spring.

2. Tap out shaft (1) until expansion plug (3) comes out. Set plug aside.

3. Tap shaft (1) all the way out and tag it.

4. Tag and take out fork (4).

5. Do steps 1 through 4 again for fourth and fifth speed shifter shaft (5).

6. Turn over cover assembly (6).

GO TO FRAME 4

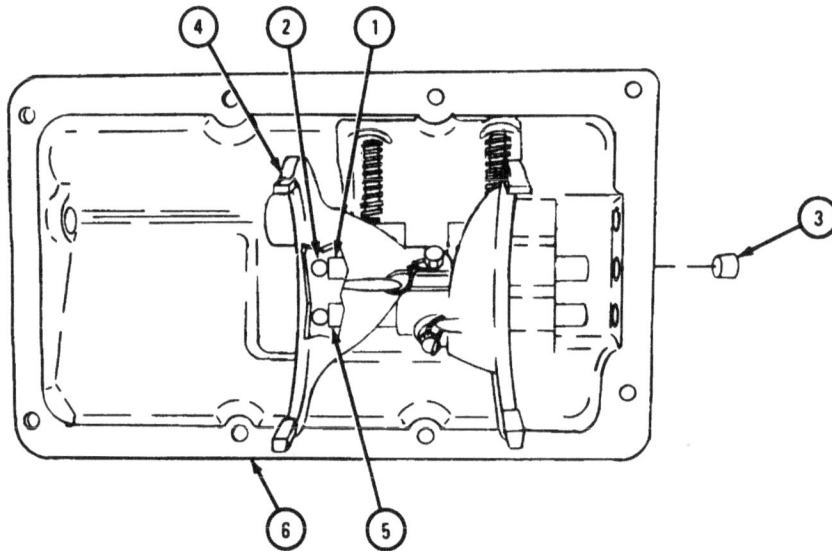

TA 087836

FRAME 4

1. Take off two interlock safety nuts (1) and two washers (2).

2. Turn over cover assembly (3).

GO TO FRAME 5

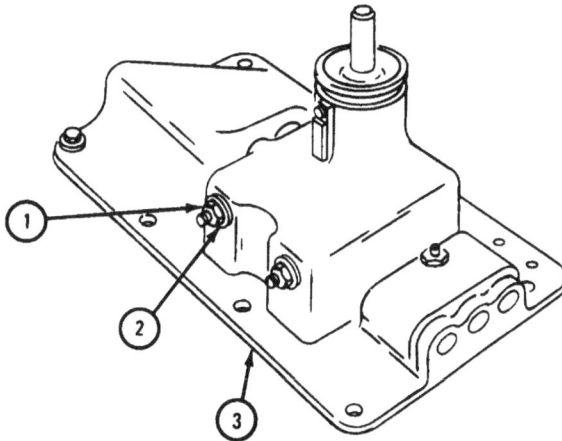

TA 087837

FRAME 5

1. Tap out two screws (1).

2. Takeout six washers (2) and two springs (3).

3. Take out plate (4).

4. Using hammer and punch, tap out and throw away first and reverse shifter shaft oil seal (5).

GO TO FRAME 6

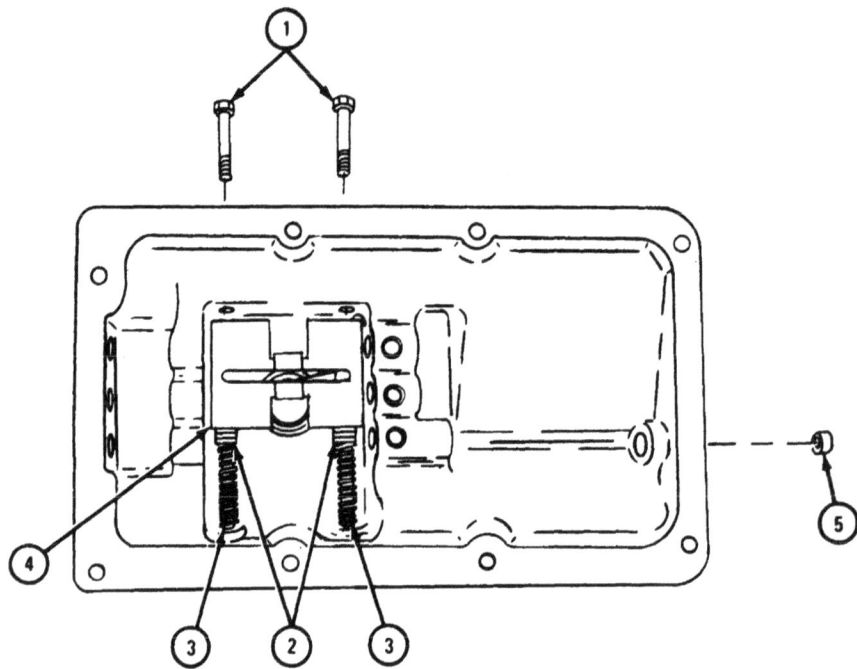

TA 087838

FRAME 6

1. Take out lockring (1), spring (2), and spring cup (3).

2. Slide out shifter shaft lever (4) through inside of shifter shaft cover (5)

END OF TASK

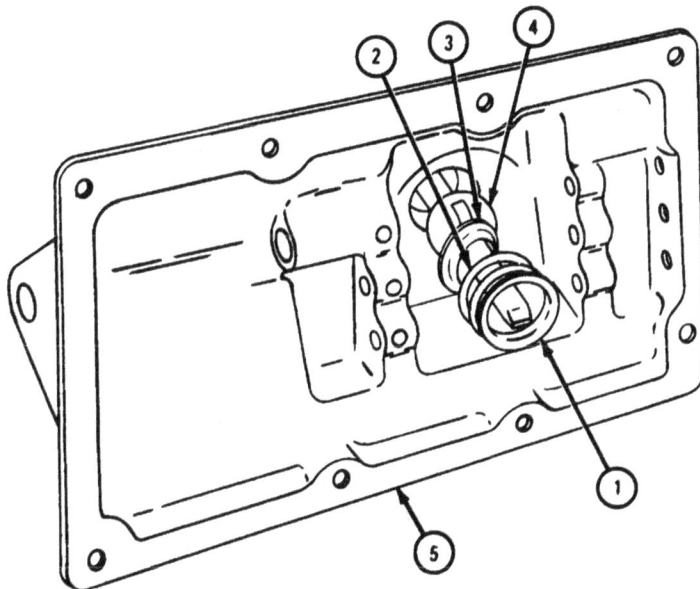

TA 087839

g. <u>Clutch Release Bearing Carrier.</u>

FRAME 1

Soldier A 1.   Take out lubrication fitting (1).

2.   Set up carrier (2) in hydraulic press as shown.

Soldier B 3.   Working under press, hold bottom of carrier (2) to keep it from falling when bearing (3) is pressed off. Tell soldier A when ready.

Soldier A 4.   Press off bearing (3).

5.   Take out two clutch bearing buttons (4).

END OF TASK

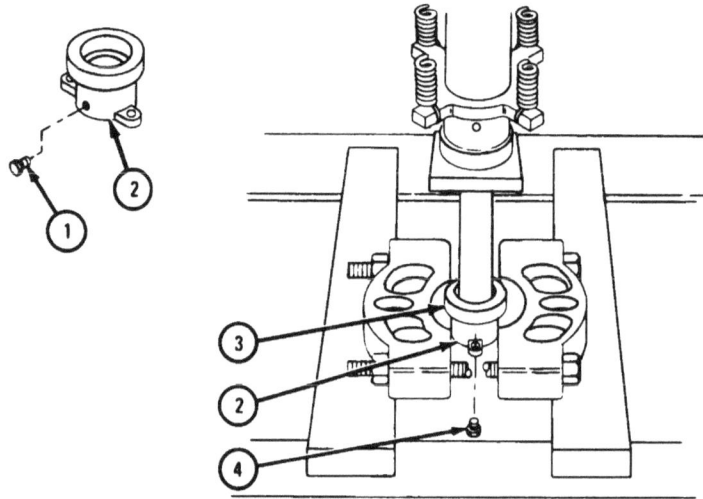

TA087840

2-6. CLEANING. This paragraph gives general instructions for cleaning the transmission parts.

a. Clean all bearing cones and cups. Refer to inspection, care, and maintenance of antifriction bearings, TM 9-214.

WARNING

Dry cleaning solvent is flammable. Do not use near an open flame.   Keep a fire extinguisher nearby when solvent is used. Use only in well-ventilated places.   Failure to do this may result in injury to personnel and damage to equipment.

Do not use more than 30 psi of air pressure for drying parts.   Eye shields must be worn when using compressed air. Eye injury can occur if eye shields are not used.

CAUTION

When scraping gasket material from surface of parts, be careful not to scratch or gouge metal surfaces.

b.   Clean all other parts with solvent.   Scrape all gasket material from surface of parts.   Rinse parts in clean solvent and dry with compressed air.

c.   Make sure all oil passages are open.   Open clogged passages with compressed air or by working a stiff wire back and forth. Flush with solvent.

2-7. GENERAL INSPECTION. This paragraph gives instructions to check for damage on the transmission case, cover, gearshafts, and gears.

CAUTION

It is easy to damage the equipment if you don't know what you are doing. Do not try to do this task unless you are experienced at it, or you have an experienced person with you.

FRAME 1

NOTE

Small chips, burrs or scratches in housing castings and transmission case can be repaired. Cracks in housing castings that do not go into screw holes or openings can be repaired. If parts are damaged in any other way, throw parts away and get new ones.

1. Check that transmission case (1) does not have any broken bolts or stripped threads. Mark them for repair.

2. Check that clutch release fork (2) is not cracked or bent.

3. Check that transmission case (1), power takeoff access cover (3), counter-shaft rear bearing cover (4), retainer cap (5), input shaft front bearing cover (6), and inspection cover (7) do not have cracks, chips, warped areas or small holes.

GO TO FRAME 2

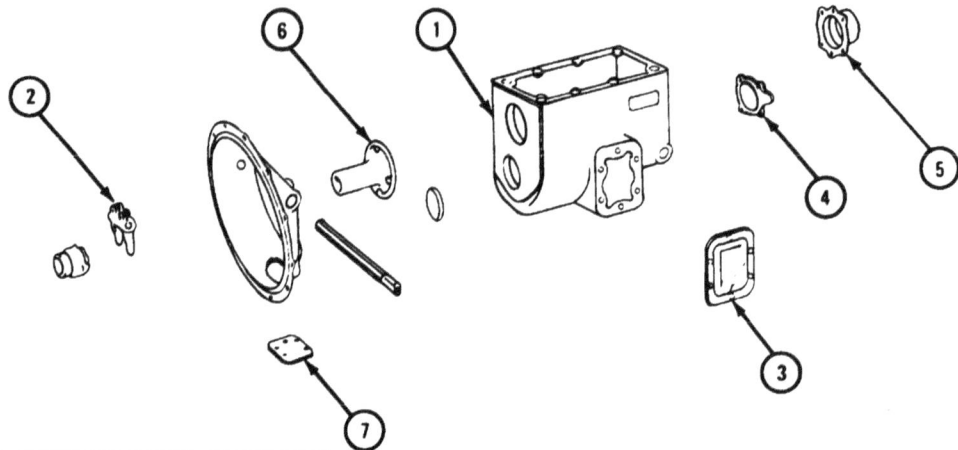

NOTE: CHECK ONLY THOSE PARTS WHICH ARE CALLED OUT IN THIS FRAME. PARTS WITHOUT CALLOUTS ARE SHOWN ONLY FOR REFERENCE PURPOSES OR ARE CHECKED IN ANOTHER FRAME.

TA 087841

FRAME 2

NOTE

Small chips, burrs or scratches in housing castings can be repaired.   Cracks in housing castings that do not go into screw holes or openings can be repaired. If parts are damaged in any other way, throw parts away and get new ones.

1.   Check that carrier (1), front bearing retainer (2), and clutch housing (3) do not have cracks, chips, warped areas or small holes.

2.   Check that lever shaft (4) is not cracked or bent.

3.   Check that all threaded parts are not stripped or crossthreaded.

GO TO FRAME 3

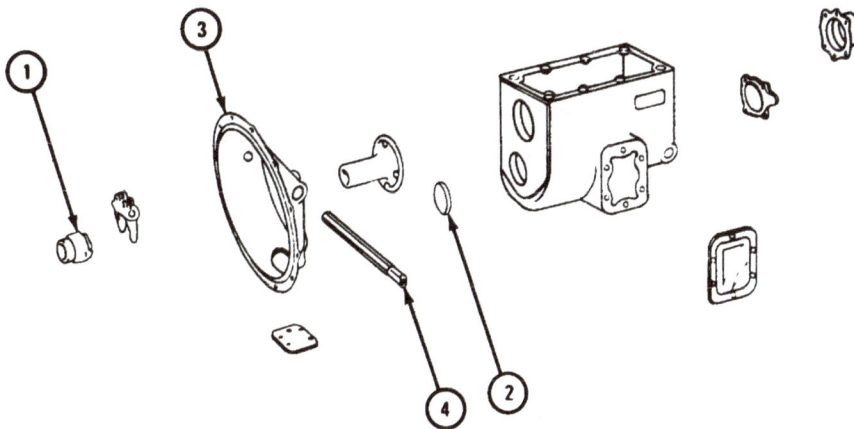

NOTE
CHECK ONLY THOSE PARTS WHICH ARE CALLED OUT IN THIS FRAME. PARTS WITHOUT CALLOUTS ARE SHOWN ONLY FOR REFERENCE PURPOSES OR ARE CHECKED IN ANOTHER FRAME.

TA 087842

FRAME 3

NOTE

Small chips, burrs or scratches on shafts, gears, synchronizers, and synchronizer sleeves can be repaired. If parts
are damaged in any other way, throw parts away and get
new ones.

1. Check that shafts (1, 2, and 3) are not chipped or cracked.

2. Check that idler shaft (4) is not cracked or bent.

3. Check that shaft splines (5) are not chipped, cracked or twisted.

4. Check that two synchronizers (6) are not chipped or cracked and that internal
splines are not twisted.

GO TO FRAME 4

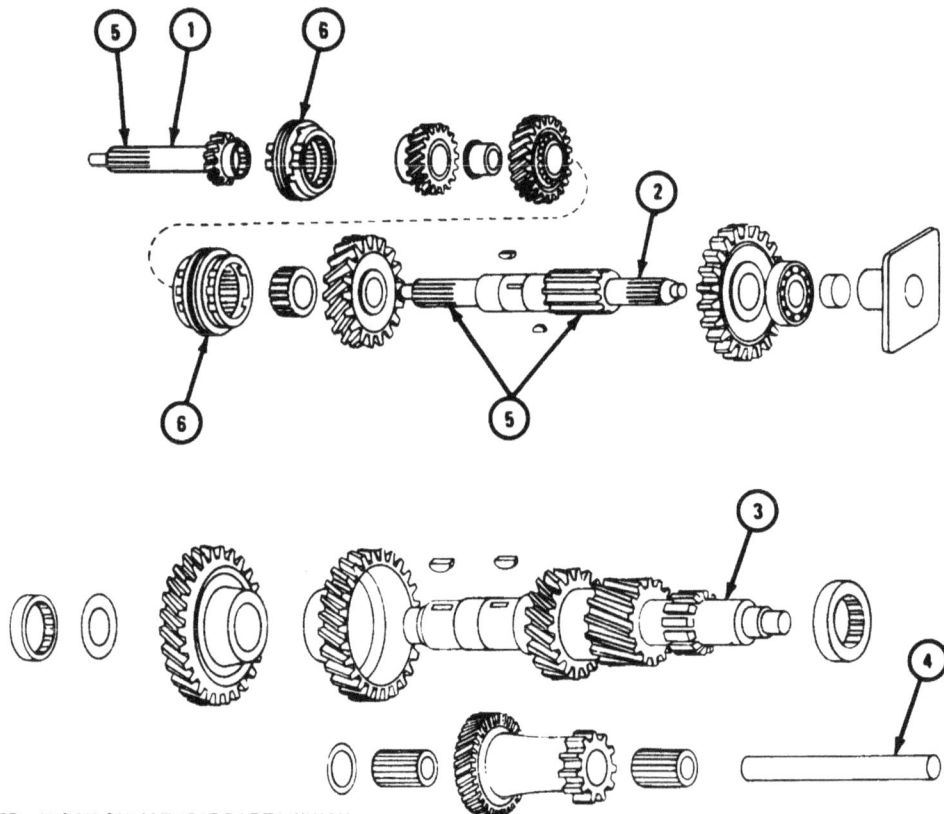

NOTE: CHECK ONLY THOSE PARTS WHICH
ARE CALLED OUT IN THIS FRAME.
PARTS WITHOUT CALLOUTS ARE
SHOWN ONLY FOR REFERENCE
PURPOSES OR ARE CHECKED IN
ANOTHER FRAME.

TA 087843

FRAME 4

NOTE

Small chips, burrs or scratches on gears, keys, and sleeves can be repaired. If parts are damaged in any other way, throw parts away and get new ones.

1. Check that four keys (1) are not damaged.

NOTE

Four gears (2) on countershaft and reverse idler gear assembly do not come off. If these gears are damaged, throw away shafts and get new ones.

2. Check that six gears (3) and two sleeves (4) have no chips, cracks, or broken teeth.

3. Check that all bearings (5) and needle bearings (6) are not damaged. Refer to TM 9-214.

GO TO FRAME 5

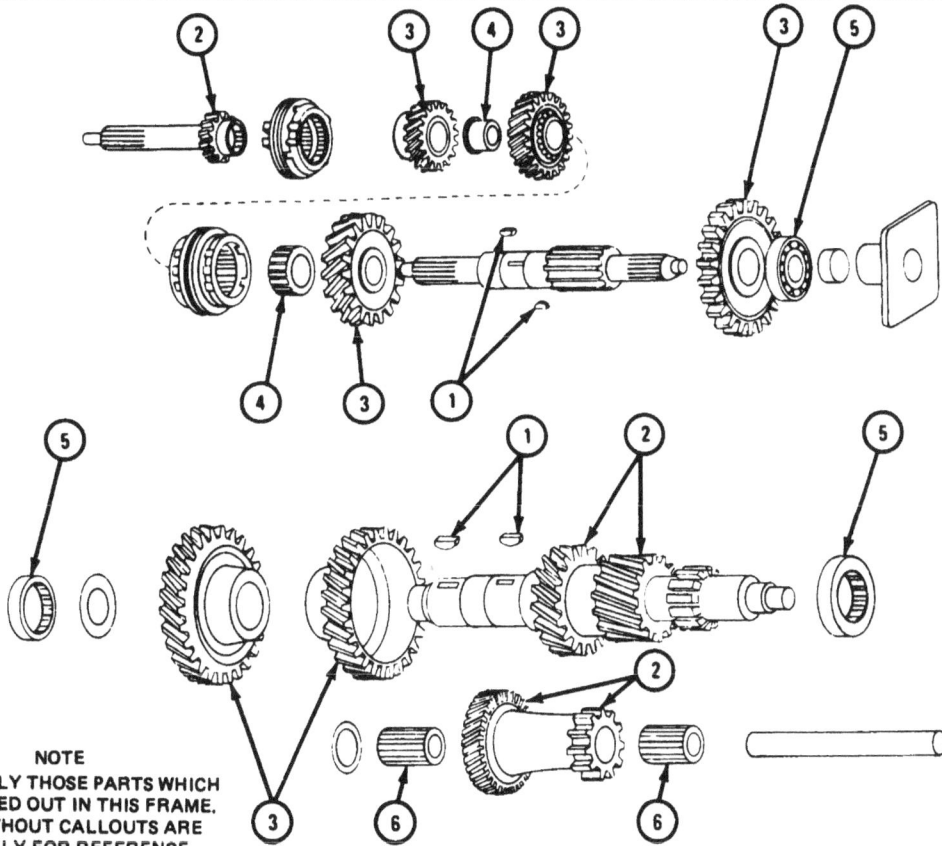

NOTE
CHECK ONLY THOSE PARTS WHICH ARE CALLED OUT IN THIS FRAME. PARTS WITHOUT CALLOUTS ARE SHOWN ONLY FOR REFERENCE PURPOSES OR ARE CHECKED IN ANOTHER FRAME.

TA 087844

FRAME 5

NOTE

Small chips, burrs or scratches on shifter forks and shifter fork shafts can be repaired. If parts are damaged in any other way, throw parts away and get new ones.

1. Check that three shifter forks (1) and three shifter shafts (2) are not cracked or bent.

2. Check that six springs (3) are not damaged in any way.

3. Check that three balls (4) have no flat spots. Check that plate (5) is not warped, cracked or damaged in any other way.

4. Check that all threaded parts are not stripped or crossthreaded.

5. Check that cover (6) has no cracks, chips, warped areas or small holes.

END OF TASK

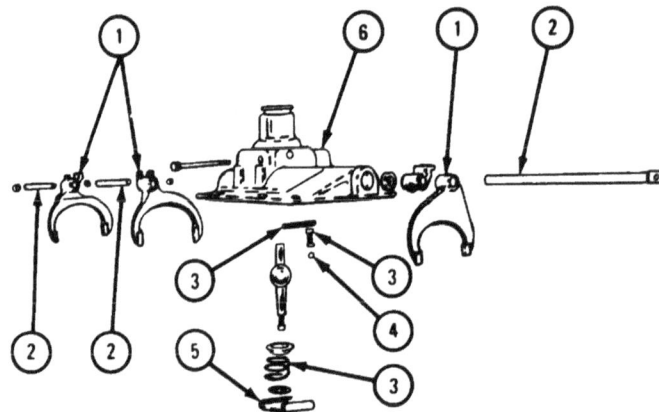

NOTE
CHECK ONLY THOSE PARTS WHICH ARE CALLED OUT IN THIS FRAME. PARTS WITHOUT CALLOUTS ARE SHOWN ONLY FOR REFERENCE PURPOSES OR ARE CHECKED IN ANOTHER FRAME.

TA 087845

2-8. WEAR LIMIT INSPECTION. The following paragraphs give instructions for checking the minimum and maximum wear limits for each subassembly to which a part or parts may be worn before a new part is needed.

    a.   Input Shaft Assembly.

FRAME 1	

NOTE

Readings must be within limits given in table 2-3. If readings are not within given limits, throw away part and get a new one.

1.   Measure bearing inside diameter (1) and out side diameter (2).

2.   Measure bearing journal diameter (3).

3.   Measure pilot diameter (4).

4.   Measure width of splines (5).

GO TO FRAME 2

NOTE: CHECK ONLY THOSE PARTS WHICH ARE CALLED OUT IN THIS FRAME. PARTS WITHOUT CALLOUTS ARE SHOWN ONLY FOR REFERENCE PURPOSES OR ARE CHECKED IN ANOTHER FRAME.

Table 2-3.   Input Shaft Assembly Wear Limits

Index Number	Item/Point of Measurement	Size and Fit of New Parts (inches)	Wear Limit (inches)
1	Bearing inside diameter	1.5748 to 1.5753	None
2	Bearing outside diameter	3.5427 to 3.5433	None
3	Bearing journal diameter	1.5748 to 1.5752	None
4	Pilot diameter	0.7465 to 0.7475	0.020
5	Spline width	0.229 to 0.231	0.015

FRAME 2

NOTE

Readings must be within limits given in table 2-4. The letter L indicates a loose fit and the letter T indicates a tight fit.   If readings are not within given limits, throw away part and get a new one.

1.   Measure input shaft pilot bearing bore (1).

2.   Measure diameter of 14 pilot roller bearings (2).

3.   Measure fit of bearing (3) on bearing journal (4).

END OF TASK

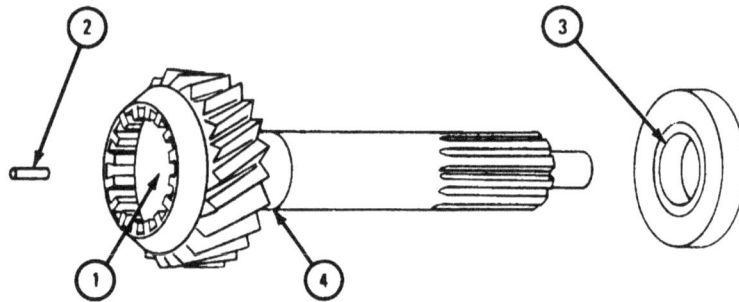

TA 087847

Table 2-4.   Input Shaft Assembly Fits and Tolerances

Index Number	Item/Point of Measurement	Size and Fit of New Parts (inches)	Wear Limit (inches)
1	Input shaft pilot bearing bore	1.7193 to 1.7198	0.0008
2	Roller bearing diameter	0.3123 to 0.3127	0.0006
3 and 4	Fit of bearing on shaft	0.005L to 0.0004T	None

b. Mainshaft Assembly.

FRAME 1

NOTE

Readings must be within limits given   in table 2-5. If readings are not within given limits,   throw away part and get a new one.

1. Measure thickness of thrust washer (1).

2. Measure fifth gear bore (2).

3. Measure sleeve outside diameter (3) and inside diameter   (4).

4. Measure third gear bore (5).

GO TO FRAME 2

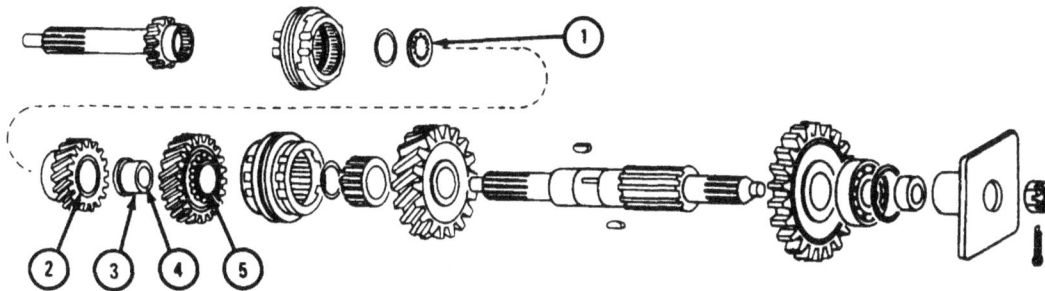

NOTE: CHECK ONLY THOSE PARTS WHICH
        ARE CALLED OUT IN THIS FRAME.
        PARTS WITHOUT CALLOUTS ARE
        SHOWN ONLY FOR REFERENCE
        PURPOSES OR ARE CHECKED IN
        ANOTHER FRAME.

TA 087848

Table 2-5.   Mainshaft Fifth and Third Gear Wear Limits

Index Number	Item/Point of Measurement	Size and Fit of New Parts (inches)	Wear Limit (inches)
1	Thrust washer thickness	0.150 to 0.152	0.010
2	Fifth gear bore	2.1250 to 2.1255	0.001
3	Sleeve outside diameter	2.1205 to 2.1210	0.001
4	Sleeve inside diameter	1.7495 to 1.750	None
5	Third gear bore	1.9680 to 1.9685	0.004

FRAME 2

NOTE

Readings must be within limits given in table 2-6. If readings are not within given limits, throw away part and get a new one.

1. Measure synchronizer sleeve bore (1).

2. Measure second gear bore (2).

3. Measure diameter of mainshaft journal (3).

4. Measure diameter of mainshaft journal (4).

5. Measure diameter of mainshaft journal (5).

GO TO FRAME 3

NOTE
CHECK ONLY THOSE PARTS WHICH ARE CALLED OUT IN THIS FRAME. PARTS WITHOUT CALLOUTS ARE SHOWN ONLY FOR REFERENCE PURPOSES OR ARE CHECKED IN ANOTHER FRAME.

TA 087849

Table 2-6. Mainshaft Journals and Second Gear Wear limits

Index Number	Item/Point of Measurement	Size and Fit of New Parts (inches)	Wear Limit (inches)
1	Synchronizer sleeve bore	1.9680 to 1.9685	None
2	Second gear bore	2.0780 to 2.0785	None
3	Mainshaft journal diameter	1.0908 to 1.0913	0.001
4	Mainshaft journal diameter	1.7500 to 1.7505	None
5	Mainshaft journal diameter	1.9640 to 1.9645	0.004

FRAME 3

NOTE

Readings must be within limits given in table 2-7. If
readings are not within given limits, throw away part
and get a new one.

1. Measure diameter of mainshaft journal (1).

2. Measure diameter of mainshaft journal (2).

3. Measure diameter of mainshaft journal (3).

4. Measure bearing inside diameter (4).

GO TO FRAME 4

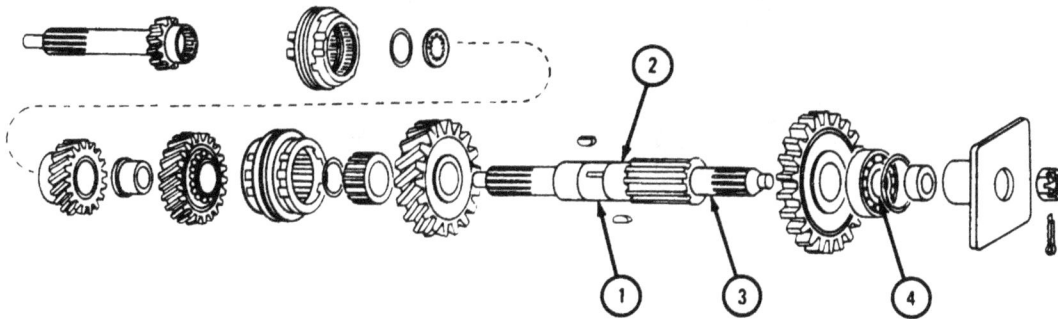

NOTE
CHECK ONLY THOSE PARTS WHICH ARE CALLED OUT IN
THIS FRAME. PARTS WITHOUT CALLOUTS ARE SHOWN
ONLY FOR REFERENCE PURPOSES OR ARE CHECKED IN
ANOTHER FRAME.

TA 087850

Table 2-7. Mainshaft Journals and Bearing Wear Limits

Index Number	Item/Point of Measurement	Size and Fit of New Parts (inches)	Wear Limit (inches)
1	Mainshaft journal diameter	1.9688 to 1.9698	None
2	Mainshaft journal diameter	2.0740 to 2.0745	None
3	Mainshaft journal diameter	1.5746 to 1.5750	None
4	Bearing inside diameter	1.5748	None

FRAME 4

NOTE

Readings must be within limits given in table 2-8. The letter L indicates a loose fit and the letter T indicates a tight fit.   If readings are not within given limits, throw away part and get a new one.

1. Measure fit of internal gear (1) on synchronizer (2).

2. Measure fit of synchronizer (2) on mainshaft splines (3).

3. Measure fit of fifth gear (4) on sleeve (5).

4. Measure fit of sleeve bore (6) on mainshaft splines (3).

5. Measure fit of third gear (7) on mainshaft journal (8).
GO TO FRAME 5

NOTE
CHECK ONLY THOSE PARTS WHICH ARE CALLED OUT IN THIS FRAME. PARTS WITHOUT CALLOUTS ARE SHOWN ONLY FOR REFERENCE PURPOSES OR ARE CHECKED IN ANOTHER FRAME.

TA 087851

Table 2-8. Mainshaft Fifth and Third Gear Fits and Tolerances

Index Number	Item/Point of Measurement	Size and Fit of New Parts (inches)	Wear Limit (inches)
1 and 2	Fit of internal gear on synchronizer	0.004 to 0.009	0.018
2 and 3	Fit of synchronizer on mainshaft splines	0.0041 to 0.0076	0.010L
4 and 5	Fit of fifth gear on sleeve	0.004L to 0.005L	0.008L
6 and 3	Fit of sleeve bore on mainshaft splines	0.000 to 0.001T	None
7 and 8	Fit of third gear on mainshaft journal	0.0035L to 0.0045L	0.010L

FRAME 5

NOTE

Readings must be within limits given in table 2-9. The letter L indicates a loose fit and the letter T indicates a tight fit.   If readings are not within given limits, throw away part and get a new one.

1. Measure fit of synchronizer (1) on sleeve splines (2).

2. Measure fit of sleeve bore (3) on mainshaft journal (4).

3. Measure fit of second gear (5) on mainshaft journal (6).

4. Measure fit of first-reverse gear (7) on mainshaft splines (8).

5. Measure fit of bearing (9) on mainshaft splines (10).

END OF TASK

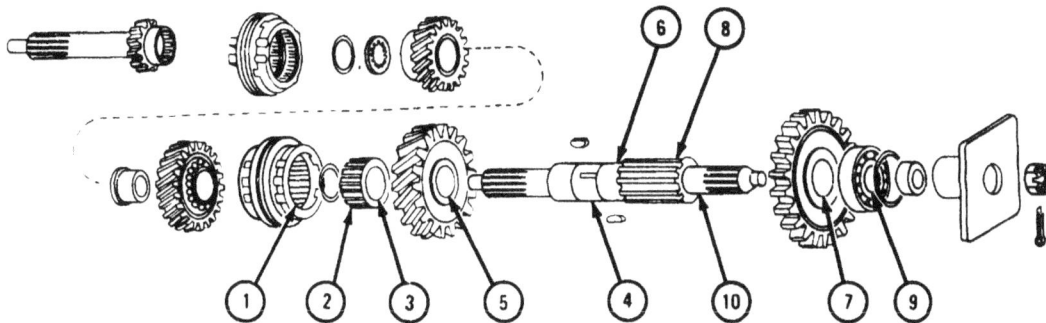

NOTE
CHECK ONLY THOSE PARTS WHICH ARE CALLED OUT IN THIS FRAME. PARTS WITHOUT CALLOUTS ARE SHOWN ONLY FOR REFERENCE PURPOSES OR ARE CHECKED IN ANOTHER FRAME.

TA 087852

Table 2-9.   Mainshaft Second and First-Reverse Gear Fits and Tolerances

Index Number	Item/Point of Measurement	Size and Fit of New Parts (inches)	Wear Limit (inches)
1 and 2	Fit of synchronizer on sleeve splines	0.004L to 0.007L	0.0106L
3 and 4	Fit of sleeve bore on main-shaft journal	0.003T to 0.018T	None
5 and 6	Fit of second gear on main-shaft journal	0.0035L to 0.0045L	0.010L
7 and 8	Fit of first-reverse gear on mainshaft splines	0.006L to 0.012L	0.016L
9 and 10	Fit of bearing on mainshaft splines	0.0002L to 0.0002T	None

c. Countershaft Assembly.

FRAME 1

NOTE

Readings must be within limits given in table 2-10. If readings are not within given limits, throw away part and get a new one.

1. Measure inside diameter of bearing (1).
2. Measure thickness of thrust washer (2)
3. Measure drive gear bore (3).
4. Measure fifth gear bore (4).
5. Measure diameter of countershaft journal (5).

GO TO FRAME 2

NOTE: CHECK ONLY THOSE PARTS WHICH ARE CALLED OUT IN THIS FRAME. PARTS WITHOUT CALLOUTS ARE SHOWN ONLY FOR REFERENCE PURPOSES OR ARE CHECKED IN ANOTHER FRAME.

TA 087853

Table 2-10. Countershaft Drive and Fifth Gear Wear Limits

Index Number	Item/Point of Measurement	Size and Fit of New Parts (inches)	Wear Limit (inches)
1	Bearing inside diameter	1.7332 to 1.7337	0.001
2	Thrust washer thickness	0.0598	0.007
3	Drive gear bore	1.995 to 2.0005	None
4	Fifth gear bore	2.2495 to 2.2505	None
5	Countershaft journal diameter	1.7317 to 1.7322	0.002

FRAME 2

NOTE

Readings must be within limits given in table 2-11. If
readings are not within given limits, throw away part
and get a new one.

1. Measure diameter of countershaft journal (1).

2. Measure diameter of countershaft journal (2).

3. Measure diameter of countershaft journal (3).

4. Measure inside diameter of bearing (4).

GO TO FRAME 3

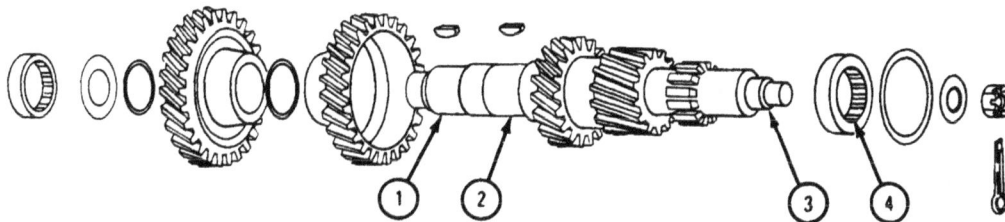

NOTE
CHECK ONLY THOSE PARTS WHICH ARE CALLED OUT IN
THIS FRAME. PARTS WITHOUT CALLOUTS ARE SHOWN
ONLY FOR REFERENCE PURPOSES OR ARE CHECKED IN
ANOTHER FRAME.

TA 087854

2-11. Countershaft Journals and Bearing Wear Limits

Index Number	Item/Point of Measurement	Size and Fit of New Parts (inches)	Wear Limit (inches)
1	Countershaft journal diameter	2.001 to 2.002	None
2	Countershaft journal diameter	2.251 to 2.252	None
3	Countershaft journal diameter	1.3788 to 1.3782	None
4	Bearing inside diameter	1.3780	None

FRAME 3

NOTE

Readings must be within limits given in table 2-12.
The letter L indicates a loose fit and the letter T
indicates a tight fit.   If readings are not within
given limits, throw away part and get a new one.

1.   Measure fit of drive gear (1) on countershaft journal (2).

2.   Measure fit of fifth gear (3) on countershaft journal (4).

3.   Measure fit of bearing (5) on countershaft journal (6).

4.   Measure fit of bearing (7) on countershaft journal (8).

END OF TASK

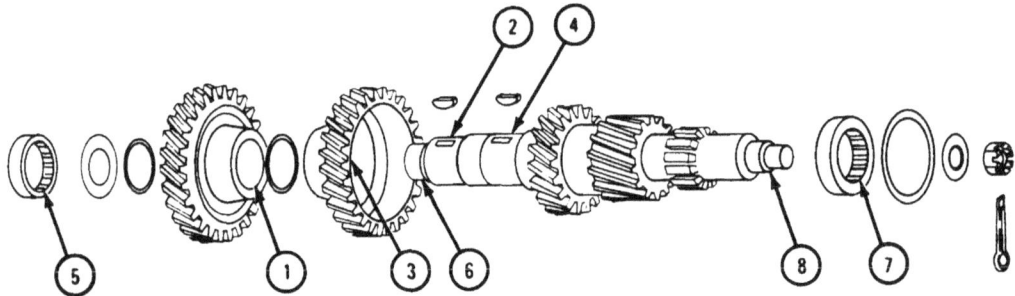

NOTE
CHECK ONLY THOSE PARTS WHICH ARE CALLED OUT IN
THIS FRAME. PARTS WITHOUT CALLOUTS ARE SHOWN
ONLY FOR REFERENCE PURPOSES OR ARE CHECKED IN
ANOTHER FRAME.

TA 087855

Table 2-12.   Countershaft Assembly Fits and Tolerances

Index Number	Item/Point of Measurement	Size and Fit of New Parts (inches)	Wear Limit (inches)
1 and 2	Fit of drive gear on counter-shaft journal	0.0005T to 0.0025T	None
3 and 4	Fit of fifth gear on counter-shaft journal	0.0005T to 0.0025T	None
5 and 6	Fit of bearing on counter-shaft journal	0.0005L to 0.0015L	None
7 and 8	Fit of bearing on counter-shaft journal	0.0002L to 0.0002T	None

d. Reverse Idler Gear Assembly.

FRAME 1

NOTE

Readings must be within limits given in table 2-13. If readings are not within given limits, throw away part and get a new one.

1. Measure diameter of shaft (1).
2. Measure thickness of two thrust washers (2).
3. Measure inside diameter (3) and outside diameter (4) of two bearings.
4. Measure gear bore (5).

GO TO FRAME 2

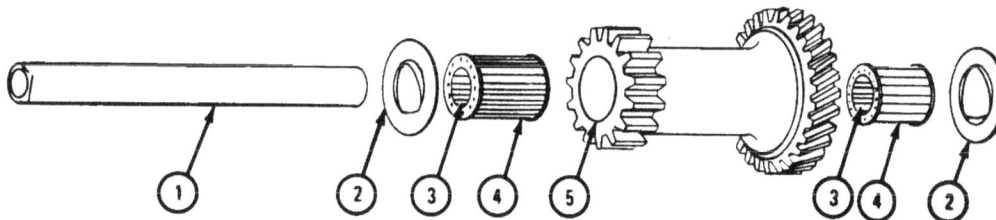

TA 087856

Table 2-13. Reverse Idler Gear Assembly Wear Limits

Index Number	Item/Point of Measurement	Size and Fit of New Parts (inches)	Wear Limit (inches)
1	Shaft diameter	1.010 to 1.015	0.0015
2	Thrust washer thickness	0.091 to 0.093	None
3	Bearing inside diameter	1.0155	None
4	Bearing outside diameter	1.5000	None
5	Gear bore	1.5005 to 1.5015	0.007

FRAME 2

NOTE

Readings must be within limits given in table 2-14. The letter L indicates a loose fit. If readings are not within given limits, throw away part and get a new one.

1. Measure fit of two bearings (1) on shaft (2).

2. Measure fit of two bearings (1) in gear bore (3).

END OF TASK

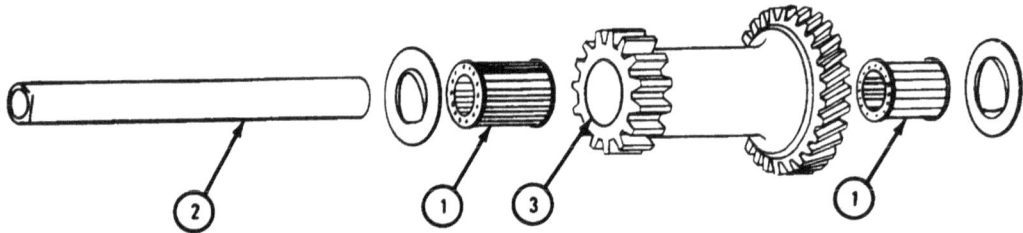

NOTE
CHECK ONLY THOSE PARTS WHICH ARE CALLED OUT IN
THIS FRAME. PARTS WITHOUT CALLOUTS ARE SHOWN
ONLY FOR REFERENCE PURPOSES OR ARE CHECKED IN
ANOTHER FRAME.

TA 087857

Table 2-14. Reverse Idler Gear Assembly Fits and Tolerances

Index Number	Item/Point of Measurement	Size and Fit of New Parts (inches)	Wear Limit (inches)
1 and 2	Fit of bearing on shaft	0.005L to 0.0015L	None
1 and 3	Fit of bearing in gear bore	0.0005L to 0.0015L	0.010

2-9. REPAIR.   This paragraph gives instructions to repair the transmission case, cover, gear shafts, and gears.

    a.   Smooth out any chips, scratches or burrs on gear shafts and gears with a honing stone.

    b.   Weld cracks and small holes in housing and cover castings. Refer to TM 9-237.

    c.   Drill out any bolts or studs broken off in tapped holes.

    d.   Drill out threaded holes that are stripped or out-of-round to the next larger size and retap them. When putting transmission together, use a bolt or stud the size of the newly tapped hole.

2-10. ASSEMBLY OF SUBASSEMBLIES. The following paragraphs from dust and assemble the transmission subassemblies.

NOTE

Keep all parts clean and protected from dust and dirt. Coat all bearings with multipurpose lubricant during assembly. Coat all oil seals with engine lubricating oil during assembly. Coat shafts and bores of gears with white lead pigment during assembly. Use new seals and snaprings during assembly.

a. Input Shaft Assembly.

FRAME 1

1. Set up input shaft (1) in hydraulic press as shown.
2. Press bearing (2) into place.
3. Take input shaft (1) out of press.

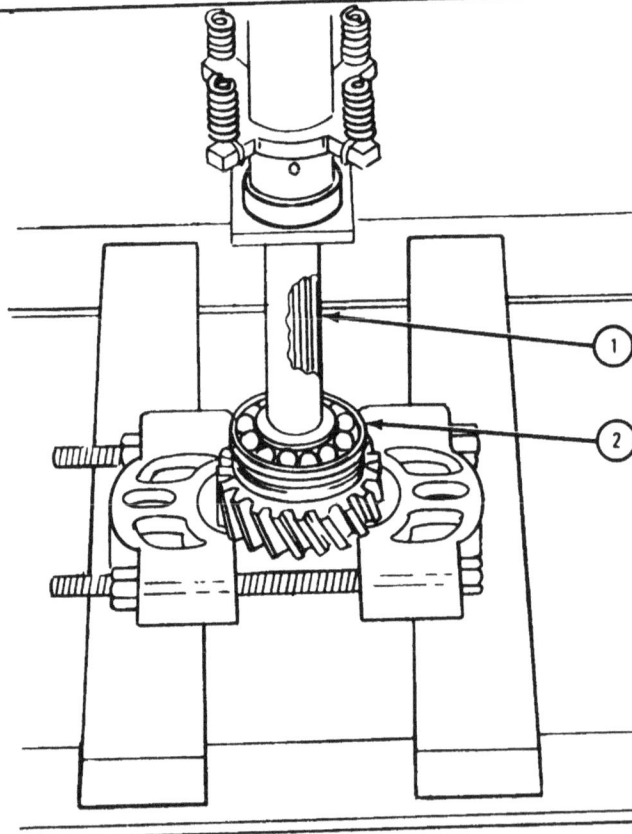

GO TO FRAME 2

TA 088100

FRAME 2

1. Put on snapring (1) as shown.

END OF TASK

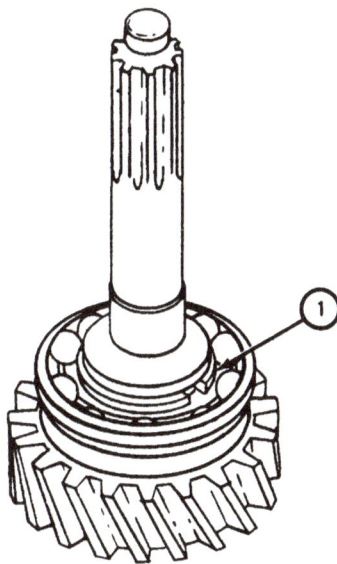

TA 087858

b. Mainshaft Assembly.

FRAME 1

1. Put in key (1).

2. Put second speed gear (2) on mainshaft (3) with collar of gear facing down.
    Aline keyway in second speed gear with key (1) on mainshaft.

GO TO FRAME 2

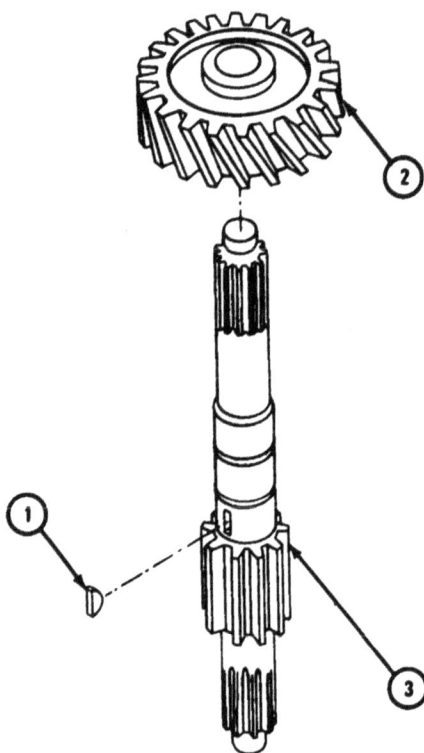

TA 087859

FRAME 2

1.   Set up hydraulic press as shown.
2.   Press second speed gear (1) into place.
3.   Take mainshaft assembly (2) out of press.
GO TO FRAME 3

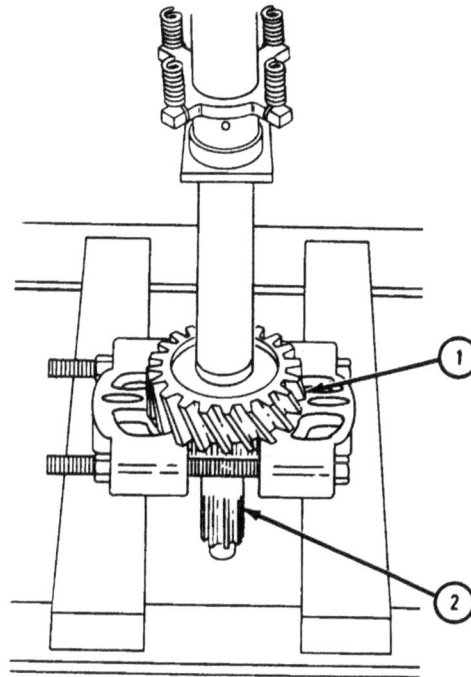

TA 087860

FRAME 3

1. Put in key (1).

2. Aline keyway in second and third speed synchronizer sleeve (2) with key (1) on mainshaft (3) and put on second and third speed synchronizer sleeve.

GO TO FRAME 4

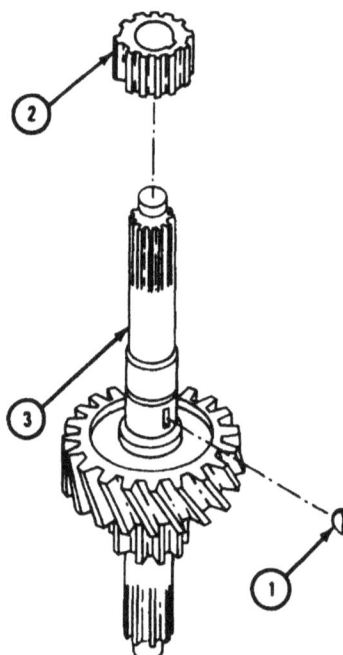

TA 087861

FRAME 4

1. Set up hydraulic press as shown.

2. Press second and third speed synchronizer sleeve (1)into place on mainshaft (2).

3. Take mainshaft (2) out of press.

GO TO FRAME 5

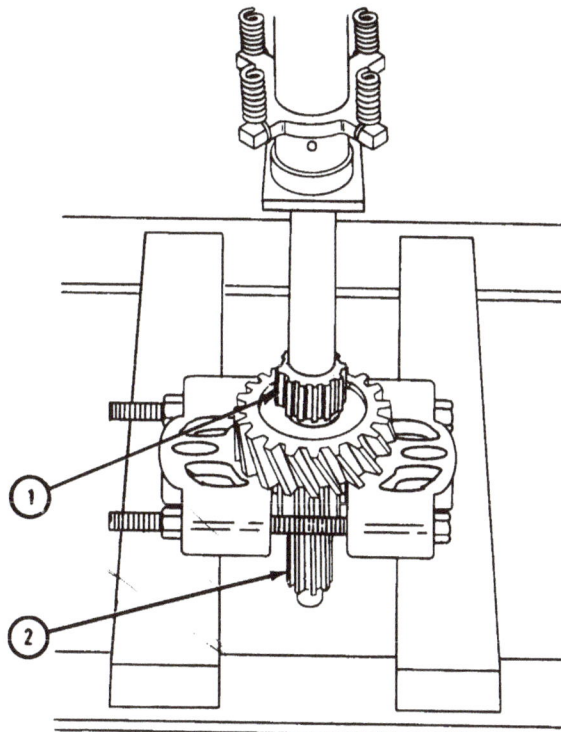

TA 087862

FRAME 5

1. Put on snapring (1).
2. Slide on second and third gear synchronizer (2).

GO TO FRAME 6

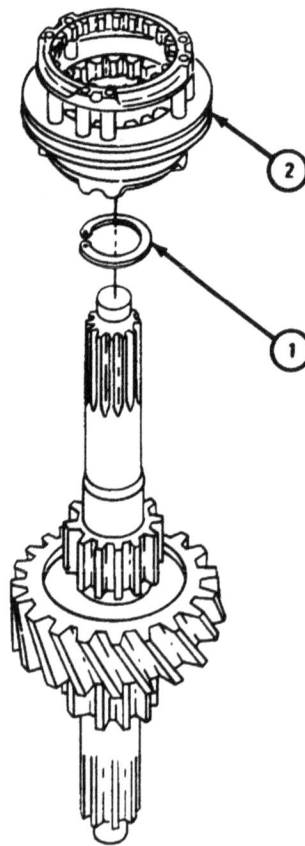

TA 087863

FRAME 6

1. Put third speed gear (1) on mainshaft (2). Set up hydraulic press as shown.
2. Press third speed gear (1) into place.

GO TO FRAME 7

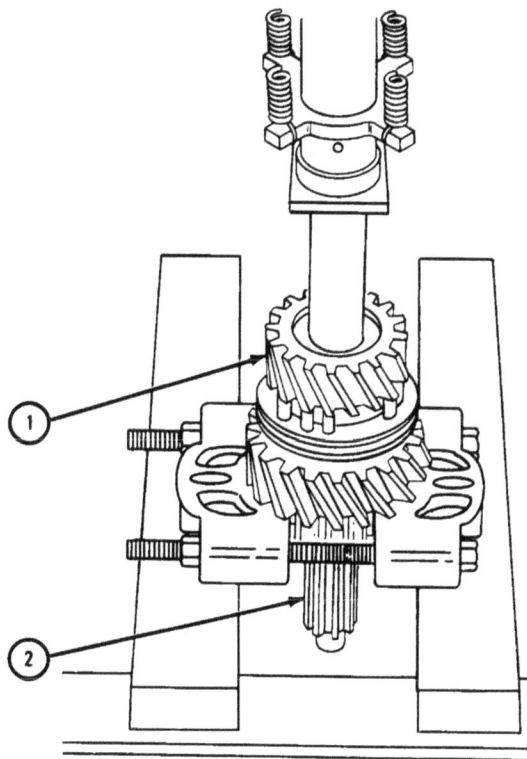

TA 087864

FRAME 7

1. Set up hydraulic press as shown.

2. Press on fifth speed gear sleeve (1).

GO TO FRAME 8

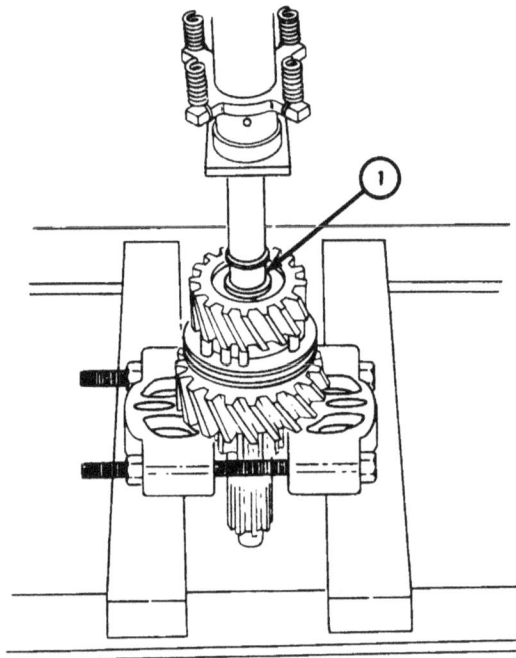

TA 087865

FRAME 8

1. Put on fifth speed gear (1).
2. Put on thrust washer (2).
3. Put on snapring (3).

GO TO FRAME 9

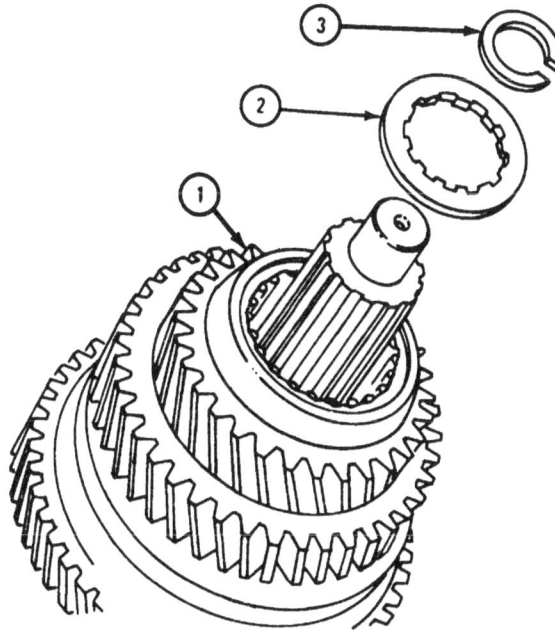

TA 087866

FRAME 9

1. Put on fourth and fifth speed synchronizer (1).
2. Take mainshaft (2) out of press.

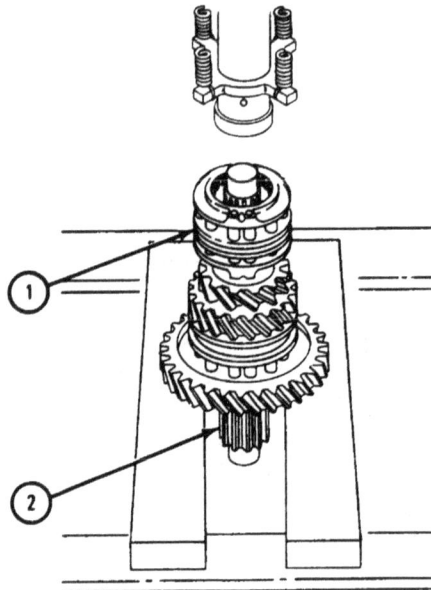

END OF TASK

TA 087867

c. <u>Countershaft Assembly.</u>

FRAME 1

1.   Put key (1) in countershaft (2).
2.   Put fourth speed gear (3) on countershaft (2) with collar of gear facing up.
     Aline keyway in fourth speed gear with key (1) in countershaft.

GO TO FRAME 2

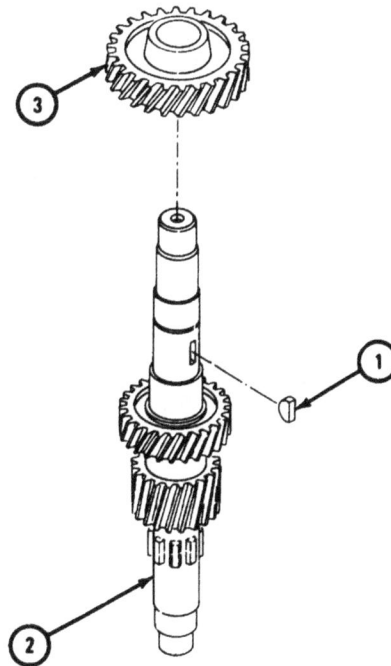

TA 087868

FRAME 2

1.   Set up hydraulic press as shown.

2.   Press fourth speed gear (1) into place.

GO TO FRAME 3

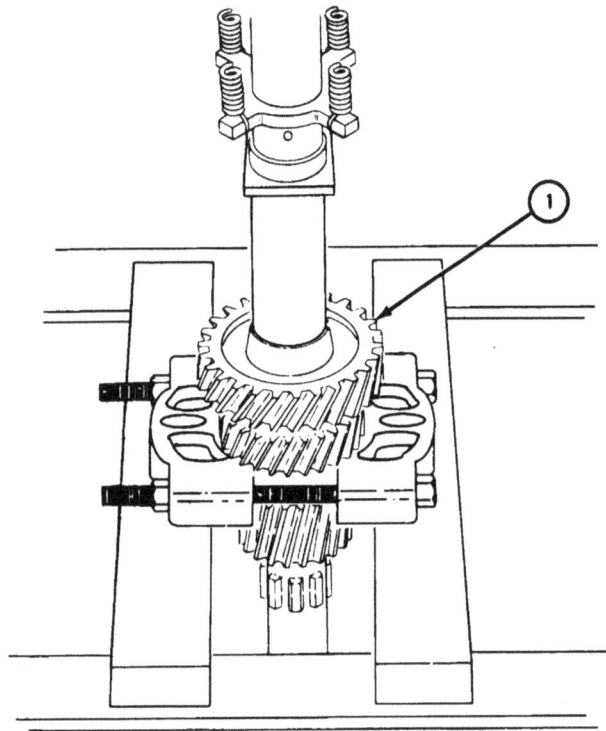

TA 087869

FRAME 3

1. Put on snapring (1).

2. Put in key (2) in countershaft (3) as shown.

3. Put drive gear (4) on countershaft (3) with collar of gear facing down.
    Aline keyway in drive gear with key (2).

GO TO FRAME 4

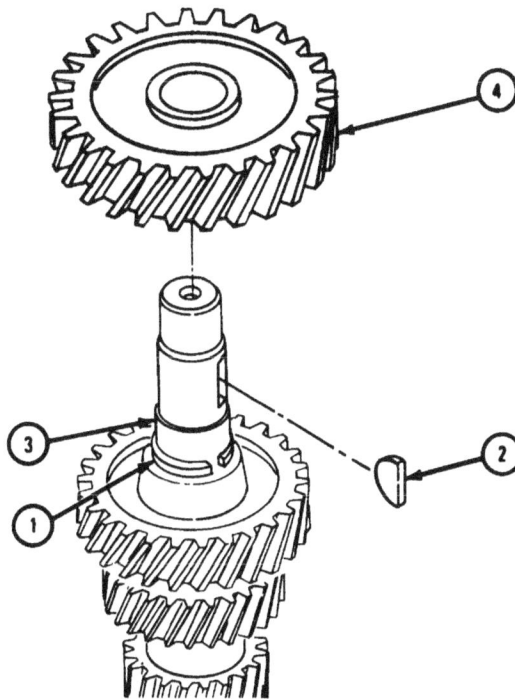

TA 087870

FRAME 4

1. Set up hydraulic press as shown.
2. Press drive gear (1) into place.

GO TO FRAME 5

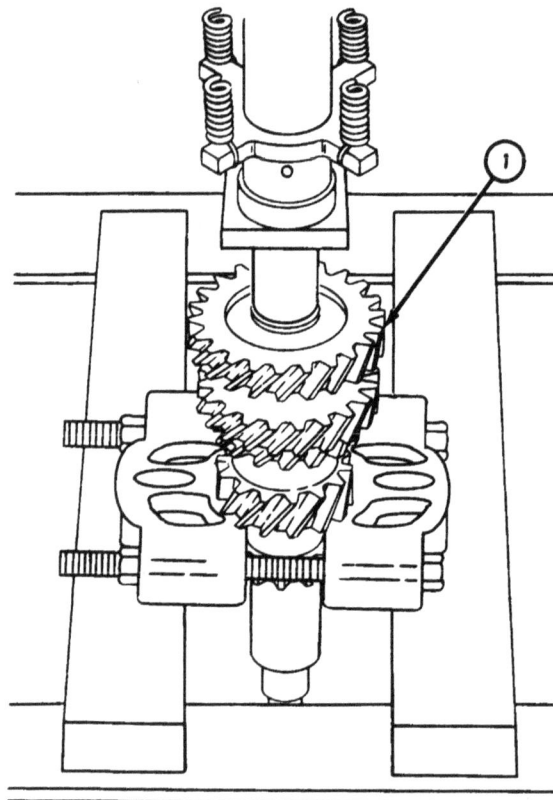

TA 087871

FRAME 5

1.   Put on snapring (1).

END OF TASK

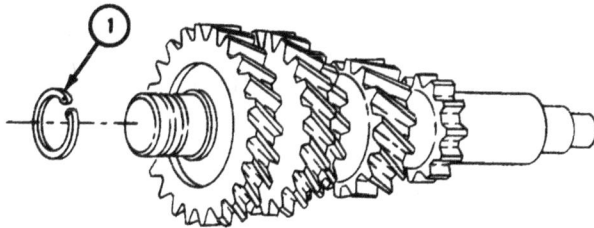

TA 087872

d. <u>Reverse Idler Gear Assembly.</u>

FRAME 1

1. Put in two bearing assemblies (1), one in each end of reverse idler gear
    assembly (2).

END OF TASK

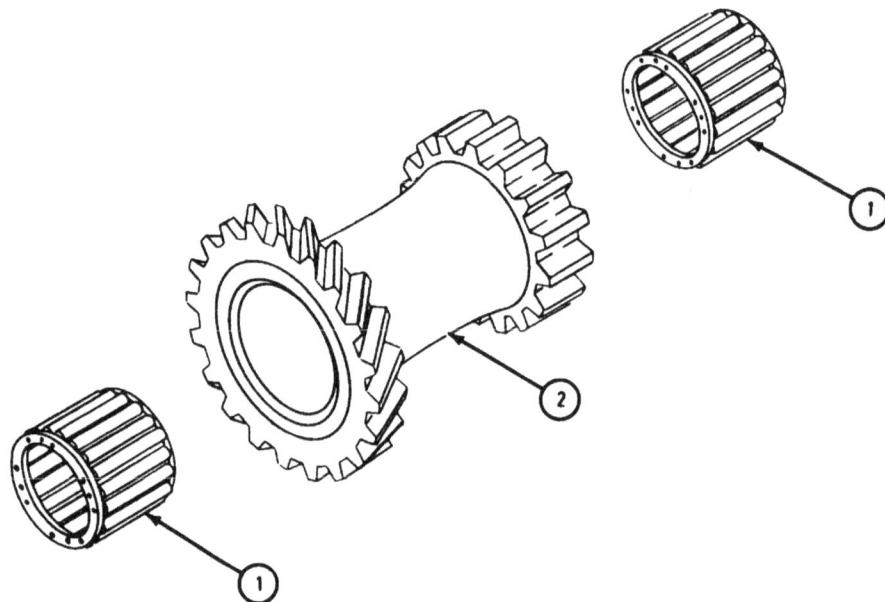

TA 087873

---

e. Mainshaft Rear Bearing Cap.

FRAME 1

1. Using hammer and wood block, tap oil seal (1) into bearing cap (2).
END OF TASK

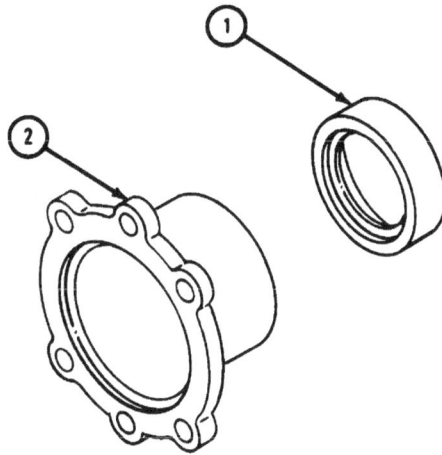

TA 087874

---

f.   Shifter Shaft Cover.

FRAME 1

1.   Put shifter shaft lever (1) in place as shown.
2.   Put on spring cup (2), spring (3), and lockring (4).
GO TO FRAME 2

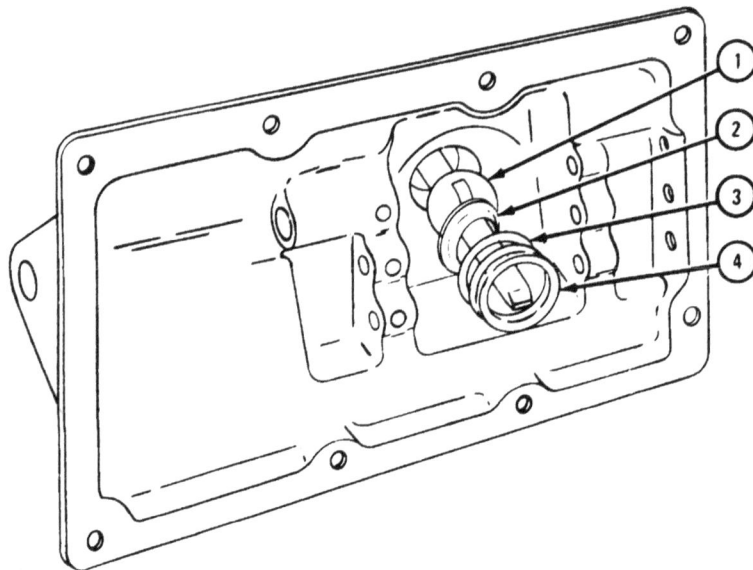

TA 087875

FRAME 2

1. Using hammer and brass drift, tap in first and reverse shifter shaft oil seal (1).

2. Put plate (2) in place, making sure that shift lever tab (3) goes through center slot of plate.

3. Put six washers (4) and two springs (5) in place as shown.

4. Put in two screws (6) through cover assembly (7), six washers springs (5).

5. Turn over cover assembly (7).

GO TO FRAME 3

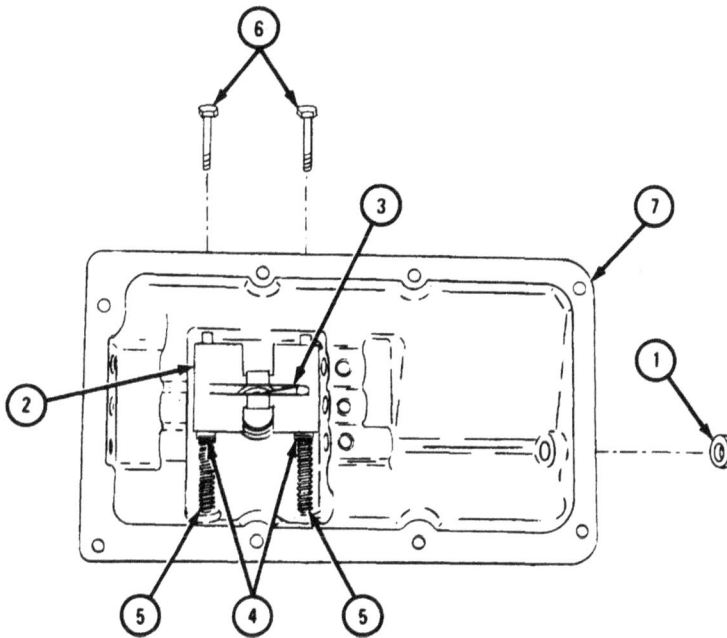

TA 087876

FRAME 3

1. Put on two washers (1) and two interlock safety nuts (2). Tighten nuts to 40 to 50 pound-feet.

2. Turn over cover assembly (3).

GO TO FRAME 4

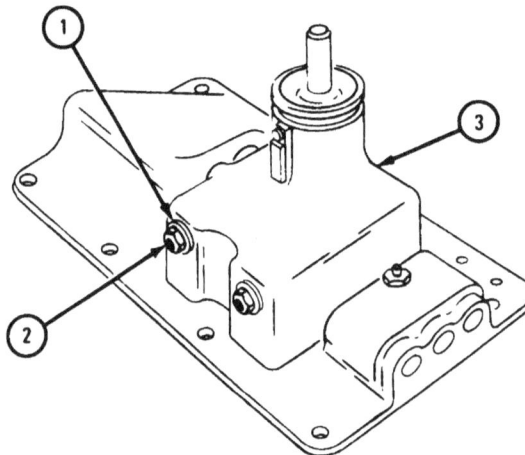

TA 087877

FRAME 4

1. Slide fourth and fifth speed shifter shaft (1) through hole in cover (2) as tagged. Leave hole (3) uncovered.

2. Slide fork (4) onto shaft (1) as shown.

3. Put spring (5) and ball (6) into hole (3). Using screwdriver, hold ball down. Slide shaft (1) over ball and into place.

4. Ane setscrew hole in fork (4) with setscrew hole in shaft (1). Make sure tab on fork sets in slot in plate (7). Put in setscrew (8).

5. Using hammer and drift, tap in expansion plug (9).

6. Do steps 1 through 5 again for second and third speed shifter shaft (10).

GO TO FRAME 5

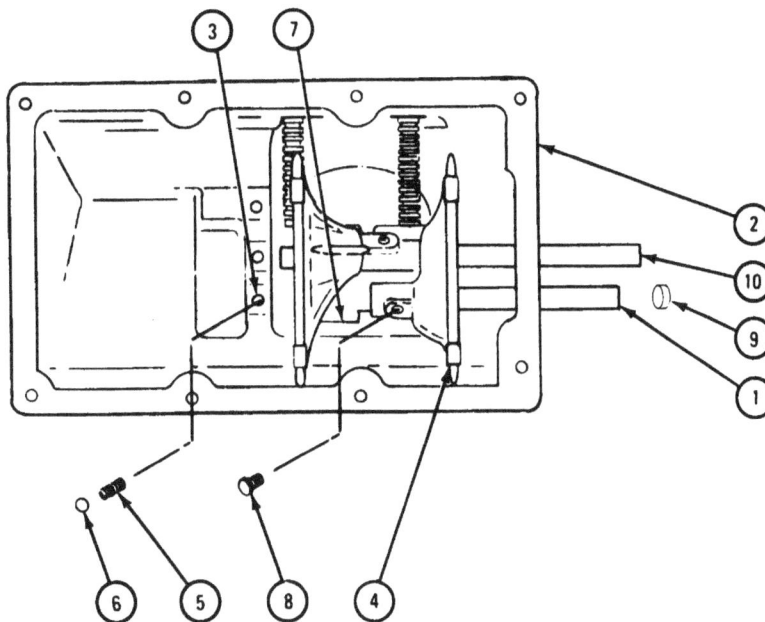

TA 087878

FRAME 5

1. Put bracket (1) into place in cover (2).

2. Slide first and reverse speed shifter shaft (3) through hole in cover (2) and bracket (1). Leave hole (4) uncovered. Make   sure that tab on bracket sets in slot in plate (5).

3. Put spring (6) and ball (7) into hole (4). Using   screwdriver, hold ball down. Slide shaft (3) over ball.

4. Loosely put in setscrew (8).

GO TO FRAME 6

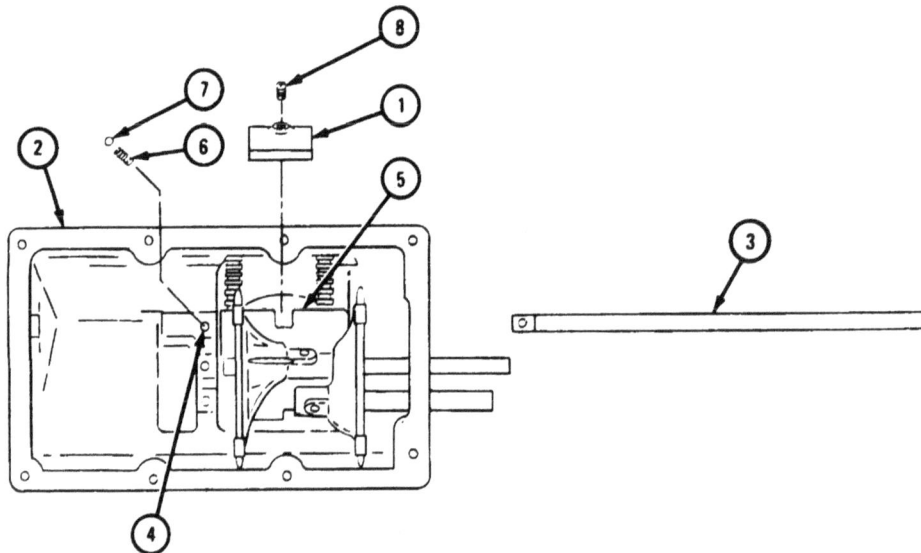

TA 087879

FRAME 6

1. Put in fork (1) and hold it in place.

2. Tap first and reverse speed shifter   shaft (2) through fork (1) and into place.

3. Aline setscrew hole in fork (1) with   setscrew hole in shaft (2). Put in setscrew (3). Tighten setscrews (3 and 4).

4. Put on four safety wires (5).

5. Using hammer and drift, tap in expansion plug (6).

END OF TASK

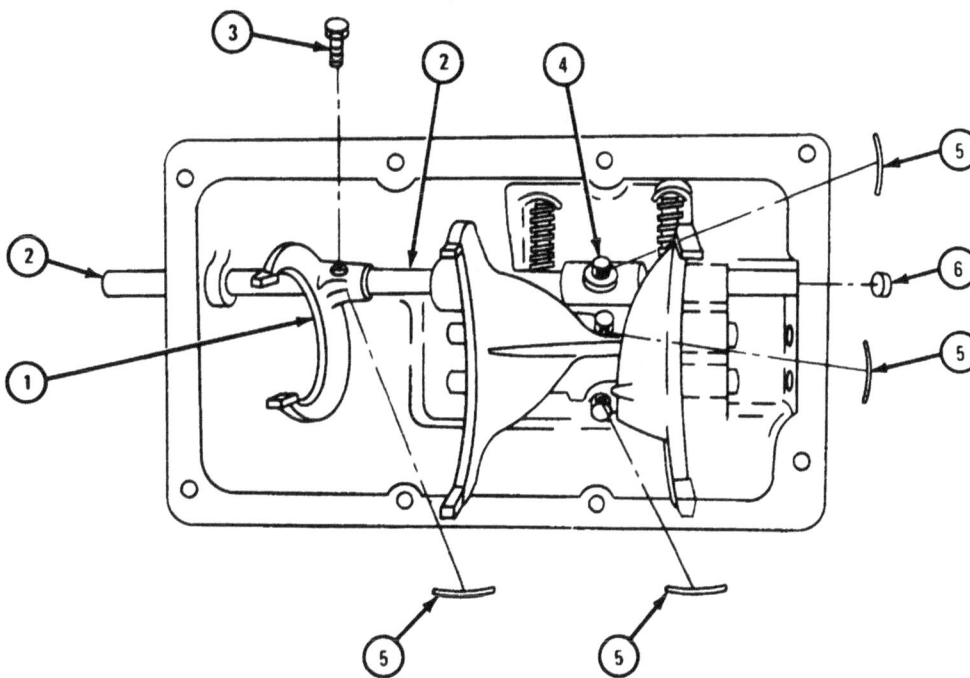

TA 087880

g. Clutch Release Bearing Carrier.

FRAME 1

1. Set up hydraulic press as shown.
2. Press bearing (1) onto carrier (2).
3. Put in lubrication fitting (3) and two clutch bearing buttons (4).
END OF TASK

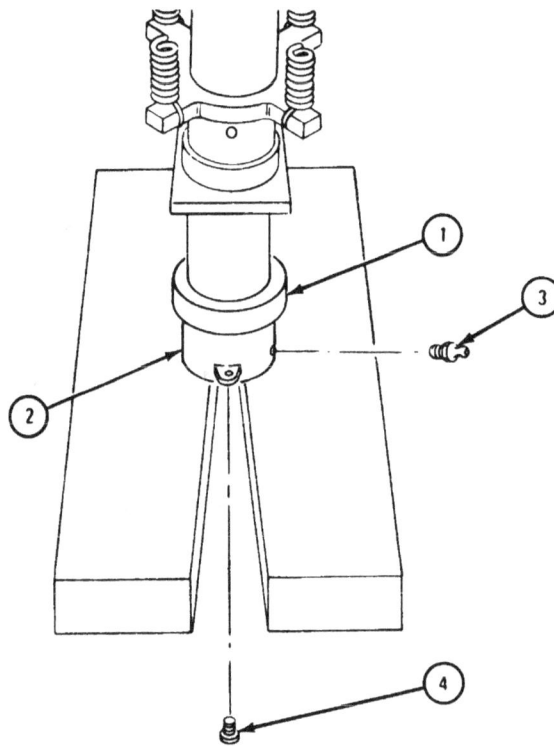

TA 087881

2-11. FINAL ASSEMBLY. The following paragraphs give instructions to assemble the transmission subassemblies into a final assembly.

NOTE

Keep all parts clean and protected from dust and dirt. Coat all oil seals, gears, and shafts with engine lubricating oil during assembly. Use new snaprings and gaskets during assembly.

a. Reverse Idler Gear Assembly.

FRAME 1	

1. Put grease on back of two thrust washers (1).

2. Put two thrust washers (1) in transmission case (2) as shown. Grease will hold two thrust washers in place.

3. Put reverse idler gear assembly (3) between two thrust washers (1) and hold it in place.

NOTE

Make sure milled side of reverse idler gear shaft (4) faces outside transmission case (2).

4. Using soft face hammer, tap in reverse idler gear shaft (4).

END OF TASK

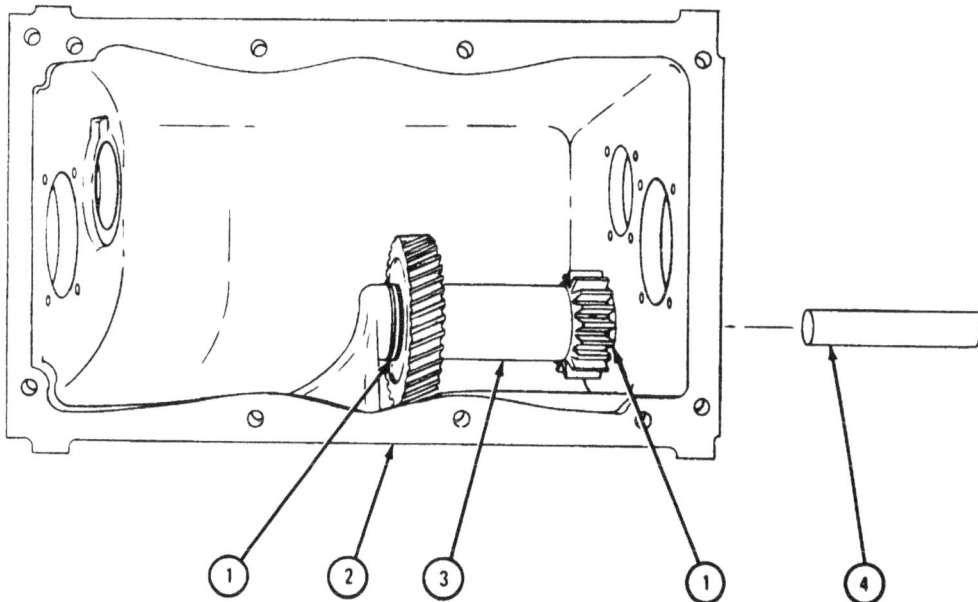

TA 087882

b.   Countershaft Assembly.

FRAME 1

1.   Using hammer and brass drift and working from inside of transmission case (1), tap in countershaft front bearing (2).

2.   Working from outside of transmission case (1) and using hammer and brass drift, tap in bearing retainer (3).

GO TO FRAME 2

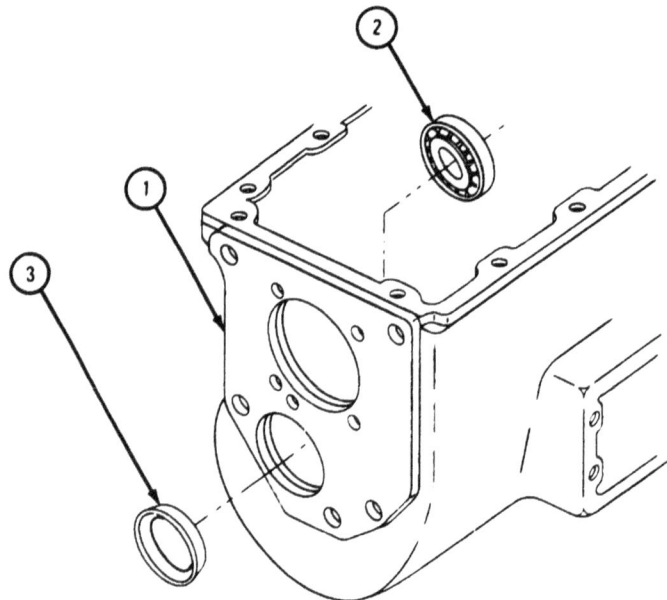

TA 087883

FRAME 2

1. Put thrust washer (1) on end of countershaft assembly (2).
2. Put countershaft assembly (2) into transmission case (3) as shown.
GO TO FRAME 3

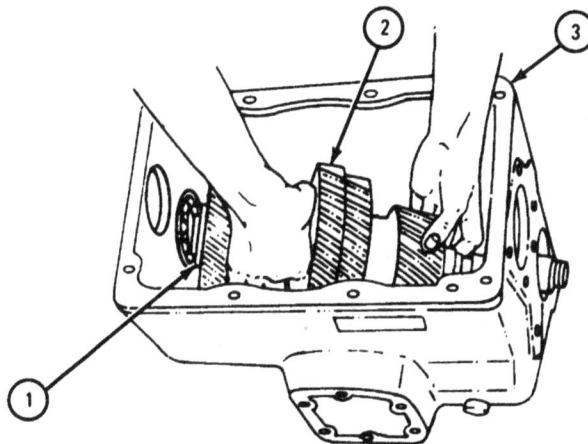

TA 087884

FRAME 3

1. Put snapring (1) on bearing (2).

2. Aline inner race of bearing (2) with countershaft (3).

3. Using hammer and brass drift, tap bearing (2) evenly into place until snap-ring (1) sets against transmission case (4).

END OF TASK

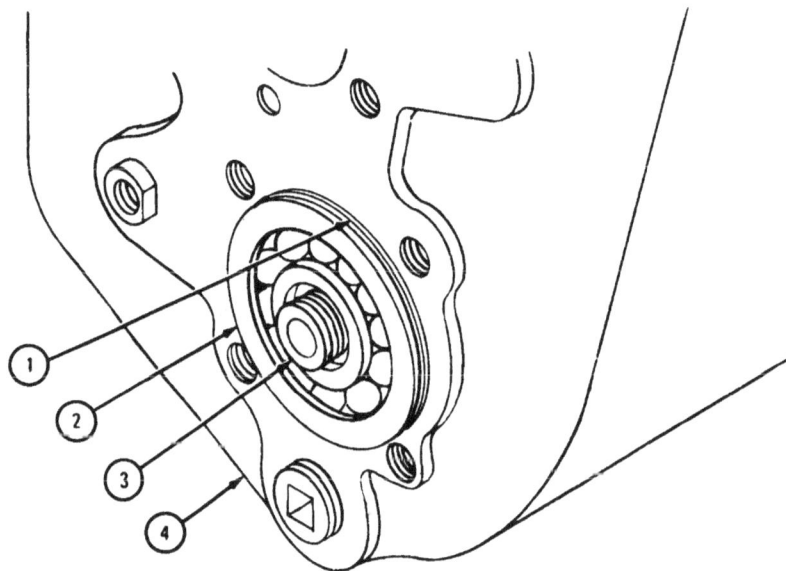

TA 087885

c.　Input Shaft Assembly.

FRAME 1

1.　Pack grease into bearing race of input shaft (1).

2.　Using rubber hammer and working from inside transmission case (2), tap input shaft (1) into place as shown.

3.　Put 14 roller bearings (3) inside bearing race of input shaft (1) as shown.

GO TO FRAME 2

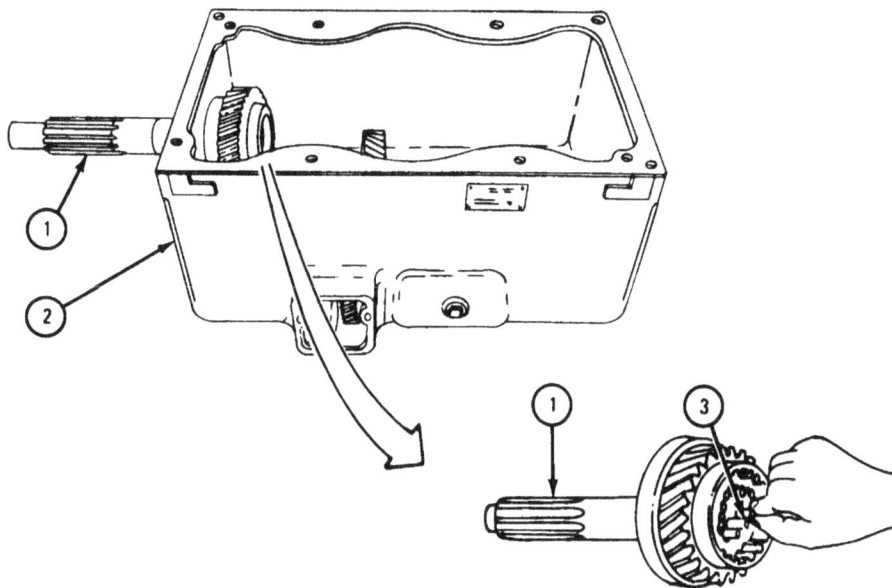

TA 087886

FRAME 2

1. Put snapring (1) on input shaft bearing (2) as shown.

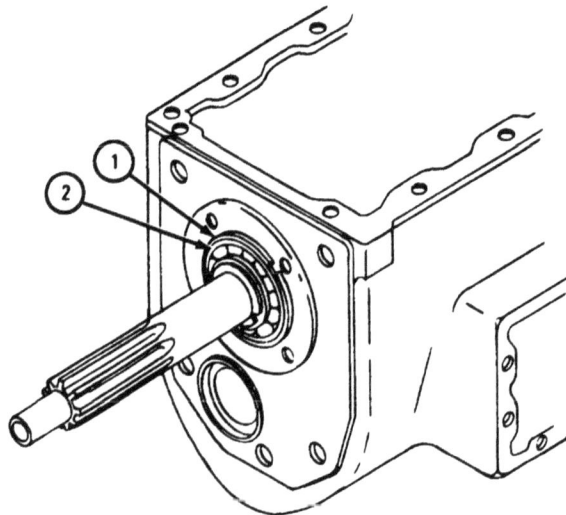

END OF TASK

TA 087887

d.   Mainshaft Assembly.

FRAME 1

1.   Put first and reverse gear (1) into back of transmission case (2) so that grooved collar on first and reverse gear faces front of transmission case.

2.   Holding both ends of mainshaft assembly (3) as shown, slide mainshaft assembly through first and reverse gear (1).

3.   Let mainshaft assembly (3) rest on countershaft assembly in transmission case (2).

GO TO FRAME 2

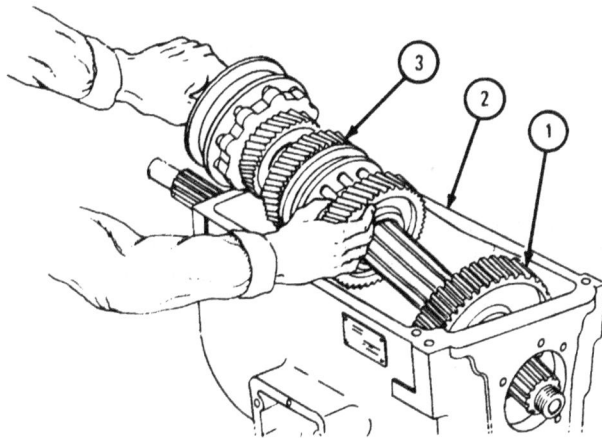

TA 087888

FRAME 2

CAUTION

Make sure that all 14 pilot roller bearings are in place
in input shaft assembly (1). If they are not in place,
transmission will be damaged.

1. Slide mainshaft assembly (2) toward input shaft assembly (1) and fit bearing
    journal (3) into middle of 14 pilot roller bearings in input shaft assembly.

2. Push mainshaft assembly (2) all the way into input shaft assembly (1).

GO TO FRAME 3

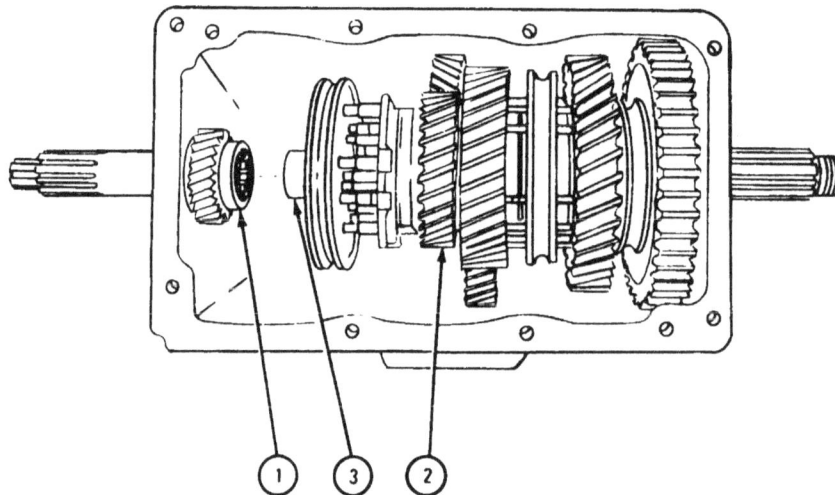

TA 087889

FRAME 3

1. Hold mainshaft assembly (1) in place. Using soft faced hammer, tap on rear
   bearing (2) until bearing is seated in transmission case (3).

END OF TASK

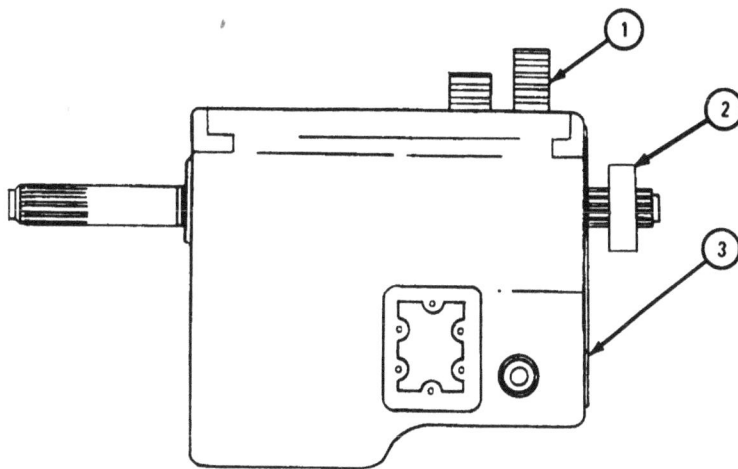

TA 087890

e. <u>Input Shaft Bearing Cap.</u>

FRAME 1

NOTE

There is an oil return channel in bearing   cap (1) and an
oil return hole in gasket (2). They must   be alined with
oil return hole in transmission case (3).

1. Put on gasket (2) and bearing cap (1), alining holes.

2. Put in four capscrews (4) and four lockwashers (5). Tighten capscrews
    evenly to 13 to 17 pound-feet.

END OF TASK

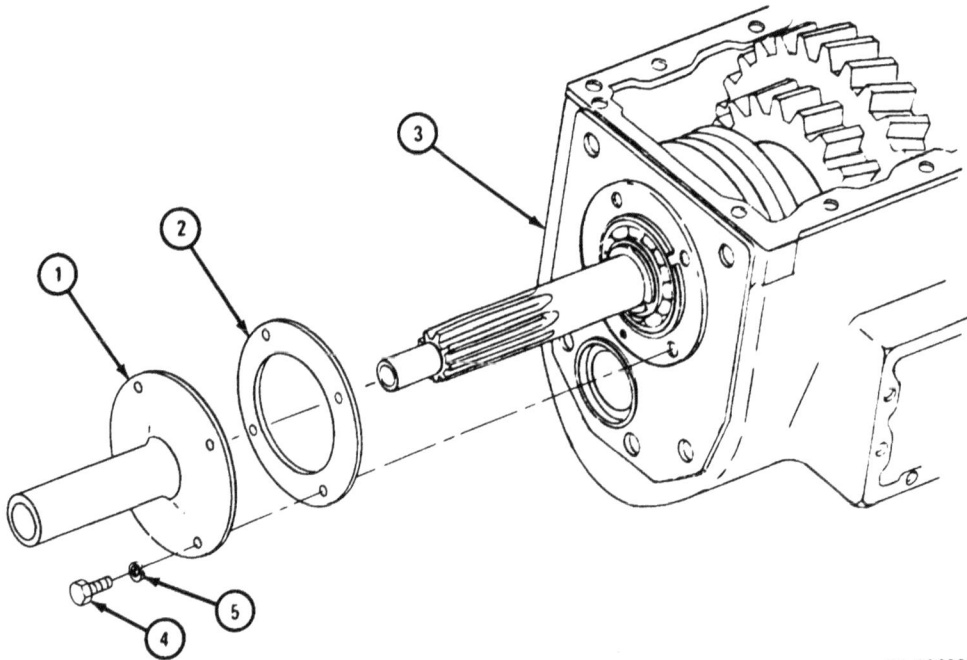

TA 087891

---

f.   Power Takeoff Access Cover (Trucks Without Power Takeoff Installed).

FRAME 1

1.   Put on gasket (1) and cover (2).

2.   Put in six capscrews (3) with six lockwashers (4). Tighten capscrews evenly to
     10 to 15 pound-feet.

END OF TASK

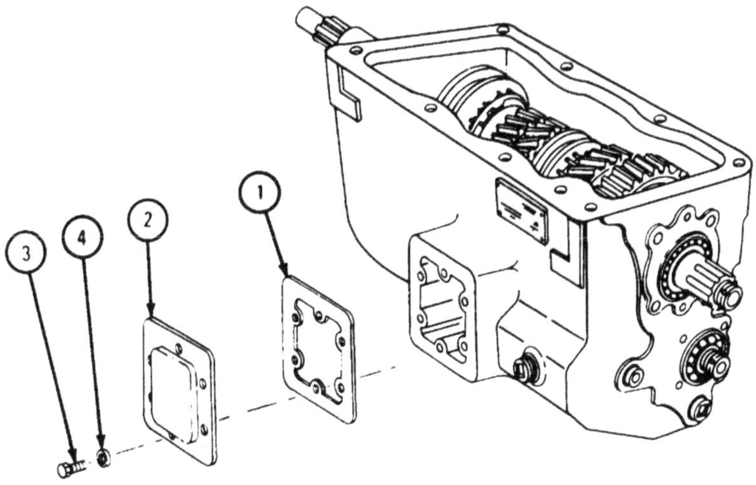

TA 087892

g. Countershaft Rear Bearing and Cover.

FRAME 1

NOTE

If synchronizer (1) and first and reverse gear (2) will not mesh, turn input shaft (3) while meshing synchronizer or gear.

1. Push second and third speed synchronizer (1) to the front to mesh.
2. Push first and reverse gear (2) all the way to rear to mesh.

GO TO FRAME 2

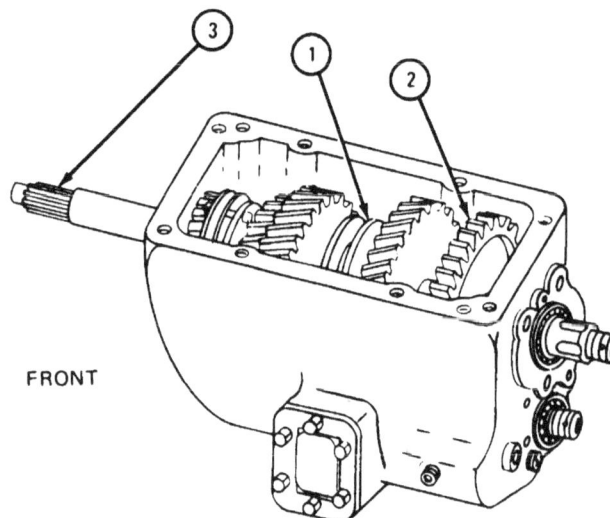

FRONT

FRAME 2

1. Put on washer (1) and slotted nut (2).   Tighten nut to 120 to 150 pound-feet.
    If slot in nut does not aline with hole in shaft (3), tighten nut until it does.

2. Put in cotter pin (4).

GO TO FRAME 3

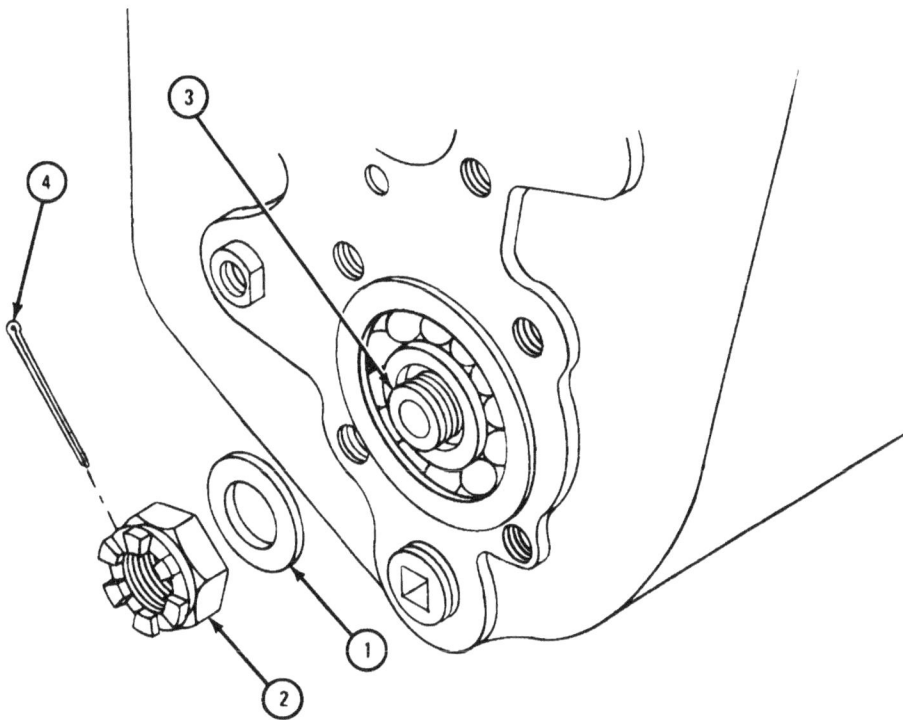

TA 087893

FRAME 3

1. Aline idler shaft (1) so that flat milled side is facing countershaft rear bearing (2).
2. Put on gasket (3) and cover (4).
3. Put in four capscrews (5) with four lockwashers (6). Evenly tighten capscrews to 25 pound-feet.
4. Put on lever (7), washer (8), and locknut (9).

END OF TASK

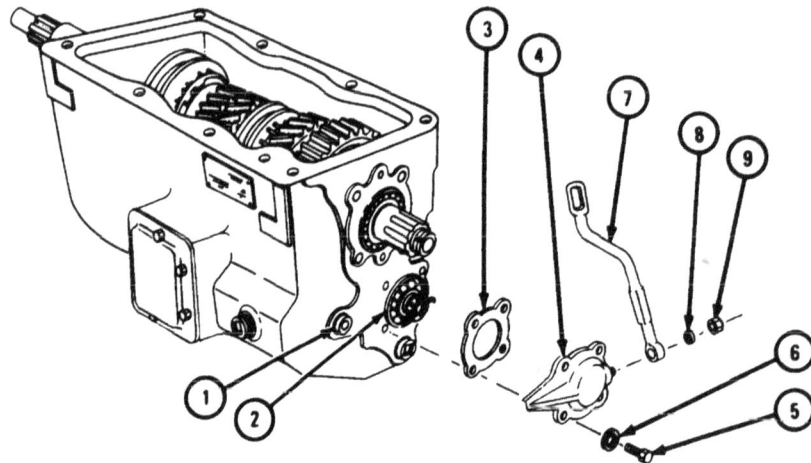

TA 087894

h. <u>Mainshaft Rear Bearing Cap.</u>

FRAME 1

1. Put on thrust washer (1), gasket (2), and cover (3). Aline all screw holes.
2. Put in four capscrews (4) with four washers (5). Evenly tighten capscrews to 25 to 32 pound-feet.

END OF TASK

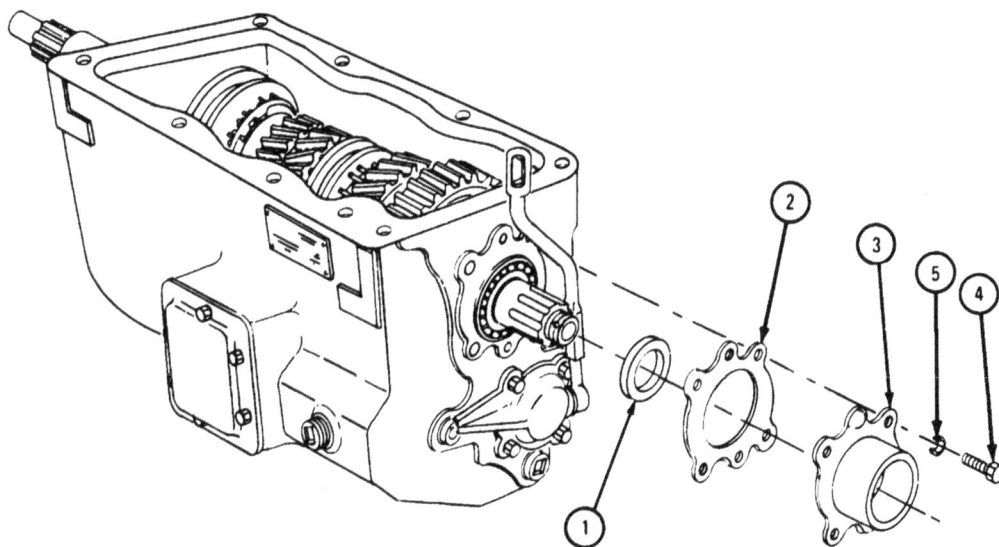

TA 087895

i. <u>Companion Flange</u>

FRAME 1

1. Using hammer and brass drift, tap on companion flange (1).

2. Put on washer (2) and slotted nut (3).   Tighten nut to 120 to 150 pound-feet.
    If slot in nut does not aline with hole in shaft (4), tighten nut until it does.

3. Put in cotter pin (5).

END OF TASK

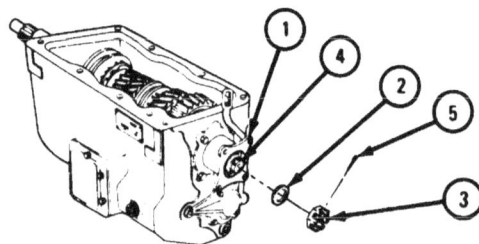

TA 087896

---

j. <u>Shifter Shaft Cover.</u>

FRAME 1

NOTE

Transmission is in neutral position when mainshaft (1)
turns and input shaft (2) does not turn.

1. Push second and third speed synchronizer (3) towards center of transmission
    to take it out of mesh.

2. Push first and reverse gear (4) to front so first and reverse gear does not
    mesh with reverse idler gear in transmission.

3. Do backlash check.   Refer to para 2-4b.

4. If backlash check is within given limits, go to frame 2.

5. If backlash check is not within given limits, disassemble transmission and do
    wear limit inspection.   Refer to para 2-4 and para 2-8.

GO TO FRAME 2

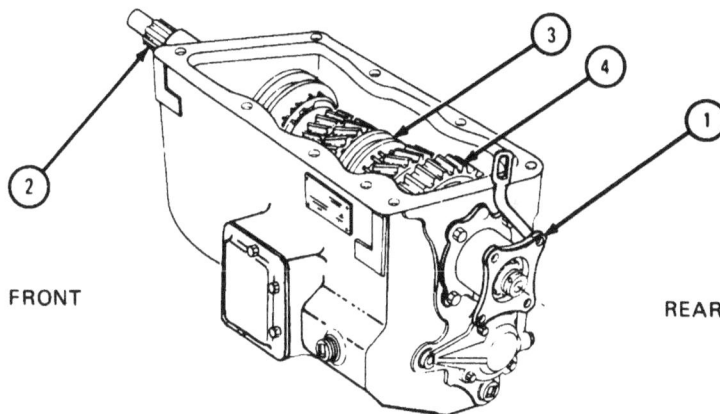

FRONT

REAR

TA 087900

FRAME 2

1. Put on gasket (1) and aline all screw holes.

2. Put shifter shaft (2) in neutral position. Refer to TM 9-2320-209-10.

3. Put on shifter shaft cover (3), making sure three forks (4) go into synchronizer grooves (5) and groove in first and reverse gear (6).

4. Put in eight capscrews (7) with eight lockwashers (8). Evenly tighten cap screws to 25 to 32 pound-feet.

END OF TASK

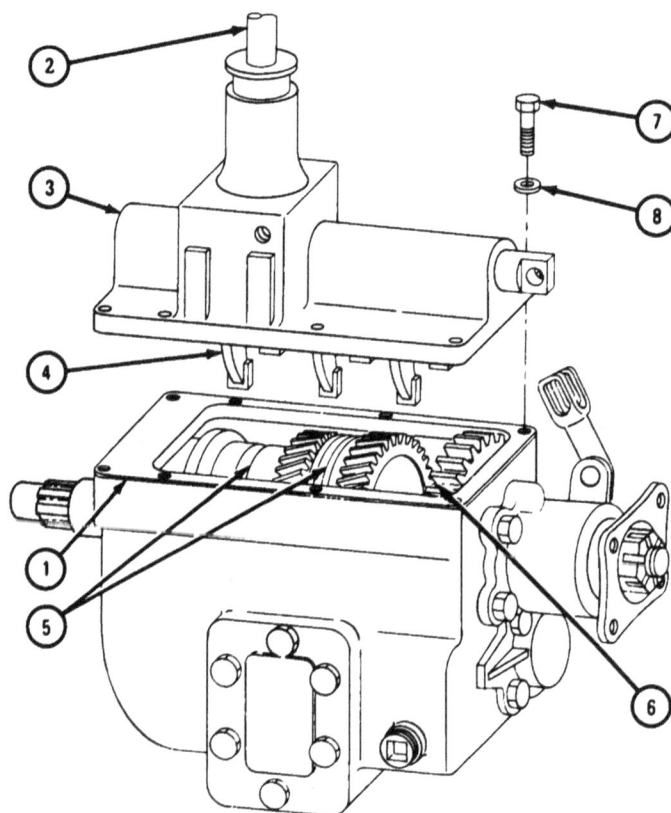

TA 087901

k. Clutch Release Assembly.

FRAME 1

1. Put on gasket (1) and cover (2) and aline all screw holes. Put in six capscrews (3).
2. Put on gasket (4).
3. Put on clutch housing (5) and aline all screw holes.
4. Put in five capscrews (6) with lockwashers (7).   Evenly tighten capscrews to 60 to 80 pound-feet.

GO TO FRAME 2

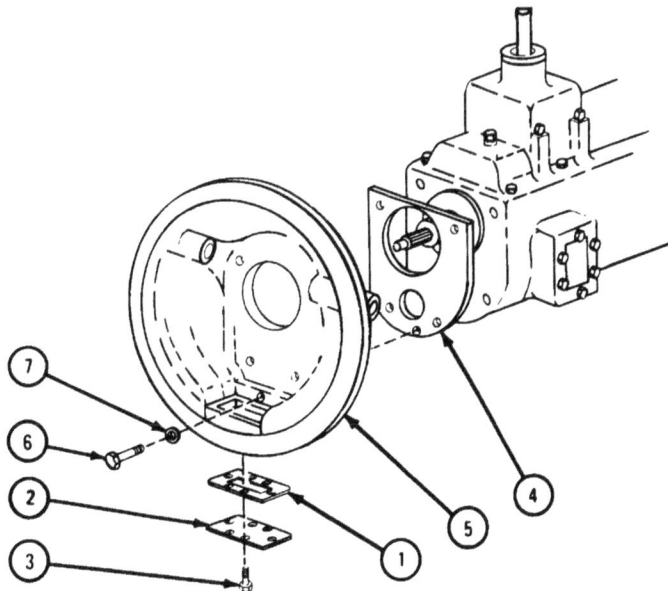

TA 087897

FRAME 2

1. Hold release fork (1) in place.

2. Put lever shaft (2) through clutch housing (3) and release fork (1) just enough so two keys (4) can be put in keyways in lever shaft.

3. Put in two keys (4). Push lever shaft (2) into clutch housing (3) until holes in two keys line up with two screw holes in release fork (1).

4. Put in two capscrews (5) with two lockwashers (6). Evenly tighten cap screws to 40 to 50 pound-feet.

GO TO FRAME 3

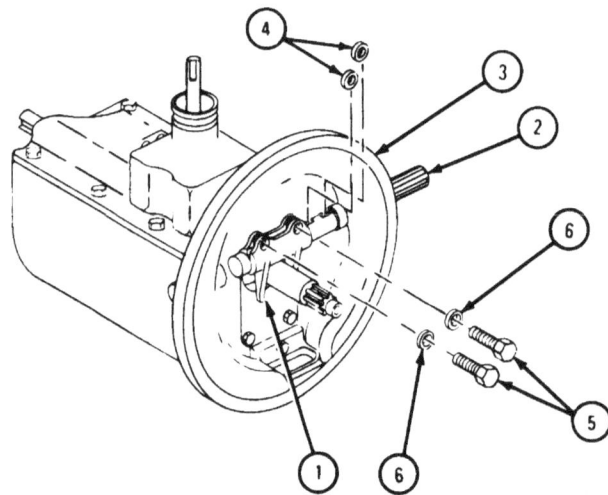

TA 087898

FRAME 3

1. Put on release bearing carrier assembly (1).
2. Hook in two support springs (2).

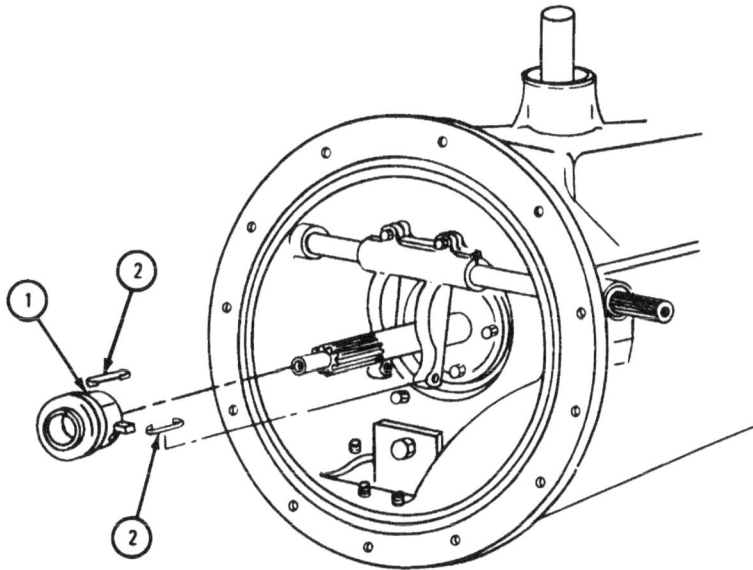

END OF TASK

TA 087899

2-12. SHIFT TEST.   The following paragraphs give instructions to test the transmission for smooth and positive shifting in all ranges after final assembly.

NOTE

Before making tests, fill transmission assembly with 1/2 pint of gear oil.   Refer to LO 9-2320-209-12/1. Attach a tag to filler plug saying transmission must be filled   after it is put back in truck.

a.   Neutral Position.

FRAME 1

1.   Put gear shift lever (1) on shifter shaft (2).

2.   Put in capscrew (3), washer (4), and nut (5).

GO TO FRAME 2

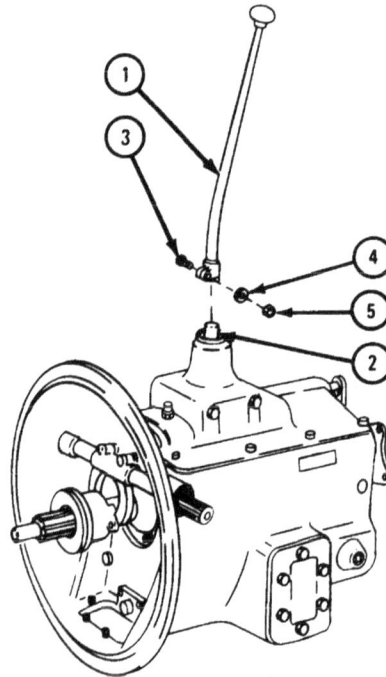

TA 102063

FRAME 2

1. Move gear shift lever (1) to N (neutral) position as shown.

2. Hold output shaft flange (2) and turn input shaft (3). Output shaft flange should not turn.

3. If output shaft flange (2) turns or if input shaft (3) does not turn freely, do the following:

   a. Remove shifter shaft cover.   Refer to para 2-4a.

   b. Disassemble shifter shaft cover.   Refer to para 2-5f.

   c. Check all parts for wear or damage. Refer to para 2-7.

   d. Assemble shifter shaft cover.   Refer to para 2-10f.

   e. Replace shifter shaft cover. Refer to para 2-11j.

END OF TASK

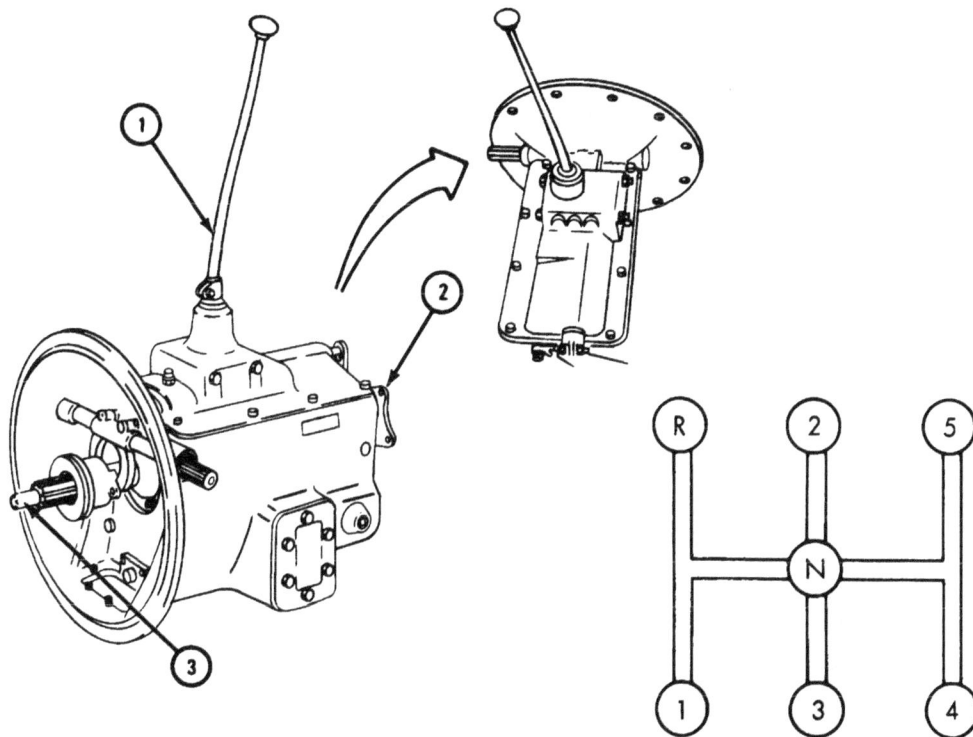

TA 102064

b. Forward Speed Positions.

NOTE

If transmission will not shift into gear easily turn
input or output shafts while shifting.

FRAME 1

1. Move gear shift lever (1) to position 1 as shown.

2. Turn input shaft (2). Output shaft flange (3) should turn.

3. Move gear shift lever (1) to positions 2, 3, 4, and 5 and do step 2 again for each position.

4. If output shaft flange (3) does not turn in each shift position, do the following

    a. Remove shifter shaft cover.   Refer to para 2-4a.

    b. Disassemble shifter shaft cover.   Refer to para 2-5f.

    c. Assemble shifter shaft cover.   Refer to para 2-10f.

    d. Replace shifter shaft cover.   Refer to para 2-11j.

END OF TASK

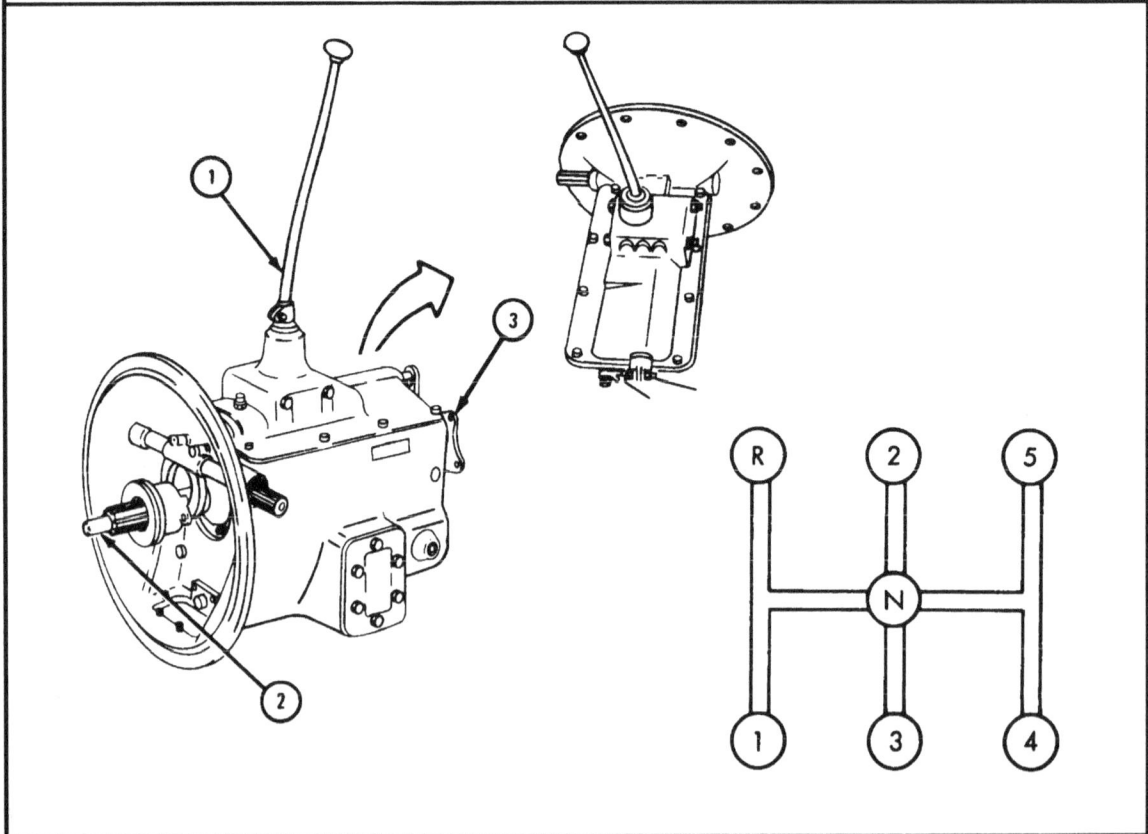

c. <u>Reverse Position.</u>

NOTE

If transmission will not shift into gear easily, turn
input or output shafts while shifting.

FRAME 1

1. Move gear shift lever (1) to R (reverse) position as shown.

2. Turn input shaft (2). Output shaft flange (3) should turn in the opposite direction
   of input shaft.

3. If output shaft flange (3) does not turn, do the following:

   a. Remove shifter shaft cover. Refer to para 2-4a.

   b. Disassemble shifter shaft cover.   Refer to para 2-5f.

   c. Assemble shifter shaft cover.   Refer to para 2-10f.

   d. Replace shifter shaft cover. Refer to para 2-11j.

GO TO FRAME 2

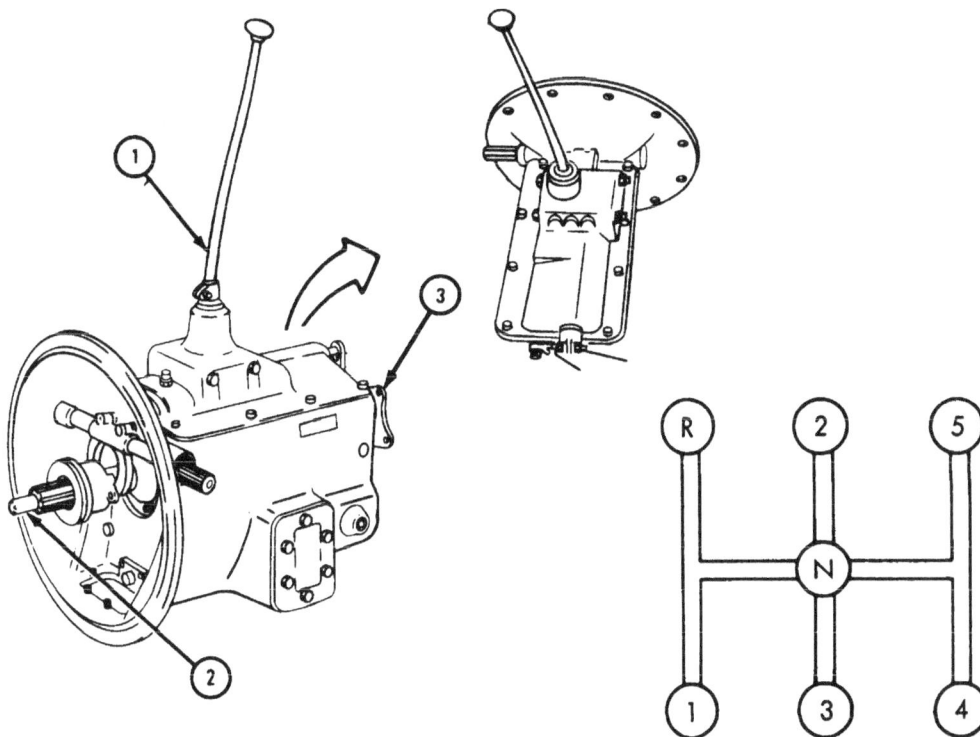

TA 102065

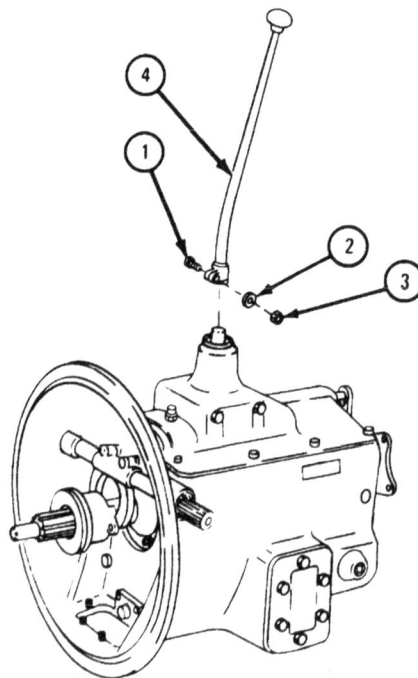TM 9-2520-246-34-1

FRAME 2

1. Take out capscrew (1), washer (2), and nut (3).
2. Take off gear shift lever (4).

END OF TASK

2-112

Section III. MAINTENANCE OF TRANSMISSION TRANSFER ASSEMBLY

NOTE

Procedures given are for model T-136-27 transmission transfer and are the same for model
T-136-21 except where noted.

TOOLS: Idler shaft front bearing cup remover and replacer, pn 7083247
Idler shaft front bearing cup remover and replacer handle, pn 7083241
Mechanical puller kit, pn 8708724
Power train rebuild tool kit, pn 7950356

SUPPLIES: Dry cleaning solvent,   type II (SD-2), Fed. Spec P-D-680
Compressed air source, 30 psi max
Multipurpose lubricant, GO 85/140, MIL-L-2105
Lubricating oil, ICE, OE/HDO 10, MIL-L-2104
Safety wire, MS20995F47
Power transfer gasket and shim set
Transfer air cylinder kit (model T-136-27)
Bearing oil seal (3)
Shifter shaft seal
Front output shaft shifter shaft seal (model T-136-21)

PERSONNEL: Two

EQUIPMENT CONDITION: Transmission transfer assembly removed from truck.

2-13. CLEANING BEFORE DISASSEMBLY. This paragraph gives instructions for cleaning
the transmission transfer assembly before disassembly. Note and scribe transmission
transfer case and covers in places that have oil soaked road mud. It is not necessary to
mark oil spots around gaskets or seals since new gaskets and seals will be put in. Scrape,
brush, and steam clean all dirt and road mud from transmission transfer assembly.

2-14. DISASSEMBLY OF TRANSMISSION TRANSFER INTO SUBASSEMBLIES. The
following paragraphs give instructions to mount and disassemble the transmission transfer
assembly into subassemblies.

a. Mounting Transmission Transfer Assembly in Stand.

FRAME 1

1. Using chain hoist, lower transmission transfer assembly (1) into stand (2).

2. Tighten four bolts (3), two on each side, to hold transmission transfer assembly (1) in place on stand (2). Take off chain hoist.

3. Use handle (4) to turn transmission transfer assembly (1) as needed to do procedures.

END OF TASK

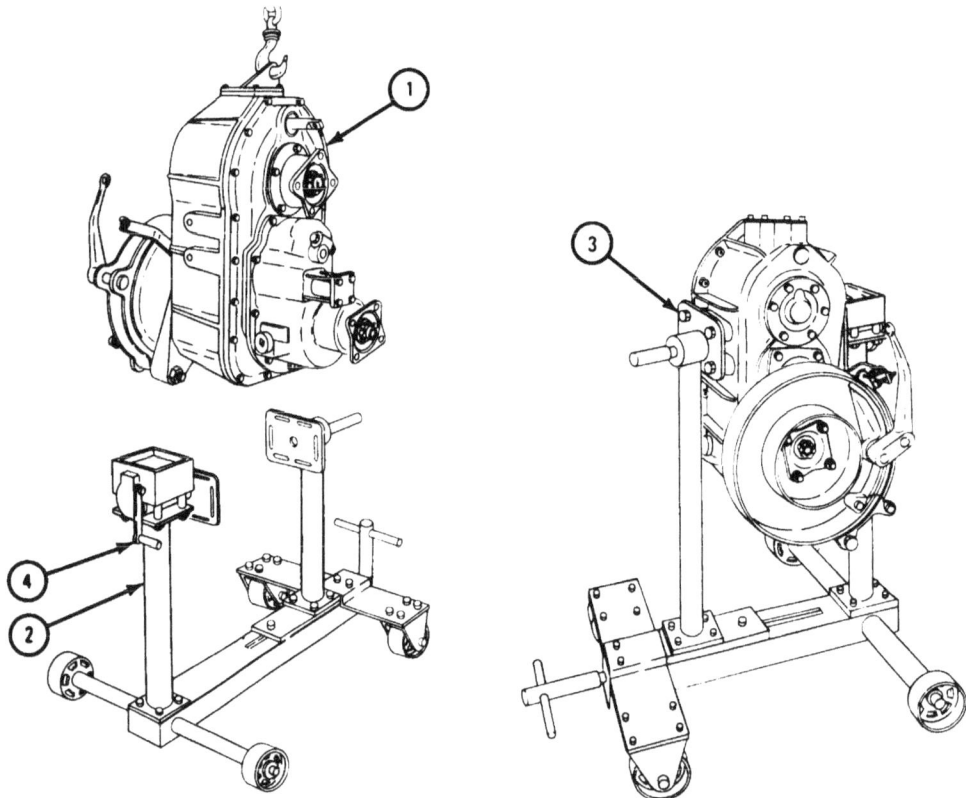

TA 101001

b. Companion Flanges.

FRAME 1	

Soldier A 1. Push in shifter shaft (1).

2. Take out two cotter pins (2 and 3).

3. Using adjustable wrench, hold companion flange (4) as shown.

Soldier B 4. Take off slotted nut (5) and washer (6).

5. Working at rear of transfer case (7), take off slotted nut (8) and washer (9).

GO TO FRAME 2

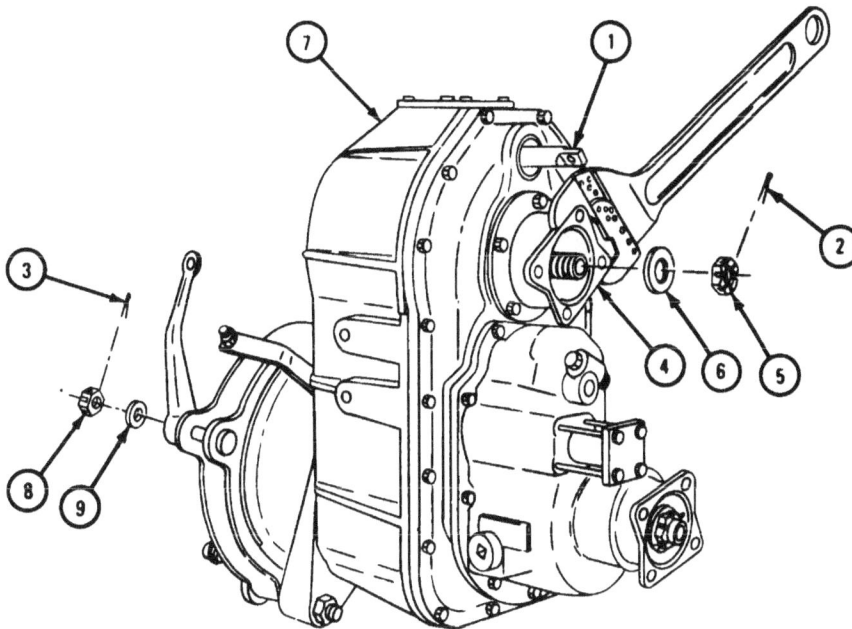

TA 087978

FRAME 2

Soldier A 1. Take out cotter pin (1).

2. Using adjustable wrench, hold companion flange   (2) as shown.

Soldier B 3. Take off slotted nut (3) and washer (4).

NOTE

If companion flanges (2 and 5) are stuck in place, use mechanical puller as shown to take them off.

4. Take off two companion flanges (2 and 5) with deflectors (6 and 7).

END OF TASK

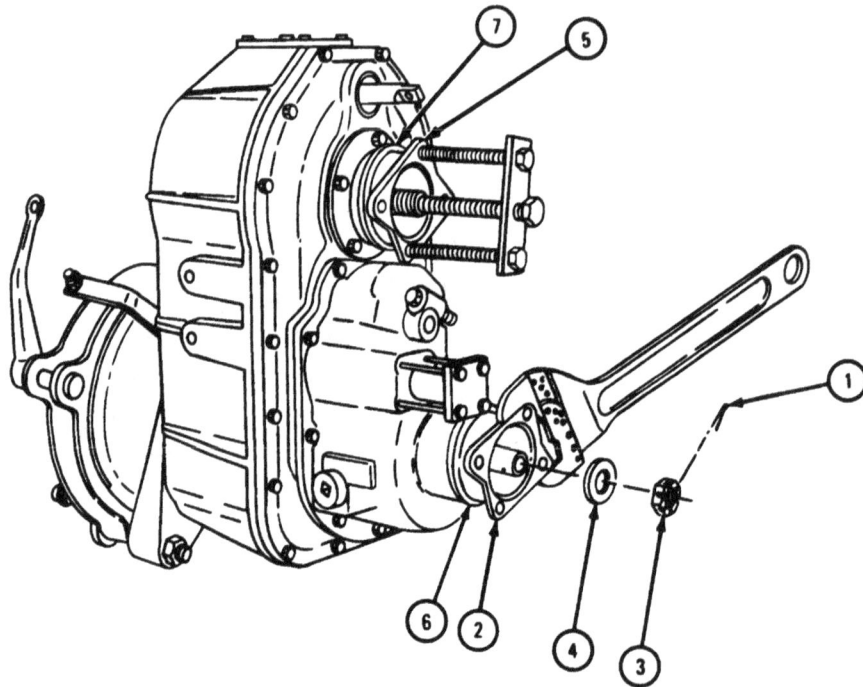

TA 087979

c. Handbrake Brake Drum and Brakeshoe Assembly.

FRAME 1

1. Loosen jamnut (1) and back off adjusting screw (2). Take off spring (3).
2. Take out jamnut (4) and anchor pin (5).
3. Pull lever assembly (6) with brakeshoe assembly (7) off handbrake brake drum (8).

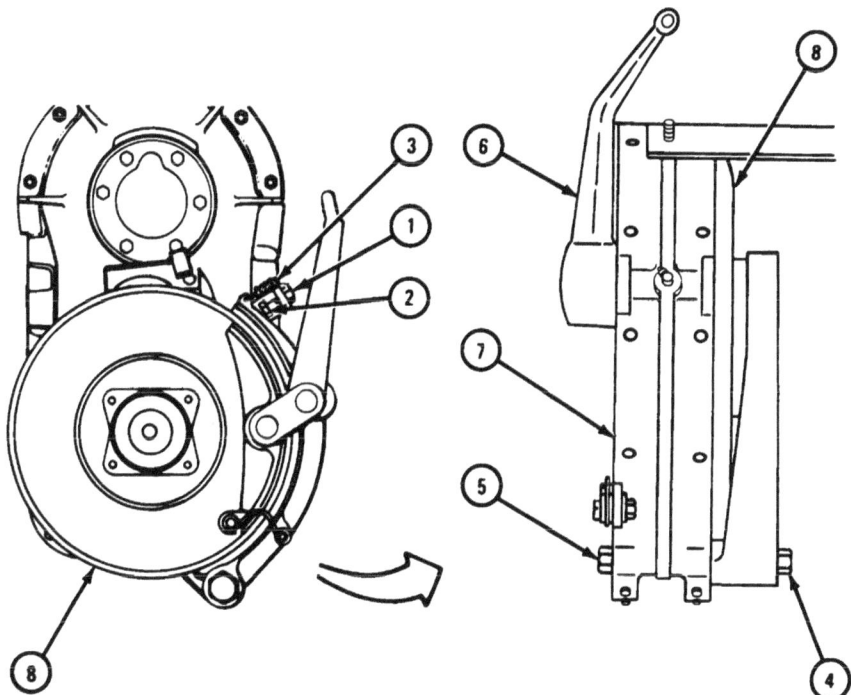

GO TO FRAME 2

TA 087980

FRAME 2

1. Pull off brake drum assembly (1).

2. Take out two capscrews (2) and two lockwashers (3).   Take off shoe stop bracket (4).

3. Take out two capscrews (5) and two lockwashers (6).   Take off brake cable bracket (7).

END OF TASK

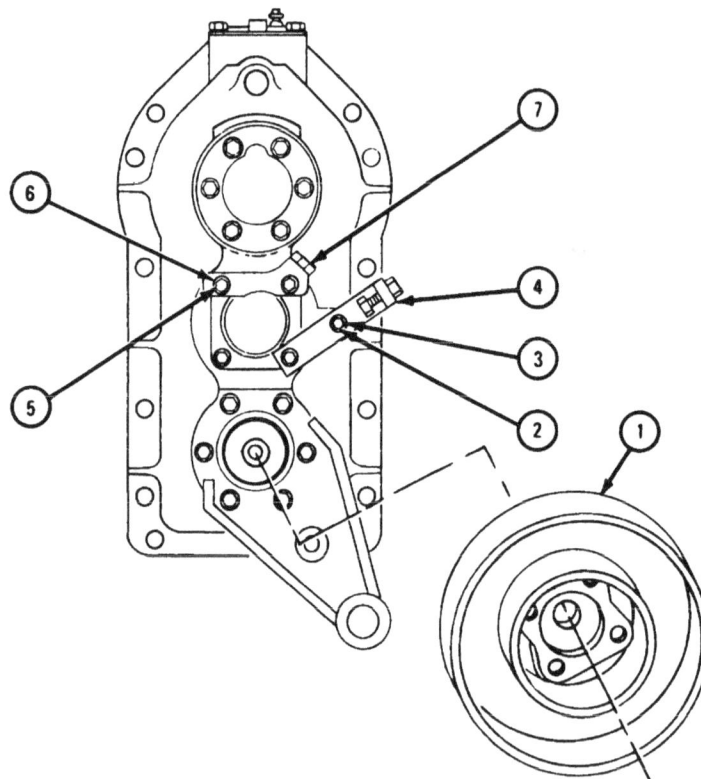

TA 087981

d. Front Output Case Assembly (Model T-136-27).

FRAME 1

1. Turn transfer assembly (1) as shown.

2. Take out 10 capscrews (2) and 10 lockwashers (3).

3. Put back two capscrews (2) into two threaded puller screw holes in flange of output shaft case (4).

4. Tighten two capscrews (2) evenly until front output shaft case (4) lifts off transfer case cover (5).

5. Carefully lift off output shaft case (4).   Sliding clutch inside output shaft case may fall out.

6. Take off gasket (6).   Take out two capscrews (2) and throw away gasket.

7. Take off shifter shaft spring (7) with two spring caps (8).

END OF TASK

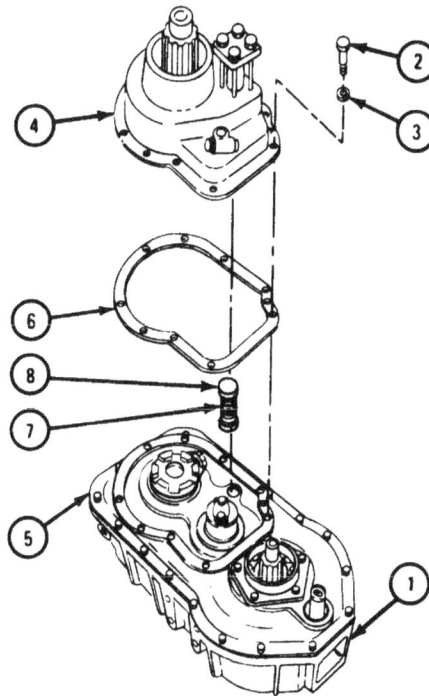

TA 087982

e. <u>Front Output Clutch (Model T-136-27).</u>

FRAME 1

1. Take off safety wire (1).
2. Loosen clutch setscrew (2).
3. Pull clutch (3) straight up and off.
END OF TASK

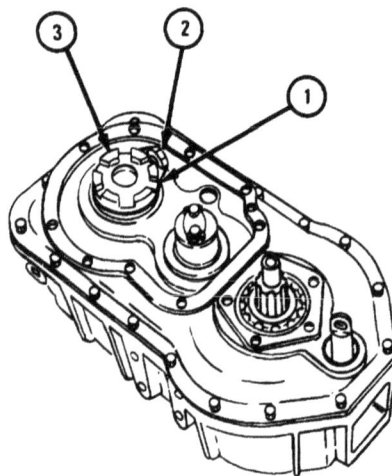

TA 087983

f.   Front Output Case Assembly (Model T-136-21).

1.   Take out 10 capscrews (1) and 10 lockwashers (2).

2.   Put back two capscrews (1) into two threaded puller screw holes in flange of output shaft case (3).

3.   Tighten two capscrews (1) evenly until front output shaft case (3) lifts off transfer case cover (4).

NOTE

When front output shaft case (3) is taken off transfer case cover (4), shifter shaft spring (5) will fall off end of shifter shaft assembly (6).

4.   Take off front output shaft case (3) with shifter shaft assembly (6) and take off and throw away gasket (7). Take out two capscrews (1).

END OF TASK

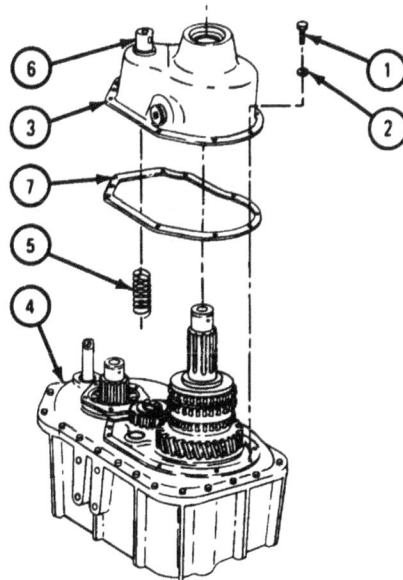

TA 087984

g. Front Output Shaft and Drive Gear (Model T-136-21).

FRAME 1

1. Bend tabs of retainer lock assembly (1) away from two capscrews (2).
2. Take out two capscrews (2).
3. Take off retainer lock assembly (1), retainer plate (3), and helical gear (4).
4. Lift out output shaft assembly (5).

END OF TASK

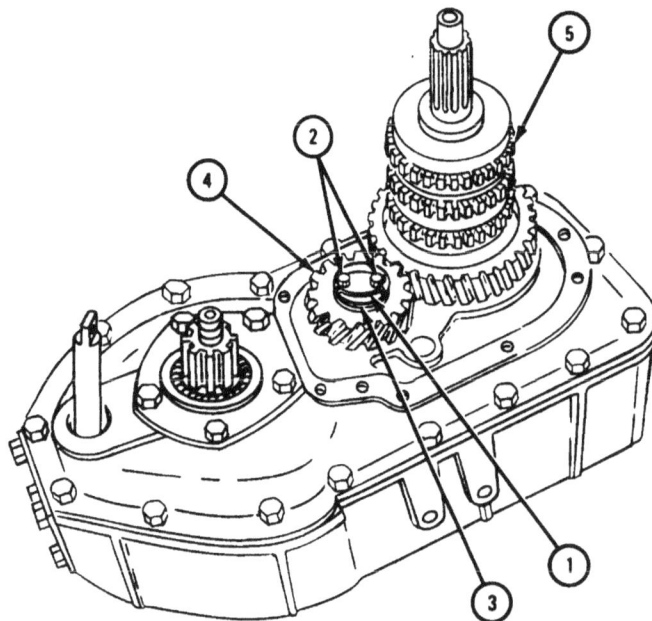

TA 087985

h. Input Shaft Front Bearing Cover.

FRAME 1

1. Take out five capscrews (1) and five lockwashers (2).

NOTE

If bearing cover (3) is stuck to transfer case cover (4), tap sides with soft faced hammer.

2. Take off bearing cover (3) and gasket (5). Throw away gasket.

3. Take off outer thrust washer (6).

4. Tap oil seal (7) out of bearing cover (3).

END OF TASK

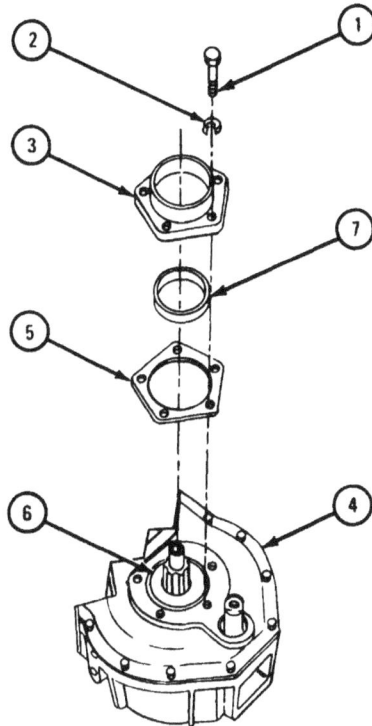

TA 087986

i. Case Cover.

FRAME 1	

NOTE

Case (1) and cover (2) are a matched set. Mark both parts so that case and cover will not be mixed up with other transmission transfer assemblies being worked on at the same time.

Do not lose or damage three taper pins (3). They are non-supply items that will have to be made.

Soldier A 1. Take out six capscrews (4) with six lockwashers (5).

2. Take out 12 capscrews (6), 12 lockwashers (7), and 12 nuts (8).

3. Put back two capscrews (4) into two puller screw holes in cover (2).

4. Tighten two capscrews (4) until cover (2) lifts off case (1).

Soldiers 5. Pry up and lift off cover (2).   Take out two capscrews (4) and throw away
A and B   gasket (9).

END OF TASK

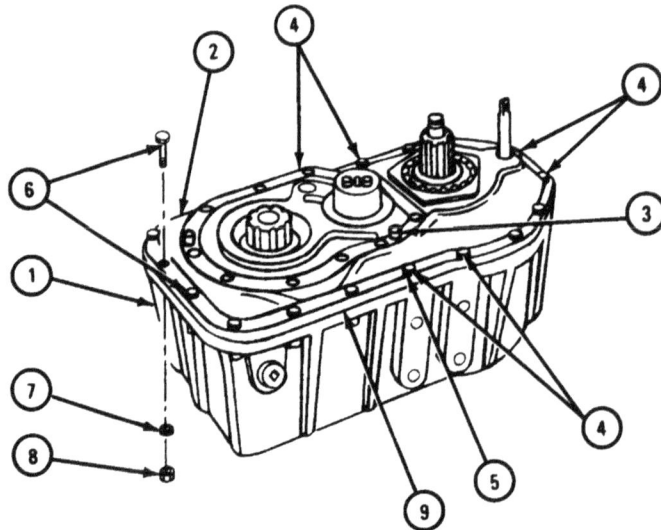

TA 087987

j.   Rear Output Shaft and Countershaft Assemblies.

FRAME 1

1.   Tilt rear output shaft assembly (1) away from countershaft assembly (2) and lift it out of transfer case (3).

2.   Tilt countershaft assembly (2) away from input shaft assembly (4) and lift it out of transfer case (3).

END OF TASK

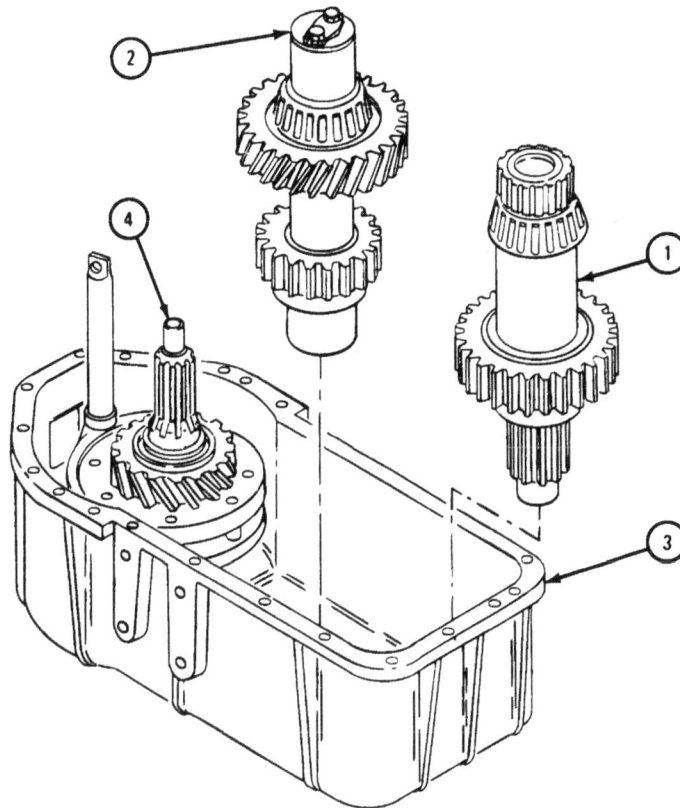

TA 087988

k. Top Cover and Shifter Shaft.

FRAME 1

1. Take out four capscrews (1) with four lockwashers (2).
2. Take off cover (3) and gasket (4). Throw away gasket.
3. Take out spring (5), plunger (6), and ball (7).
4. Take off safety wire (8).
5. Take out setscrew (9) and pull out shifter shaft (10).

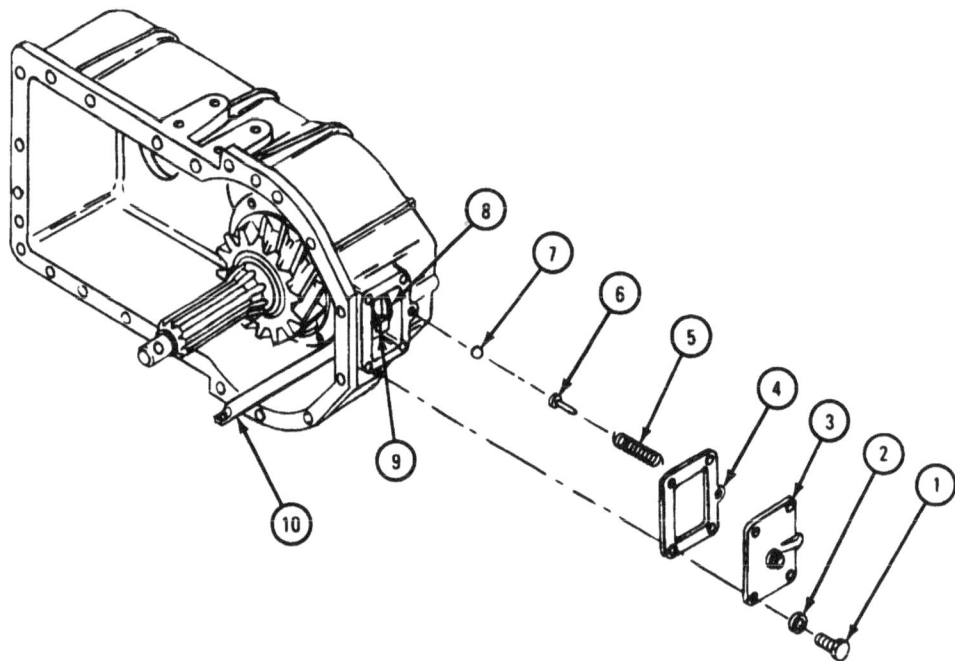

END OF TASK

TA 087989

1. Input Shaft Assembly and Shifter Fork.

FRAME 1

1. Tilt input shaft assembly (1) with shifter fork (2) toward top of transfer case (3) and lift it out. Take off shifter fork.

END OF TASK

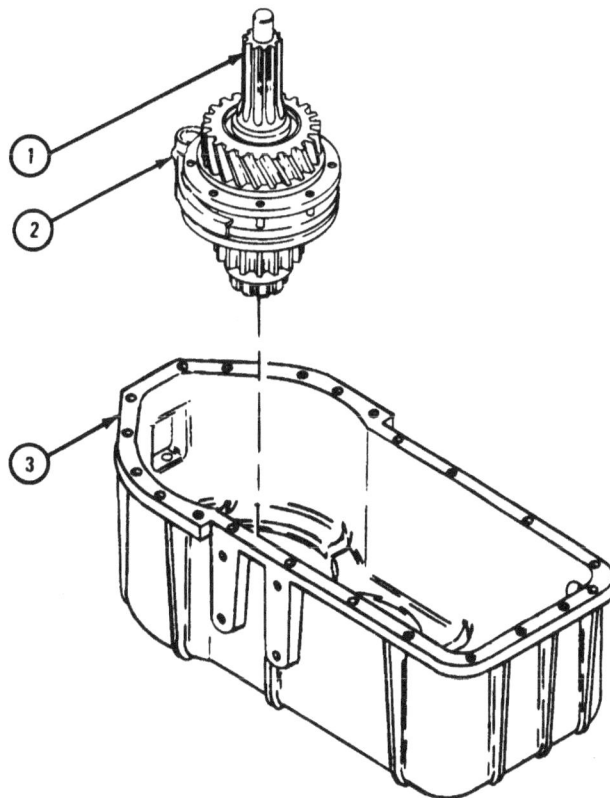

TA 087990

m. <u>Countershaft Rear Bearing Cover.</u>

FRAME 1

1. Turn transfer assembly (1) as shown.
2. Take out capscrew (2) and lockwasher (3).
3. Take off bearing cover (4), gasket (5), and shims (6).

END OF TASK

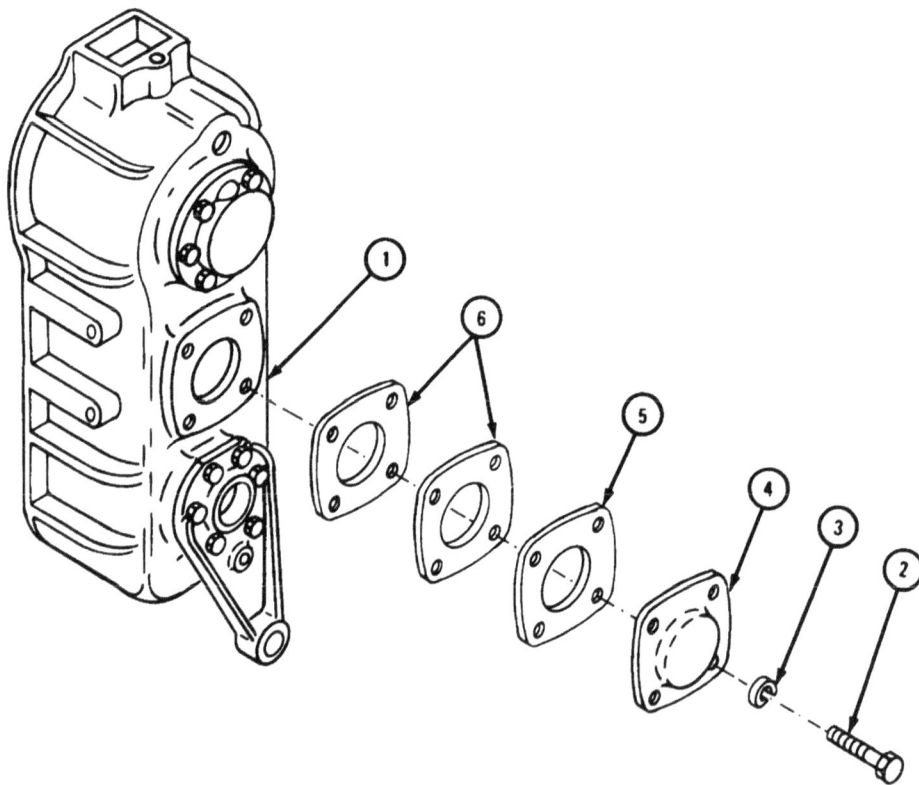

TA 087991

n. Rear Output Shaft Rear Bearing Retainer.

FRAME 1

1. Take out six capscrews (1) and six lockwashers (2).
2. Tap off bearing retainer (3) from dowel pin (4) and take off shims (5).
END OF TASK

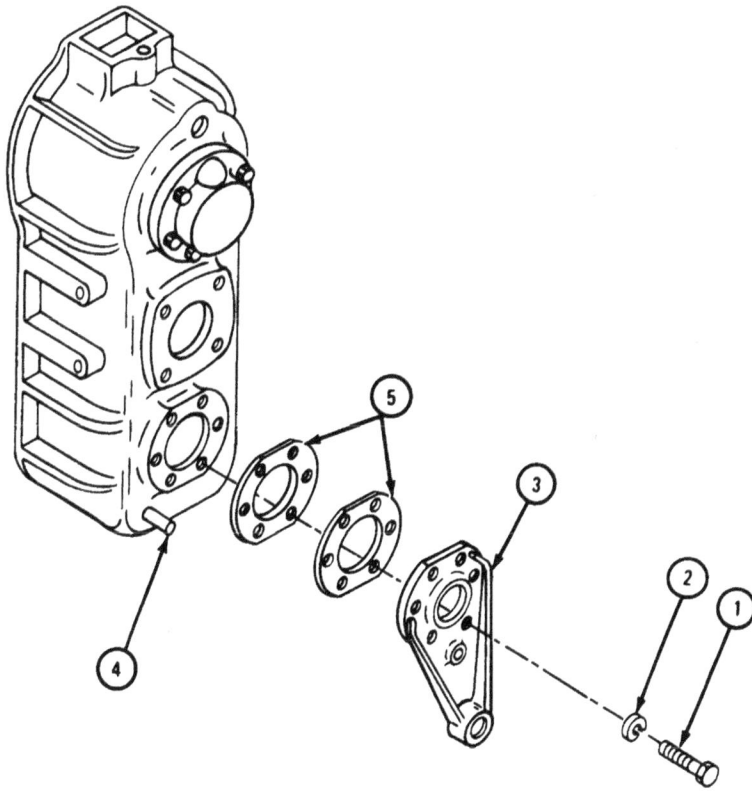

TA 087992

o. <u>Input Shaft Rear Bearing Cover (Trucks Without Transfer Power Takeoff</u>
    <u>Installed).</u>

FRAME 1

---

1. Take out six capscrews (1) and six lockwashers (2).
2. Take off bearing cover (3) and gasket (4). Throw away gasket.

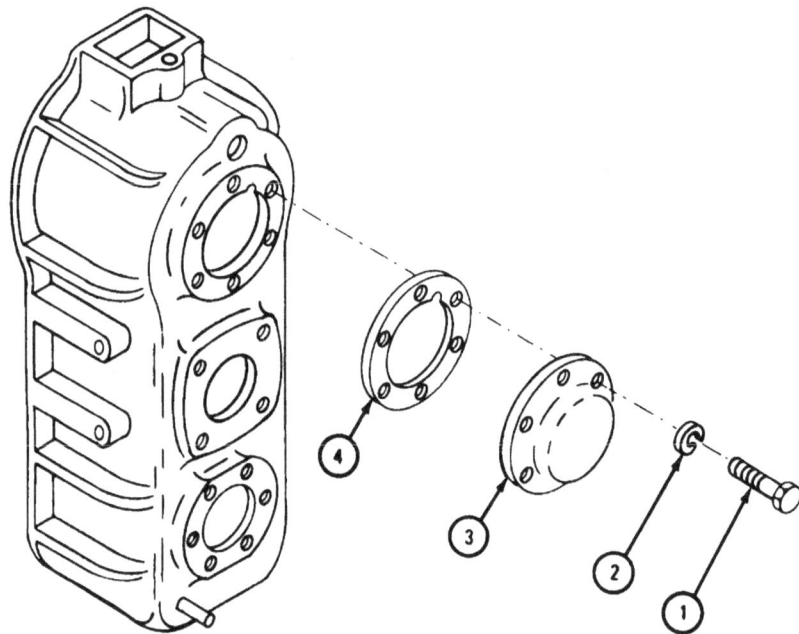

END OF TASK

TA 087993

2-15. DISASSEMBLY OF SUBASSEMBLIES. The following paragraphs give instructions to disassemble the transmission transfer subassemblies.

a. Rear Output Shaft Assembly.

FRAME 1
Soldier A 1. Set rear output shaft (1) in hydraulic press as shown.
Soldier B 2. Working from under press, hold bottom of rear output shaft (1) to keep it from falling when it is pressed out of front roller bearing (2). Tell soldier A when ready.
Soldier A 3. Press out rear output shaft (1).
GO TO FRAME 2

MODEL T-136-27   MODEL T-136-21

TA 087994

FRAME 2

Soldier A 1.   Set up rear output shaft (1) in hydraulic press as shown.

NOTE

Scribe top and bottom of driven gear (2) so that it will be
put back the same way.

Soldier B 2.   Working from under press, hold bottom of rear output shaft (1) to keep it
from falling when it is pressed out of driven gear (2) and rear bearing
(3). Tell soldier A when ready.

Soldier A 3.   Press out rear output shaft (1).

GO TO FRAME 3

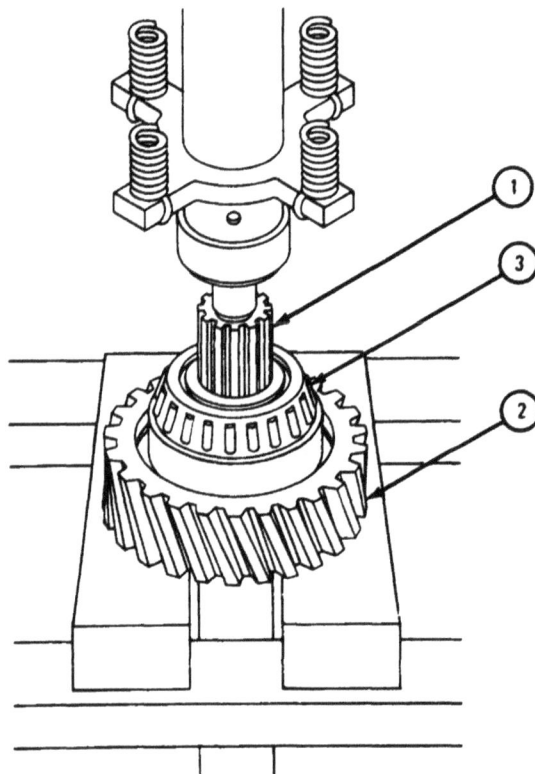

TA 087995

FRAME 3

1.   Take out key (1).
END OF TASK

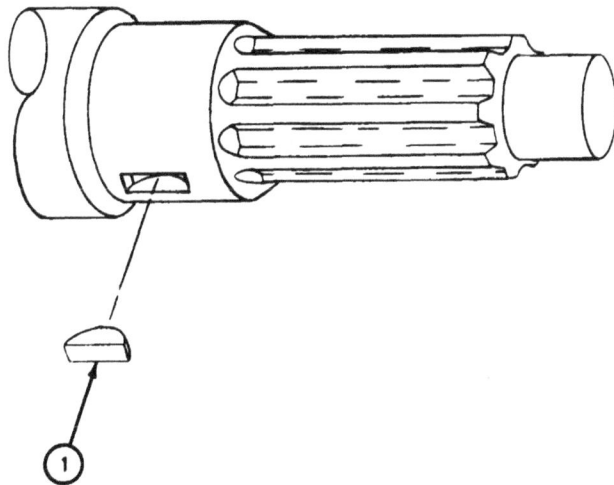

TA 087996

b. <u>Countershaft Assembly.</u>

NOTE

If working on transmission transfer model T-136-27, do frame 1.   If working on transmission transfer model T-136-21, go to frame 2.

FRAME 1

1.   Bend ears on retainer lock assembly (1) away from two capscrews (2).

2.   Take out two capscrews (2).

3.   Take off retainer lock assembly (1) and retainer   plate (3).

4.   Take off sleeve (4).

GO TO FRAME 2

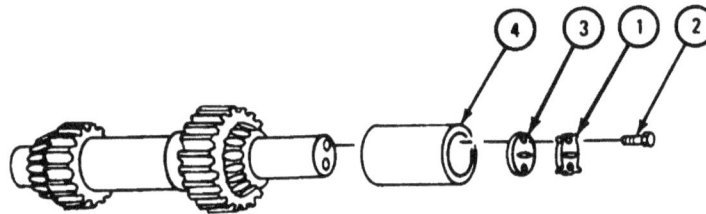

TA 087997

FRAME 2

1. Bend ears on retainer lock assembly (1) away from two capscrews (2).
2. Take out two capscrews (2).
3. Take off retainer lock assembly (1) and retainer plate (3).

GO TO FRAME 3

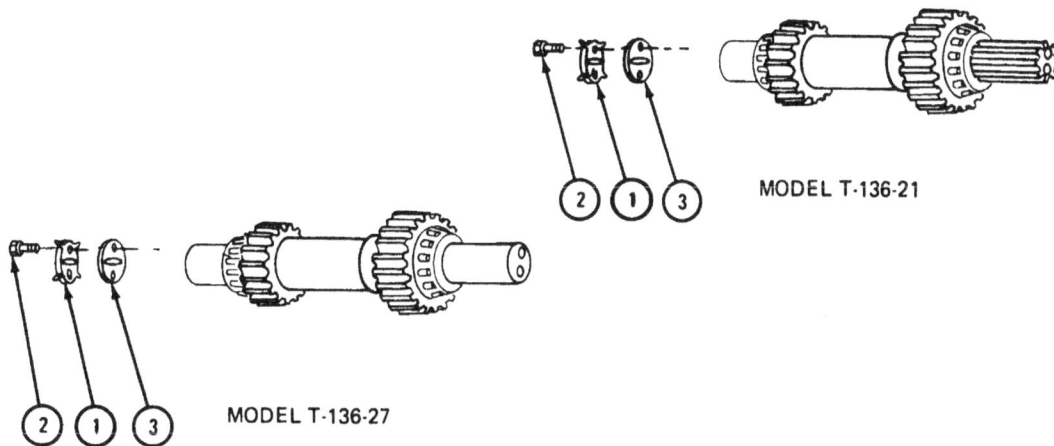

MODEL T-136-21

MODEL T-136-27

TA 087998

FRAME 3

Soldier A 1.   Set up countershaft (1) in hydraulic press as shown.

NOTE

Scribe top and bottom of high range gear (2) so that it will be put back the same way.

Soldier B 2.   Working from under press, hold bottom of countershaft (1) to keep it from falling when high range gear (2) and rear bearing (3) are pressed off. Tell soldier A when ready.

Soldier A 3.   Press off high range gear (2) and rear bearing (3).

GO TO FRAME 4

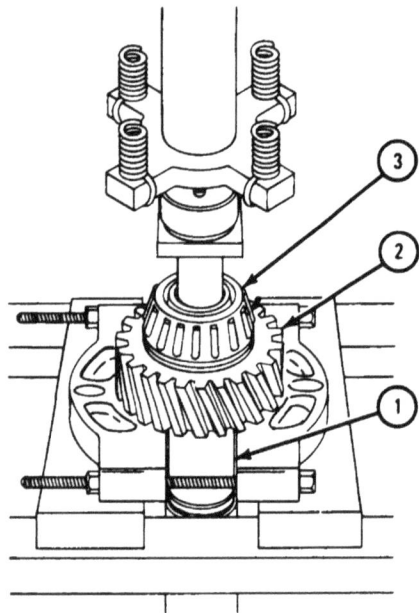

MODEL T-136-27

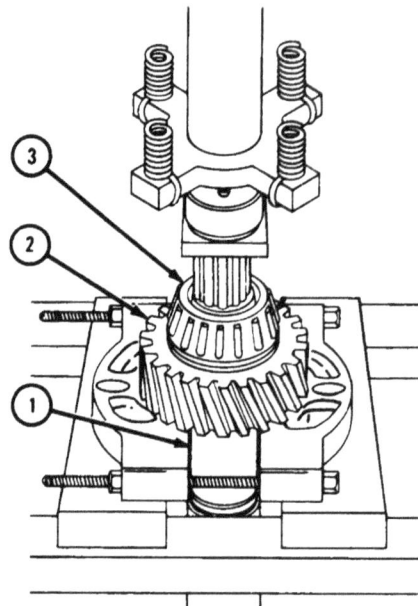

MODEL T-136-21

TA 087999

FRAME 4

Soldier A 1.   Set up countershaft (1) in hydraulic press as shown.

NOTE

Scribe top and bottom of low range gear (2) so that it will be
put back the same way.

Soldier B 2.   Working under press, hold end of countershaft (1) to keep it from falling
when countershaft is pressed out. Tell soldier A when ready.

Soldier A 3.   Press countershaft (1) out of low range gear (2) and front bearing (3).

GO TO FRAME 5

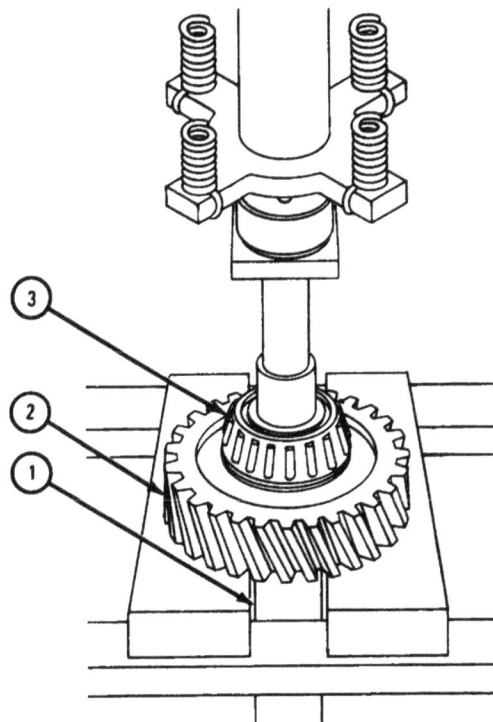

TA 088000

TM 9-2520-246-34-1

FRAME 5

1. Take two keys (1) out of countershaft (2).
END OF TASK

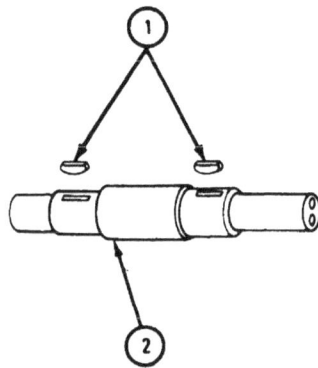

MODEL T-136-27

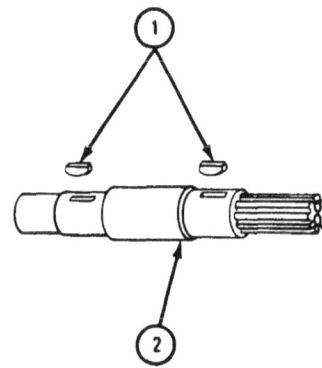

MODEL T-136-21

TA 088001

2-138

c. <u>Input Shaft Assembly.</u>

FRAME 1

1. Slide off thrust washer (1), low range gear (2), and synchronizer (3).
2. Take off snapring (4).
GO TO FRAME 2

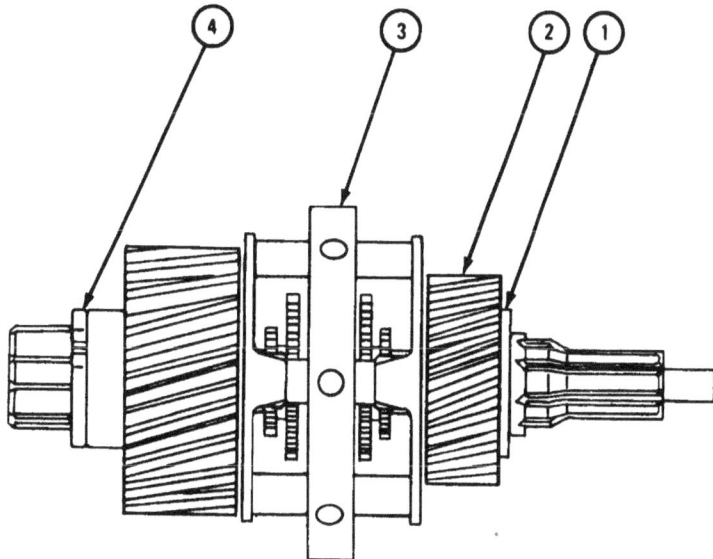

TA 088002

FRAME 2

Soldier A 1.   Set up input shaft (1) in hydraulic press as shown.

NOTE

Scribe top and bottom of gear (2) so that it will be put back the same way.

Soldier B 2.   Working from under press, hold bottom of input shaft (1) to keep it from falling when input shaft is pressed out of gear (2) and bearing (3). Tell soldier A when ready.

Soldier A 3.   Press out input shaft (1).

GO TO FRAME 3

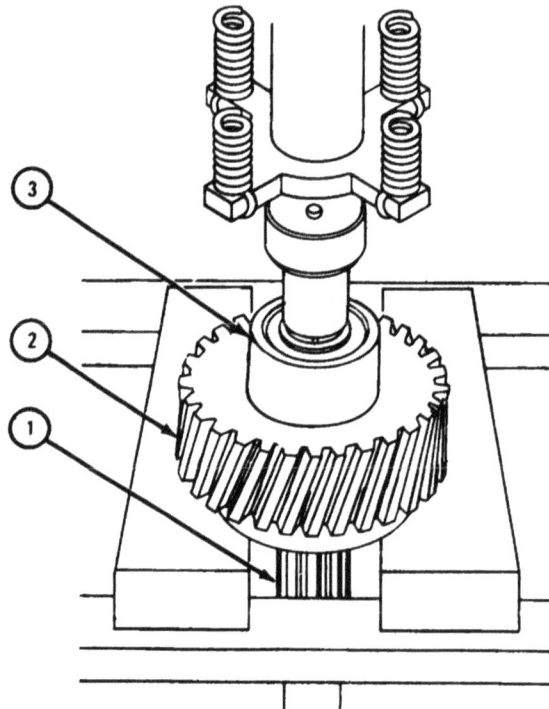

TA 088003

FRAME 3

1. Using hammer and brass drift, tap out inner bearing (1) and spacer (2).
2. Turn over gear (3) and tap out outer bearing (4).
GO TO FRAME 4

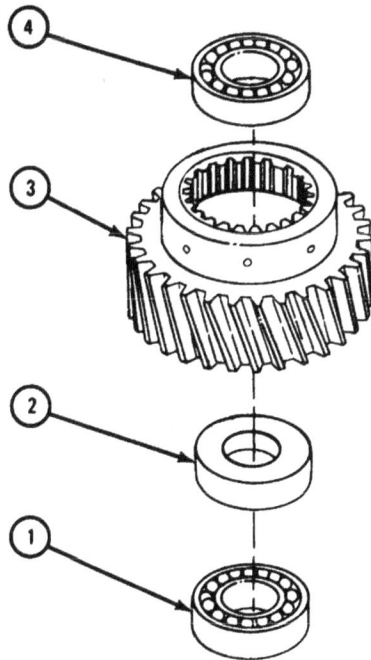

TA 088004

FRAME 4

1. Using hammer and brass drift, tap out inner bearing (1) and spacer (2).
2. Turn over gear (3) and tap out outer bearing (4).

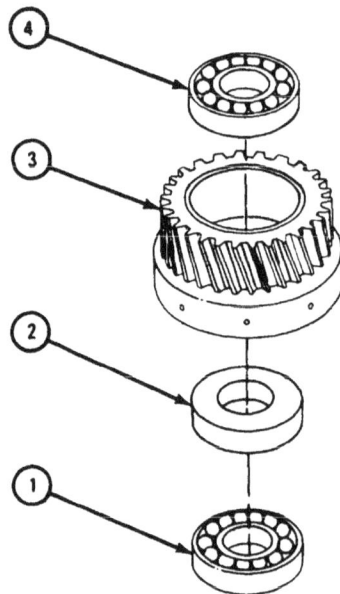

END OF TASK

TA 088005

d.   Front Output Case Assembly (Model T-136-27).

FRAME 1

1.   Bend back ears on four key washers (1).

2.   Take out four capscrews (2). Take off air tube cover (3) and brass gasket (4).

3.   Slide off air tube (5).

4.   Take off nut (6) and washer (7).   Take off piston (8) and second washer (7).

5.   Take off brass gasket (9).

GO TO FRAME 2

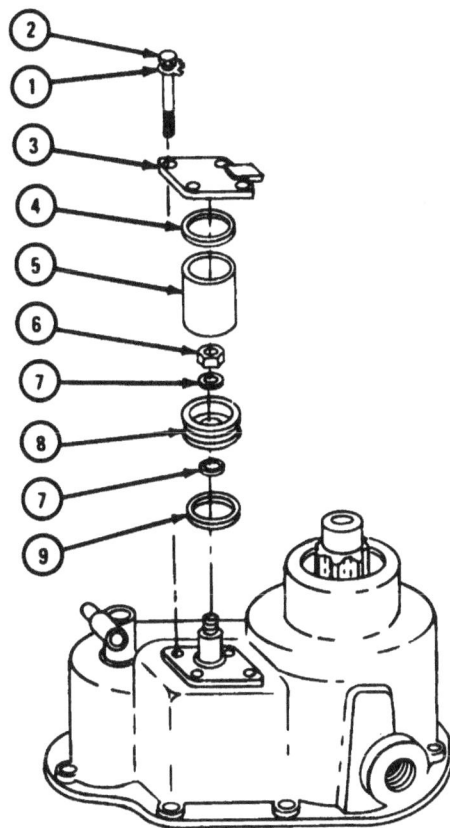

TA 088006

FRAME 2

1. Set up output shaft cover assembly (1) in hydraulic press as shown.
2. Press out front output shaft (2).
3. Take out shifter shaft (3), sliding clutch (4), and front output shaft (2).
GO TO FRAME 3

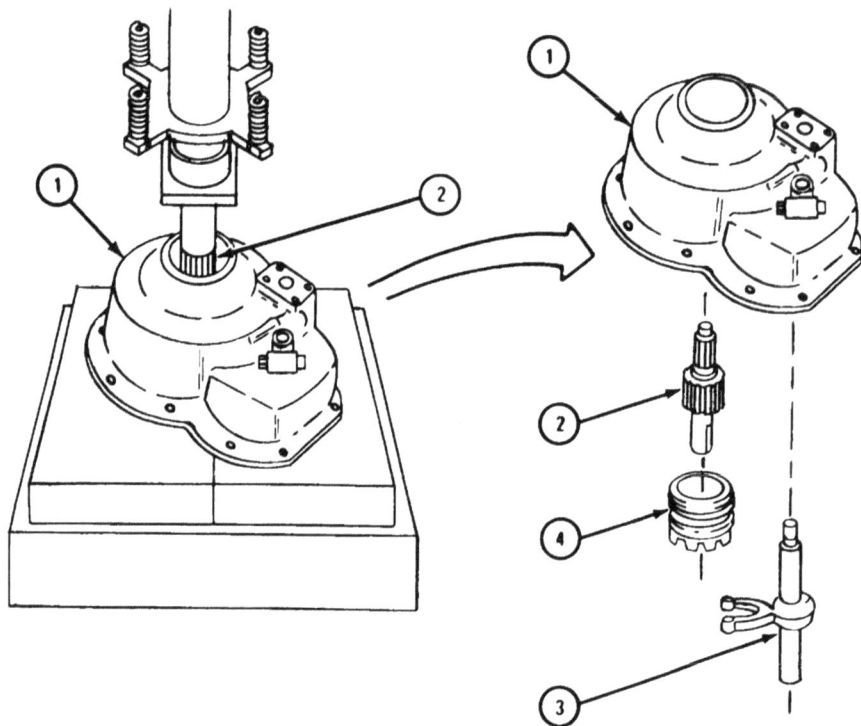

TA 088007

FRAME 3

1. Working from inside of output shaft case (1), drive out seal (2).
2. Take out snap ring (3) and thrust washer (4).

GO TO FRAME 4

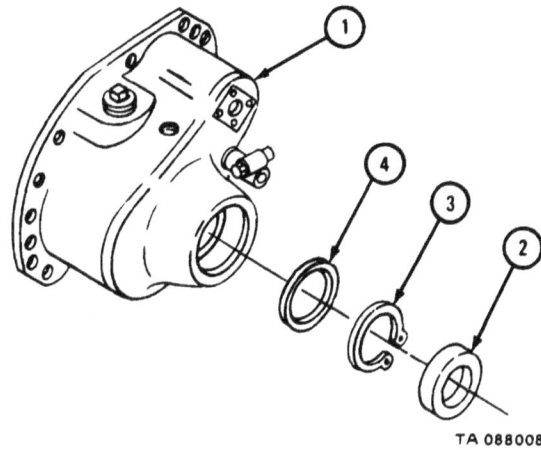

TA 088008

FRAME 4

1. Take out snapring (1).

2. Working at rear side of output shaft case (2), take out snapring (3).

3. Using bearing remover and replacer, drive out output shaft bearing (4).

4. Take out filler plug (5).

5. Take off speedometer adapter (6).

END OF TASK

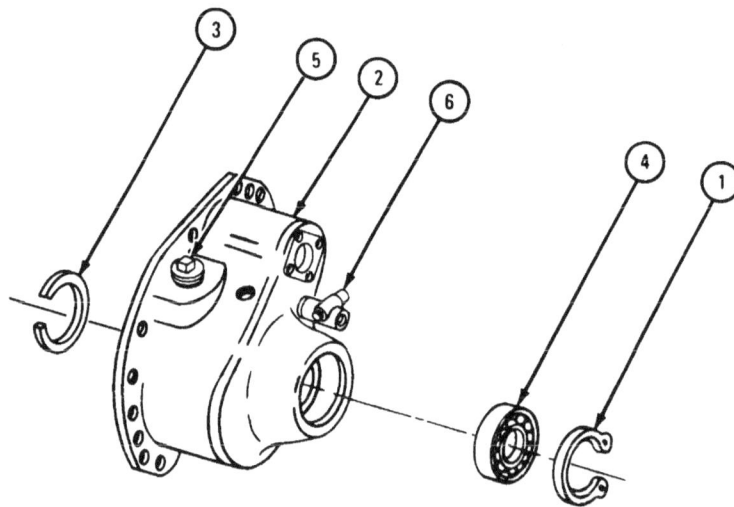

TA 088009

e. Shifter Shaft Assembly (Model T-136-27).

FRAME 1

1. Cut and take off safety wire (1).
2. Take out setscrew (2).
3. Slide shifter shaft fork (3) off shifter shaft (4).

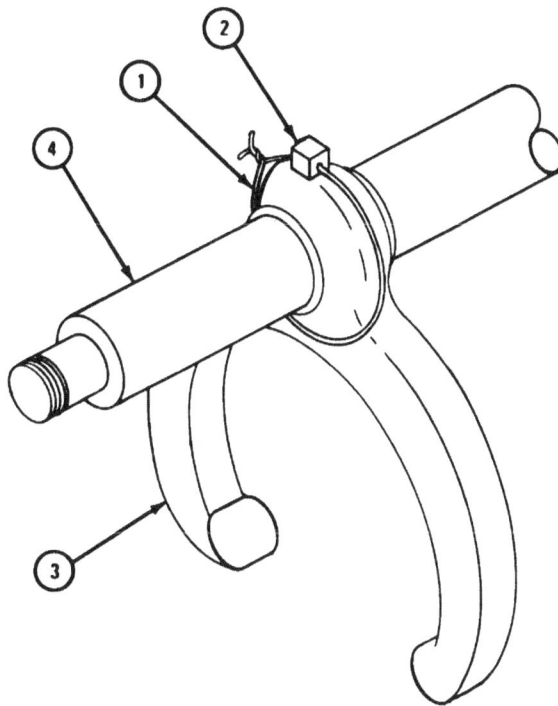

END OF TASK

TA 088010

f. Front Output Case Assembly (Model T-136-21).

FRAME 1

1. Working from front of front output shaft case (1), tap out shifter shaft (2) and take off clutch collar (3).
2. Take off safety wire (4).
3. Take out setscrew (5) and take shifter fork (6) off shifter shaft (2).
4. Take out speedometer adapter (7).

GO TO FRAME 2

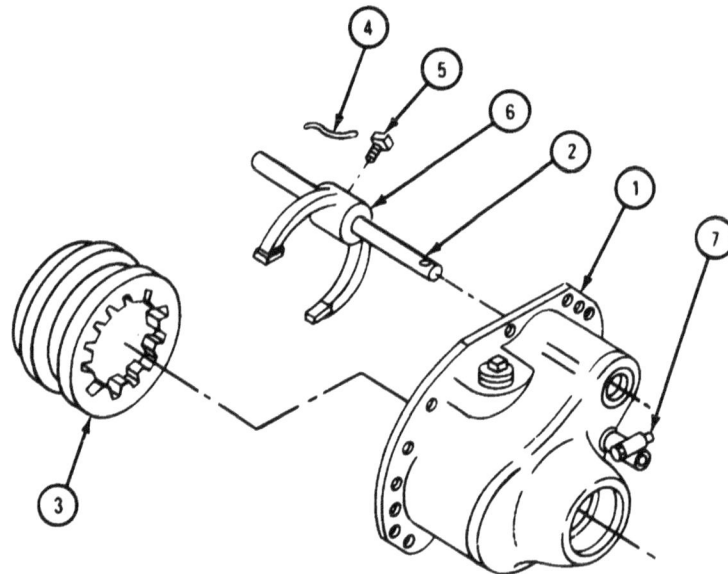

TA 088011

FRAME 2

1. Working from inside of output shaft case (1), drive out seal (2) and seal (3).
2. Take out snapring (4) and thrust washer (5).

GO TO FRAME 3

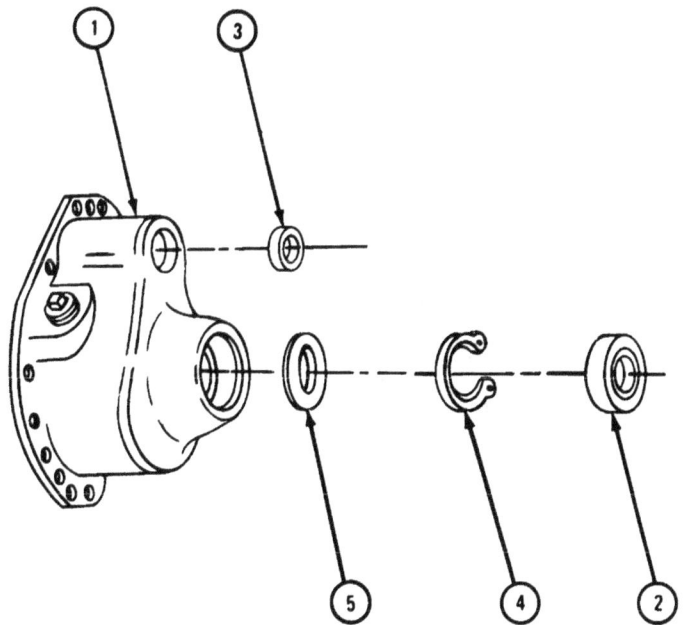

TA 088012

FRAME 3

1. Take out snapring (1).
2. Working at rear side of output shaft case (2), take out snapring (3).
3. Using bearing remover and replacer, drive out output shaft bearing (4).
4. Take out filler plug (5).

END OF TASK

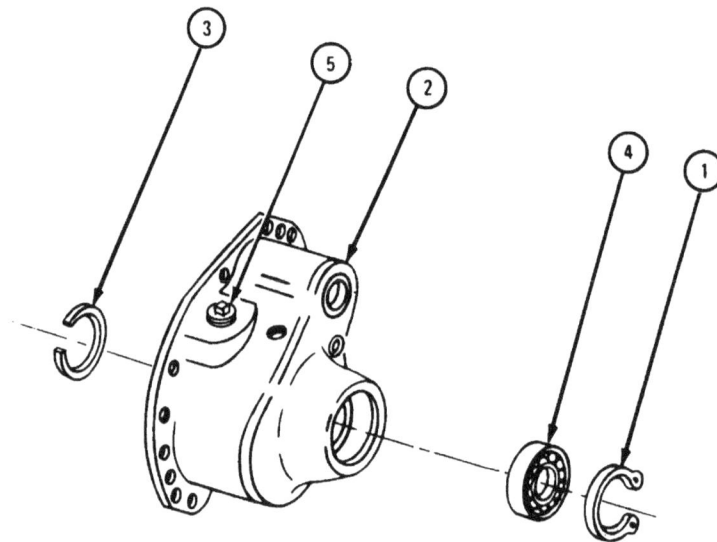

TA 088013

g. <u>Front Output Shaft (Model T-136-21).</u>

FRAME 1

1. Take off snap ring (1) and thrust washer (2).
2. Take off front outer race (3) with front sprag unit (4).
3. Take off two spacers (5).
4. Take off rear outer race (6) with rear sprag unit (7).

GO TO FRAME 2

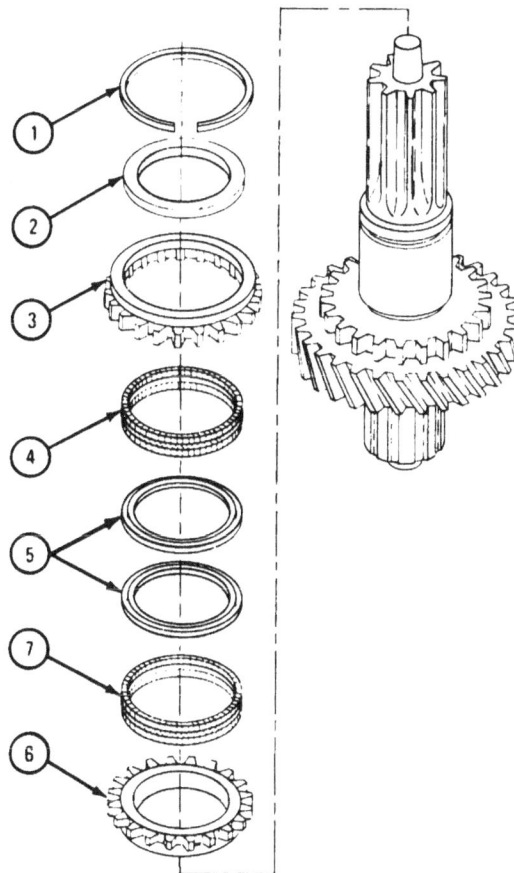

TA 088014

FRAME 2

1. Take off snapring (1).

Soldier A 2. Set up front output shaft (2) in hydraulic press as shown.

NOTE

Scribe top and bottom of transmission gear (3) so that it will be put back the same way.

Soldier B 3. Working under hydraulic press, hold end of front output shaft (2) to keep it from falling when front output shaft is pressed out of transmission gear (3) and inner race (4). Tell soldier A when ready.

Soldier A 4. Press out front output shaft (2).

GO TO FRAME 3

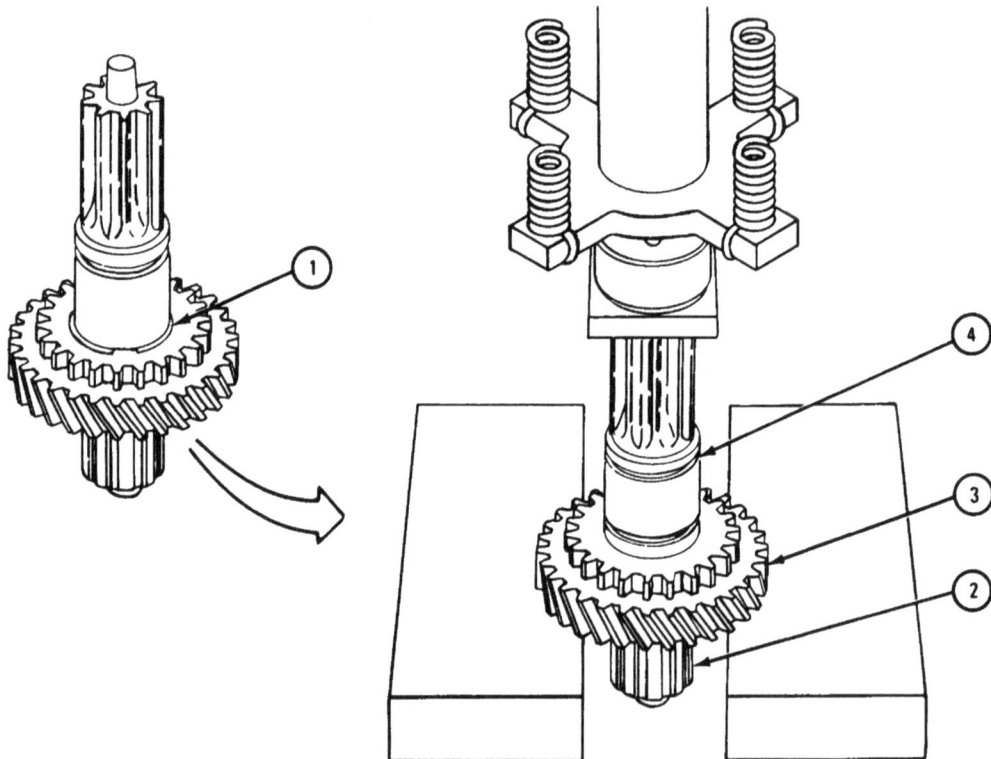

TA 088015

FRAME 3

1. Take two keys (1) out of front output shaft (2).
GO TO FRAME 4

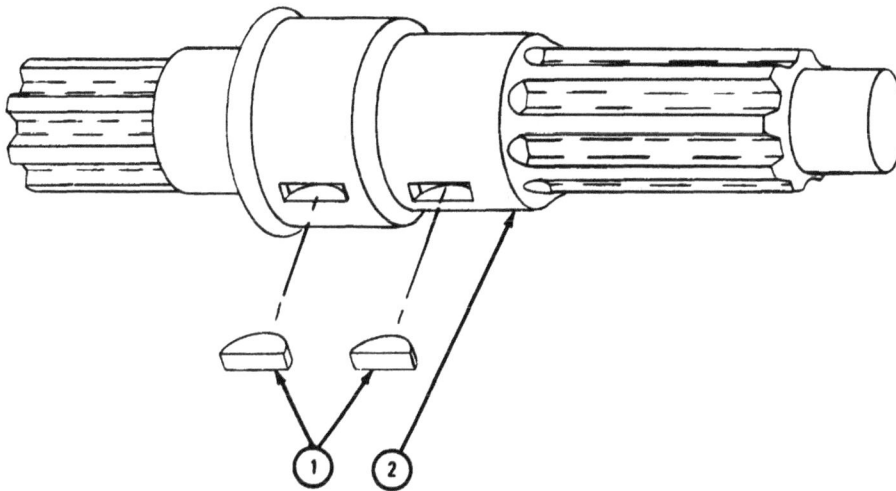

TA 088016

FRAME 4

1. Take front sprag unit (1) out of front outer race (2).
2. Take off spring (3).
3. Do steps 1 and 2 again for rear sprag unit.

END OF TASK

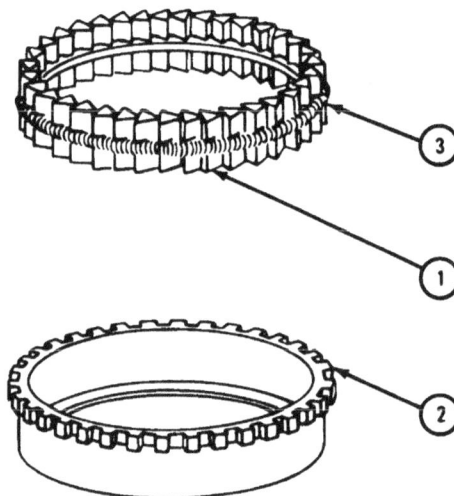

TA 088017

h. <u>Case Cover.</u>

FRAME 1

1. Take out snapring (1).

2. Using bearing cap remover and replacer and working from rear side of cover (2), drive out input shaft front bearing (3).

3. Using 12-inch long, 1/2-inch round stock and working from rear side of cover (2), drive out shifter shaft oil seal (4).

GO TO FRAME 2

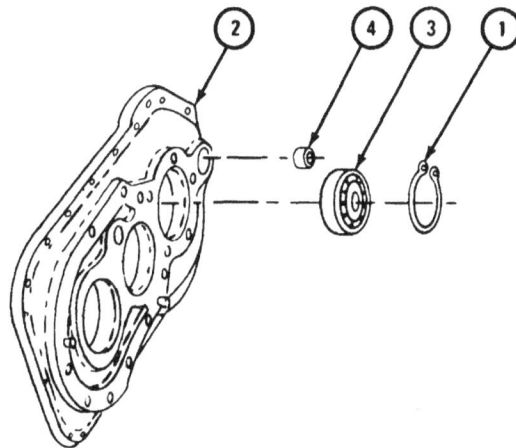

FRAME 2

1. Tap out seal (1).

2. Takeout two snaprings (2).

3. Using bearing remover and replacer and working from rear side of cover (3), drive out countershaft bearing cup (4) and rear output bearing cup (5).

END OF TASK

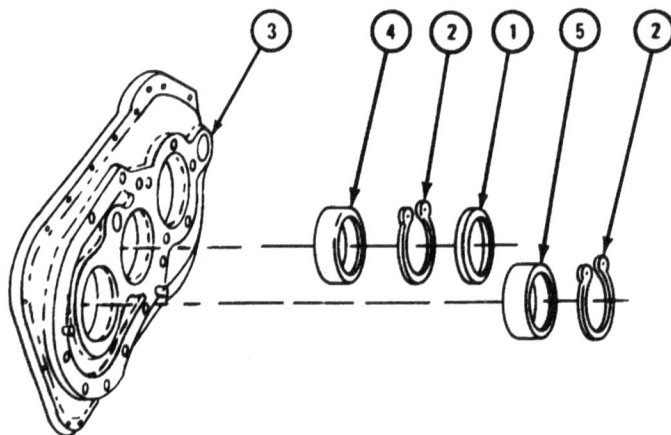

TA 088019

i. Transfer Case Housing.

FRAME 1

1. Using bearing remover and replacer and working from inside transfer case (1), drive out rear output shaft bearing cup (2) and countershaft bearing cup (3).

2. Take out magnetic drain plug (4) and drain plug (5).

3. Using 1/2-inch pin punch and working from inside transfer case (1), drive out dowel pin (6) and shifter shaft plug (7).

END OF TASK

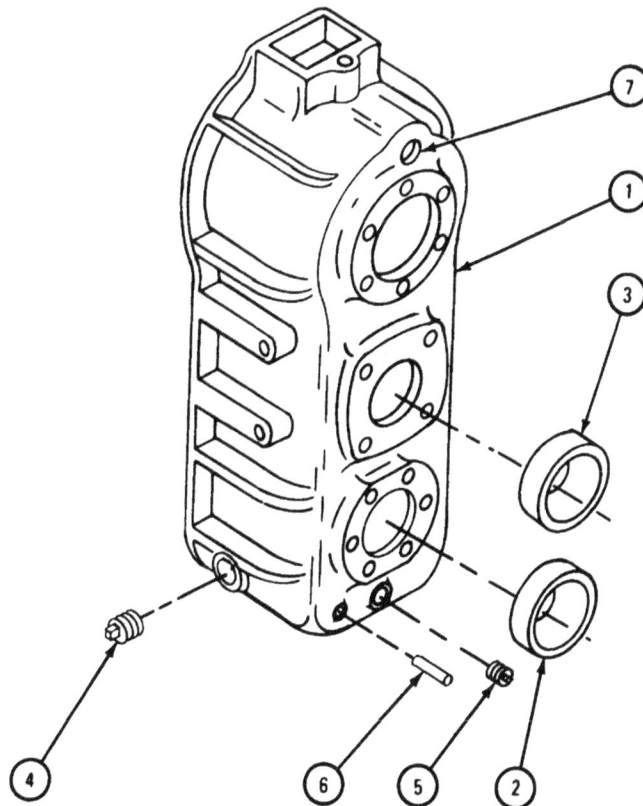

TA 088020

j. Handbrake Shoe Assembly.

FRAME 1

1. Take off two snap rings (1).
2. Take off stabilizer spring (2).
3. Take off four washers (3).

GO TO FRAME 2

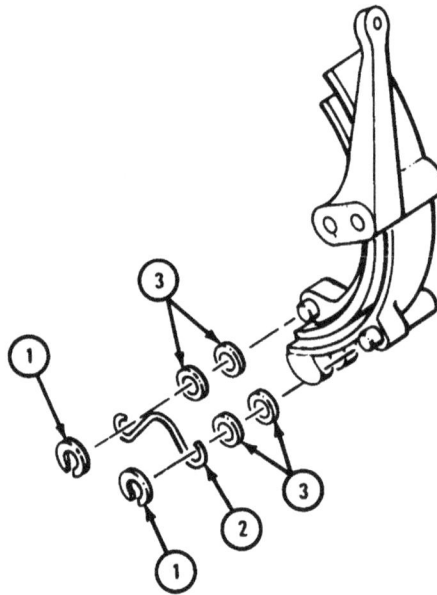

TA 088021

FRAME 2

1. Take off two retaining rings (1 and 2).
2. Lift outer brakeshoe (3) off lever pin (4).
3. Lift inner brakeshoe (5) off lever pin (6).
END OF TASK

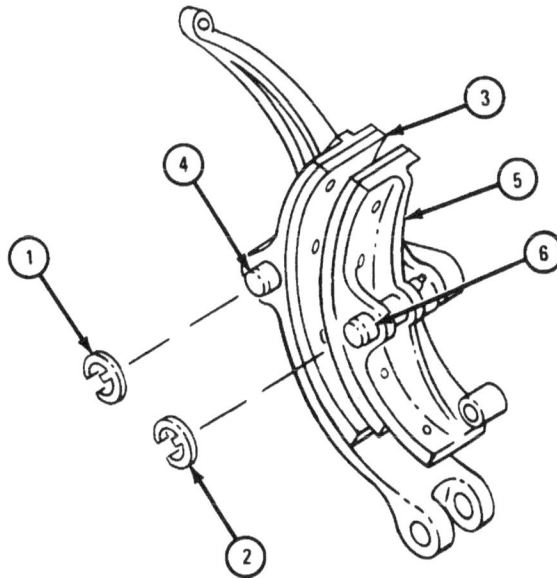

TA 088022

TM 9-2520-246-34-1

k. Handbrake Drum Assembly.

FRAME 1

1. Using hammer and drift, tap off shield (1).
2. Take off companion flange (2) and brake drum shield (3).
3. Take out four capscrews (4).
END OF TASK

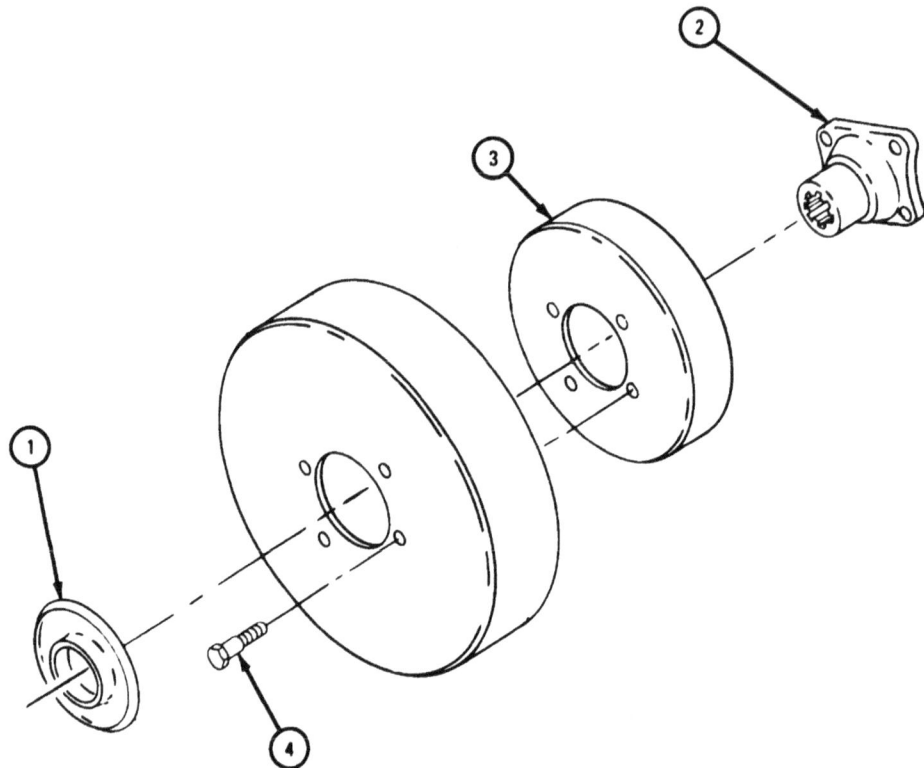

TA 088023

2-160

1. Companion Flanges and Deflectors.

FRAME 1

1. Using brass hammer, tap two deflectors (1) off two companion flanges (2).
END OF TASK

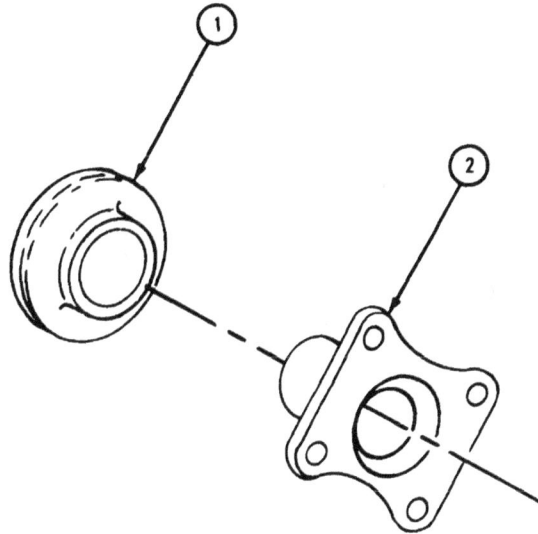

TA 088024

m. <u>Rear Output Shaft Rear Bearing Retainer.</u>

FRAME 1

1.   Using hammer and brass drift, drive out seal (1) from bearing retainer (2).
END OF TASK

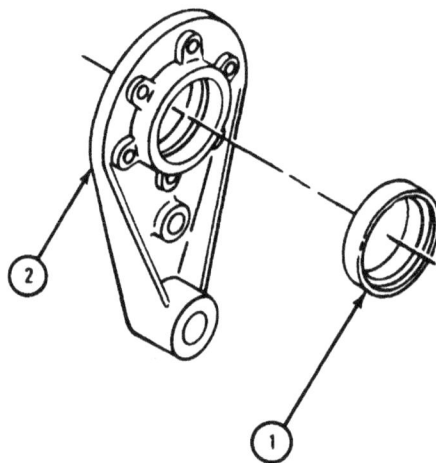

TA 088025

2-16. CLEANING.   This paragraph gives general instructions for cleaning the transmission transfer parts.

a.   Clean all bearing cones and cups. Refer to inspection, care, and maintenance of antifriction bearings, TM 9–214.

WARNING

Dry cleaning solvent is flammable. Do not use near an open flame.   Keep a fire extinguisher nearby when solvent is used.   Use only in well-ventilated places.   Failure to do this may result in injury to personnel and damage to equipment.

Do not use more than 30 psi of air pressure for drying parts.   Eye shields must be worn when using compressed air.   Eye injury can occur if eye shields are not worn.

CAUTION

When scraping gasket material from surface of parts, be careful not to scratch or gouge metal surfaces.

b.   Clean all other parts with solvent.   Scrape all gasket material from surface of parts.   Rinse parts in clean solvent and dry with compressed air.

c.   Make sure that all oil passages are open. Open clogged passages with compressed air or by working a stiff wire back and forth. Flush with solvent.

2-17. GENERAL INSPECTION.   The following paragraphs give instructions to check for damage on the transmission transfer case, covers, gearshafts, and gears.

CAUTION

It is easy to damage the equipment if you don't know what you are doing. D   o not try to do this task unless you are experienced at it, or you have an experienced person with you.

NOTE

Small chips, burrs or scratches on gears and gearshafts can be repaired. If parts are damaged in any other way, throw parts away and get new ones.

a. Front Output Shaft Assembly (Model T-136-21).

FRAME 1

1. Check that shaft splines (1) are not chipped, cracked or twisted.

2. Check that two shafts (2) are not chipped or cracked.

3. Check that gear (3) is not chipped or cracked and that gear teeth are not damaged.

4. Check that inner race (4) is not chipped or cracked.

5. Check that key (5) is not chipped or cracked.

6. Check that two sprag springs (6) are not worn, kinked or twisted.

GO TO FRAME 2

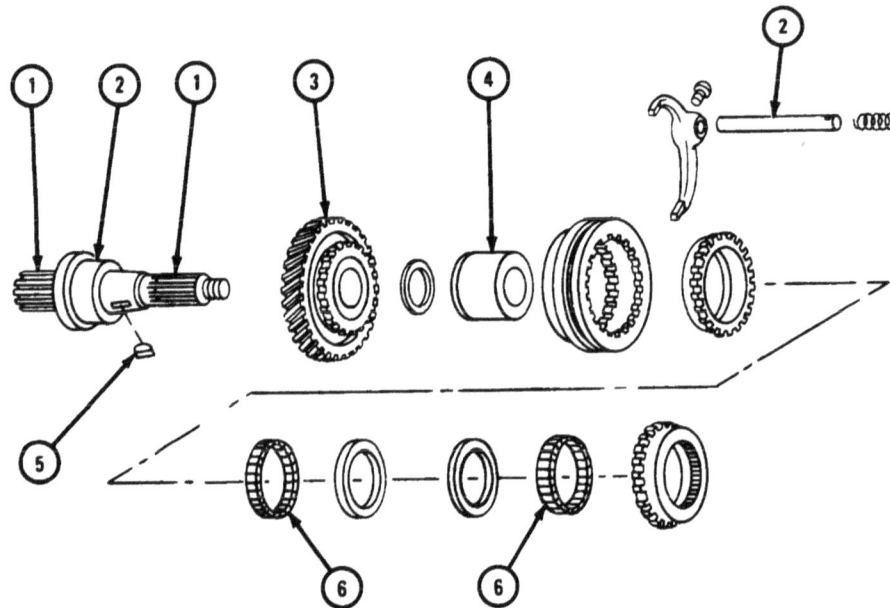

NOTE: CHECK ONLY THOSE PARTS WHICH
ARE CALLED OUT IN THIS FRAME.
PARTS WITHOUT CALLOUTS ARE
SHOWN ONLY FOR REFERENCE
PURPOSES OR ARE CHECKED IN
ANOTHER FRAME.

TA 088026

FRAME 2

1. Check that fork (1) is not cracked or bent.
2. Check that spring (2) is not weak or damaged.
3. Check that outer races (3) are not chipped or cracked and that splines are not chipped, cracked or twisted.
4. Check that collar (4) is not chipped or cracked and that splines are not chipped, cracked or twisted.

END OF TASK

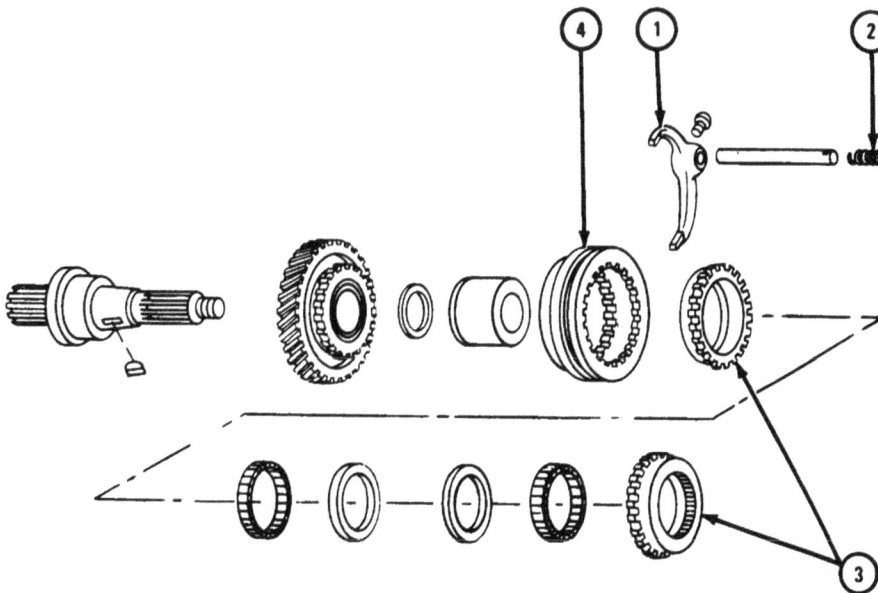

NOTE
CHECK ONLY THOSE PARTS WHICH ARE CALLED OUT IN THIS FRAME. PARTS WITHOUT CALLOUTS ARE SHOWN ONLY FOR REFERENCE PURPOSES OR ARE CHECKED IN ANOTHER FRAME.

TA 088027

b. Front Output Shaft Assembly (Model T-136-27).

FRAME 1

1. Check bearing (1). Refer to inspection, and maintenance care of antifriction bearings, TM 9-214.

2. Check that front output shaft splines (2) are not chipped, cracked or twisted.

3. Check that shaft (3) is not chipped or cracked.

4. Check that sliding clutch (4) is not chipped or cracked and that internal splines are not twisted or burred.

5. Check that threads (5) are not stripped or crossthreaded.

END OF TASK

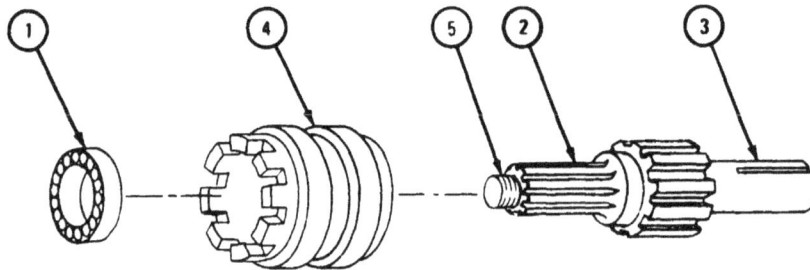

TA 088028

c.   Countershaft Assembly.

FRAME 1

1.   Check that gears (1) have no chips, cracks or damaged teeth. Check that
     keys (2) are not chipped or cracked.

GO TO FRAME 2

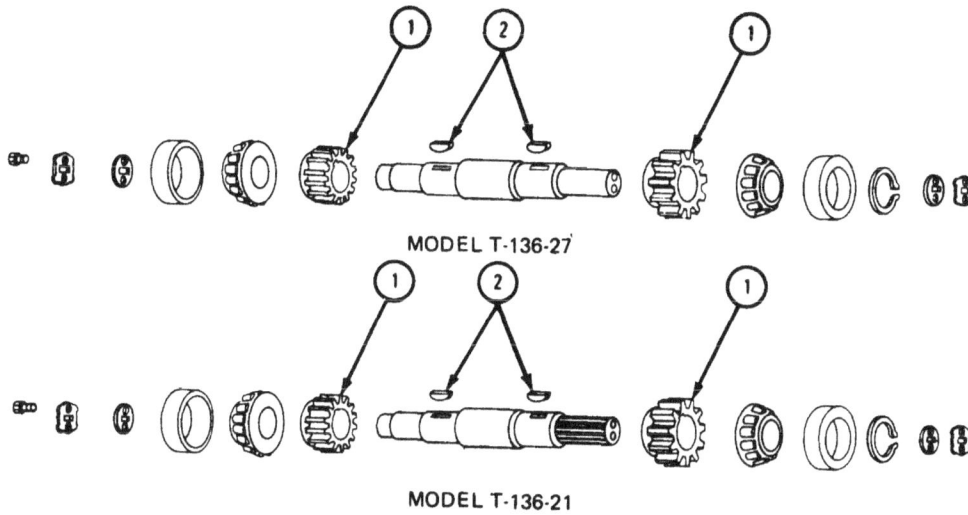

MODEL T-136-27

MODEL T-136-21

FRAME 2

1.   Check all bearing cones (1) and bearing cups (2). Refer to inspection,
     care and maintenance of antifriction bearings, TM 9-214.

2.   Check that shafts (3) are not chipped or cracked.

3.   Check that shaft splines (4) are not chipped, cracked, twisted or burred.

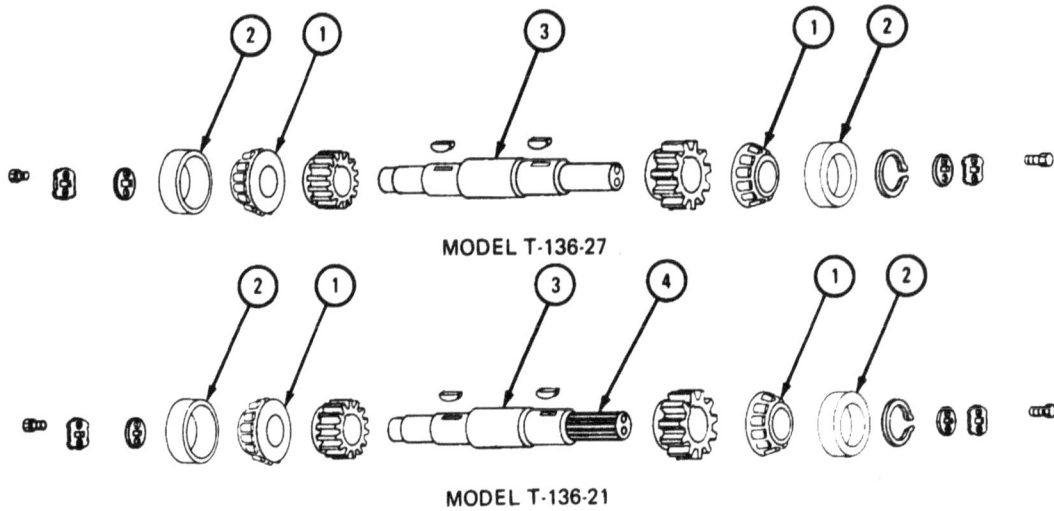

END OF TASK

MODEL T-136-27

MODEL T-136-21

NOTE
CHECK ONLY THOSE PARTS WHICH ARE CALLED OUT IN
THIS FRAME. PARTS WITHOUT CALLOUTS ARE SHOWN
ONLY FOR REFERENCE PURPOSES OR ARE CHECKED IN
ANOTHER FRAME.

TA 088030

d. Input Shaft Assembly.

FRAME 1

1. Check all bearings (1). Refer to inspection, care and maintenance of antifriction bearings, TM 9-214.

2. Check that shaft (2) is not chipped or cracked.

3. Check that shaft splines (3) are not chipped, cracked, twisted or burred.

4. Check that synchronizer (4) is not chipped, cracked and that internal splines are not twisted.

5. Check that gears (5) have no chips, cracks or damaged teeth.

6. Check that all threaded parts are not stripped or crossthreaded.

END OF TASK

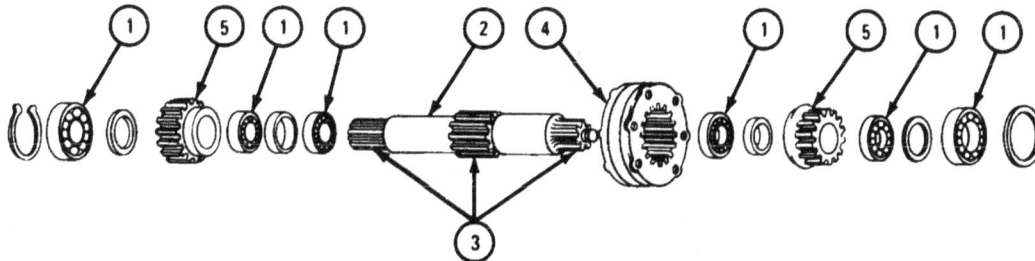

NOTE
CHECK ONLY THOSE PARTS WHICH ARE CALLED OUT IN THIS FRAME. PARTS WITHOUT CALLOUTS ARE SHOWN ONLY FOR REFERENCE PURPOSES OR ARE CHECKED IN ANOTHER FRAME.

TA 088031

e.   <u>Rear Output Shaft Assembly.</u>

FRAME 1

1.   Check all bearing cones (1) and bearing cups (2) . Refer to inspection, care and maintenance of antifriction bearings, TM 9-214.

2.   Check that rear output shaft (3) is not chipped or cracked.

3.   Check that shaft splines (4) are not chipped, cracked, twisted or burred.

4.   Check that sliding clutch (5) is not chipped or cracked and that internal splines are not twisted.

5.   Check that threads (6) are not stripped or crossthreaded.

6.   Check that gear (7) has no chips or cracks and that gear teeth are not damaged.

7.   Check that key (8) is not chipped or cracked.

END OF TASK

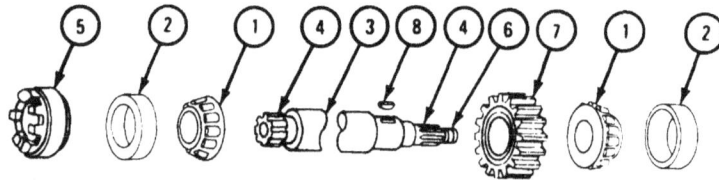

TRANSFER MODEL T-136-27

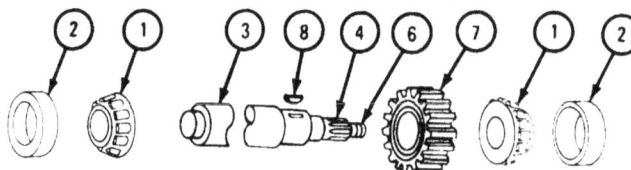

TRANSFER MODEL T-136-21

TA 088032

---

f. Transfer Case and Covers.

FRAME 1

1. Check that transfer case (1) and cover (2) have no chips, cracks or small holes.

2. Check that bearing covers (3) have no chips, cracks, small holes or worn screw holes.

3. Check that access cover (4) has no cracks, small holes or worn screw holes.

4. Check that output shaft case (5) has no chips, cracks or small holes.

5. Check that ball (6) has no flat spots.

6. Check that spring (7) is not weak or broken.

7. Check that companion flange (8) has no cracks, worn screw holes or worn or twisted splines.

8. Check that deflector (9) has no cracks or bends.

9. Check that all threads are not stripped or crossthreaded.

END OF TASK

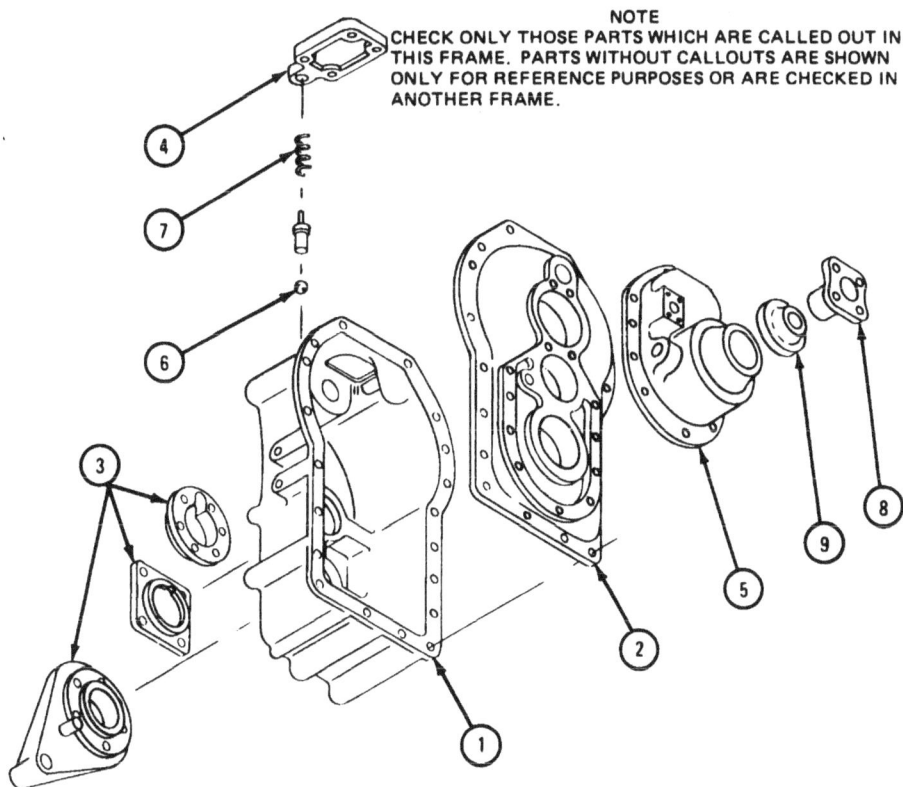

NOTE
CHECK ONLY THOSE PARTS WHICH ARE CALLED OUT IN THIS FRAME. PARTS WITHOUT CALLOUTS ARE SHOWN ONLY FOR REFERENCE PURPOSES OR ARE CHECKED IN ANOTHER FRAME.

g. Handbrake Assembly.

FRAME 1

1. Check that companion flange (1) has no cracks, worn screw holes or worn or twisted splines. Throw away damaged companion flange.
2. Check that brake drum shield (2) has no cracks or bends.
3. Check that brake drum (3) has no cracks, heavy scoring or signs of overheating.
4. Check that deflector (4) has no cracks or bends.

GO TO FRAME 2

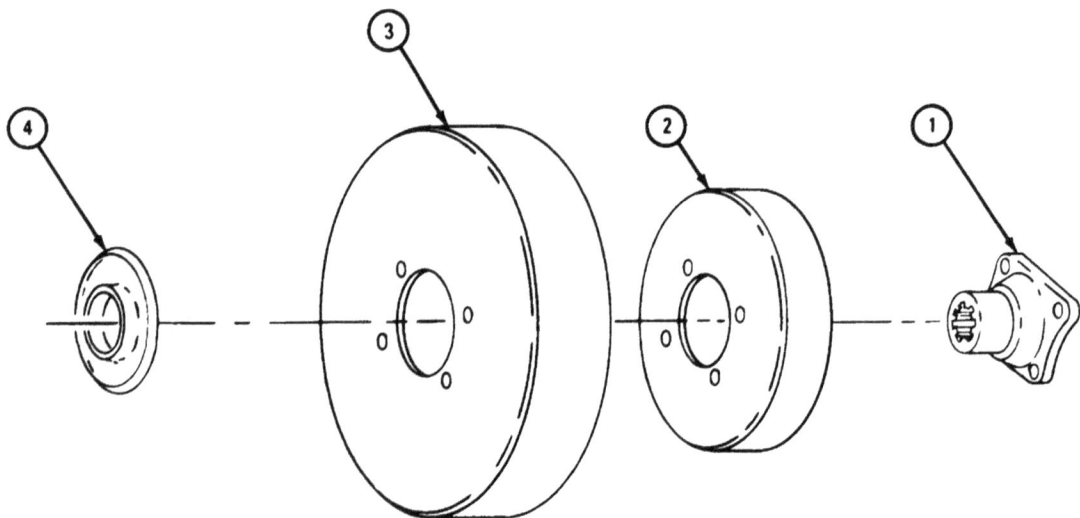

TA 088034

FRAME 2

1. Check that two brakeshoe linings (1) are not worn more than 1/16 inch from top of brake lining rivets. If linings are worn, put on new ones. Refer to TM 9-2320-209-34.

2. Check that two brakeshoe faces (2) and webs (3) have no cracks. If brakeshoes are cracked or worn, get new ones.

3. Check that handbrake lever (4) has no cracks. If lever is cracked, weld it. Refer to TM 9-237.

4. Check that all other parts have no wear and damage. Throw away damaged parts and get new ones in their place.

END OF TASK

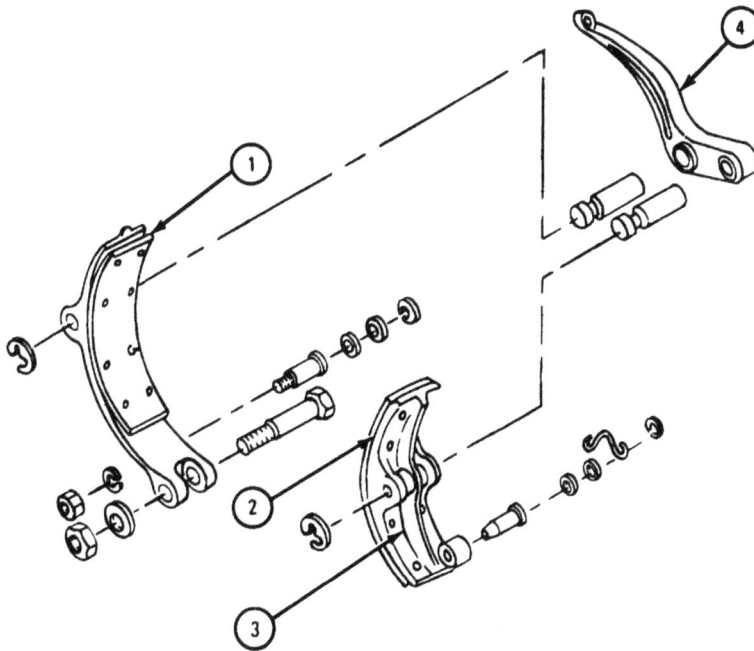

NOTE
CHECK ONLY THOSE PARTS WHICH ARE CALLED OUT IN
THIS FRAME. PARTS WITHOUT CALLOUTS ARE SHOWN
ONLY FOR REFERENCE PURPOSES OR ARE CHECKED IN
ANOTHER FRAME.

TA 088035

h. Air Cylinder Assembly (Model T-136-27).

FRAME 1

NOTE

Air cylinder parts cannot be repaired. If any parts are
damaged, throw them away and get new ones in their place.

1. Check that air cylinder cover (1) has no cracks and warpage.

2. Check that bore of air cylinder (2) has no pitting or rust and that it is not
out-of-round.

3. Check that piston (3) has no pitting or rust and that it is not out-of-round.

4. Check that piston seal (4) has no cracked, dry or split rubber.

5. Check that nut (5) and two washers (6) are not damaged.

END OF TASK

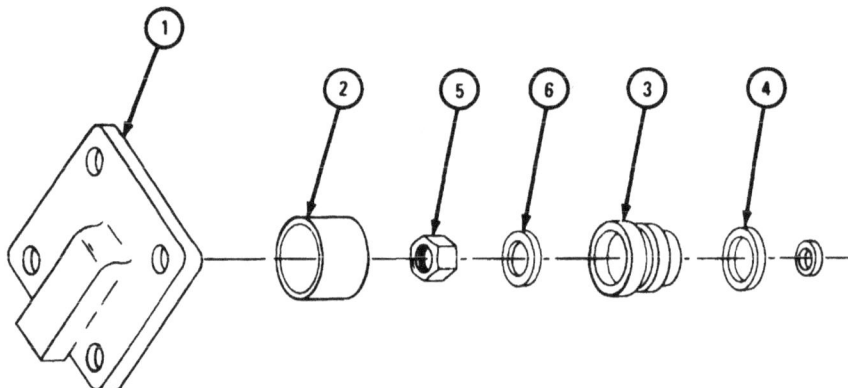

NOTE
CHECK ONLY THOSE PARTS WHICH ARE CALLED OUT IN
THIS FRAME. PARTS WITHOUT CALLOUTS ARE SHOWN
ONLY FOR REFERENCE PURPOSES OR ARE CHECKED IN
ANOTHER FRAME.

TA 088036

2-18.   WEAR LIMIT INSPECTION (MODEL T-136-27). The following paragraphs give the minimum and maximum wear limits for each subassembly to which a part or parts may be worn before a new part is needed.

a.   Low Range Shifter Shaft Assembly.

FRAME 1

NOTE

Readings must be within limits given in table 2-15. If readings are not within given limits, throw away   part and get a new one.

1.   easure diameter of shifter shaft (1).

2.   Measure width of groove in shifter fork (2).

END OF TASK

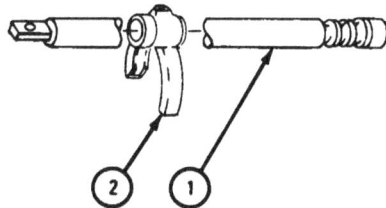

TA 088037

Table 2-15. Low Range Shifter Shaft Assembly Wear Limits

Index Number	Item/Point of Measurement	Size and Fit of New Parts (inches)	Wear Limit (inches)
1	Low range shifter shaft outside diameter	0.9945 to 0.9955	None
2	Shifter fork groove width	0.712 to 0.720	0.7400

b. <u>Air Cylinder Shifter Shaft Assembly.</u>

FRAME 1

1. Check that shifter shaft (1) is not bent or cracked. Throw away bent or cracked shifter shaft.

2. Check that shifter fork (2) is not bent or cracked. Throw away bent or cracked fork.

NOTE

Readings must be within limits given in table 2-16. If readings are not within given limits, throw away part and get a new one.

3. Measure diameter of shifter shaft (1).

4. Measure width of two shifter fork pads (3).

5. Measure to check that shifter fork pads (3) are perpendicular to shifter fork bore (4).

6. Measure height of spring (5).

END OF TASK

TA 088038

Table 2-16. Air Cylinder Shifter Shaft Assembly Wear Limits

Index Number	Item/Point of Measurement	Size and Fit of New Parts (inches)	Wear Limit (inches)
1	Shifter shaft outside diameter	0.8715 to 0.8725	0.8685
3	Shifter fork pad width	0.562 to 0.572	0.5550
3 and 4	Shifter fork perpendicularity of pad to bore	0.005	0.030
5	Free height of spring	2.64 to 2.85	2.55 minimum

c.   Front Output Shaft Assembly.

FRAME 1	

NOTE

Readings must be within limits given in table 2-17. If readings are not within given limits, throw away part and get a
new one.

1.   Measure inner bearing race (1) and outer bearing race (2).

2.   Measure clutch inside diameter (3).

3.   Measure clutch groove width (4).

4.   Measure clutch jaw taper (5).

GO TO FRAME 2

NOTE
CHECK ONLY THOSE PARTS WHICH ARE CALLED OUT IN
THIS FRAME. PARTS WITHOUT CALLOUTS ARE SHOWN
ONLY FOR REFERENCE PURPOSES OR ARE CHECKED IN
ANOTHER FRAME.

TA 088039

Table 2-17.   Front Output Shaft Assembly Wear Limits

Index Number	Item/Point of Measurement	Size and Fit of New Parts (inches)	Wear Limit (inches)
1	Bearing inside diameter	1.7712 to 1.7717	None
2	Bearing outside diameter	3.3459 to 3.3465	None
3	Clutch inside diameter	2. 515 to 2.517	None
4	Sliding clutch groove width	0.630 to 0.635	None
5	Clutch jaw taper	0.006 to 0.013	0.0020

FRAME 2

NOTE

Readings must be within limits given in table 2-18.
The letter L indicates a loose fit and the letter T
indicates a tight fit.   If readings are not within
given limits,   throw away part and get a new one.

1.   Measure front diameter (1) and rear diameter (2) of output shaft.

2.   Measure fit of fork (3) in clutch groove (4).

3.   Measure fit of bearing (5) on output shaft (1).

END OF TASK

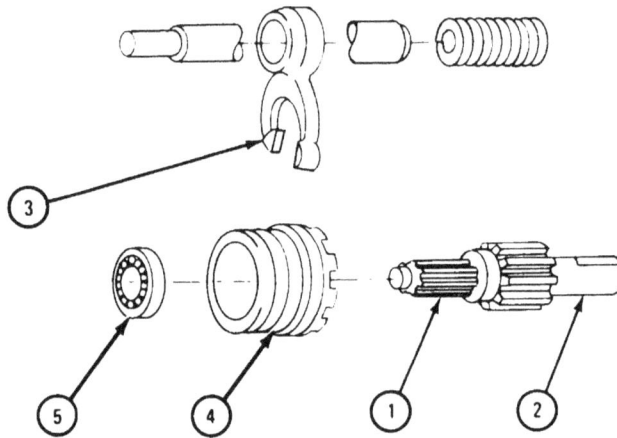

NOTE
CHECK ONLY THOSE PARTS WHICH ARE CALLED OUT IN
THIS FRAME. PARTS WITHOUT CALLOUTS ARE SHOWN
ONLY FOR REFERENCE PURPOSES OR ARE CHECKED IN
ANOTHER FRAME.

TA 088040

Table 2-18.   Front Output Shaft Assembly Fits and Tolerances

Index Number	Item/Point of Measurement	Size and Fit of New Parts (inches)	Wear Limit (inches)
1	Front output shaft front diameter	1.7716 to 1.7721	1.7716
2	Front output shaft rear diameter	1.245 to 1.246	1.2435
3 and 4	Fit of fork in clutch groove	0.058 to 0.073L	0.088L
1 and 5	Fit of bearing on output shaft	0.001T to O.0009T	None

d. Countershaft Assembly.

FRAME 1

NOTE

Readings must be within limits given in table 2-19. If readings are not within given limits, throw away part and get a new one.

1. Measure two inner bearing races (1) and two outer bearing races (2).

2. Measure two gear bores (3).

3. Measure keyway width (4) in two gears.

4. Measure width of two keys (5).

GO TO FRAME 2

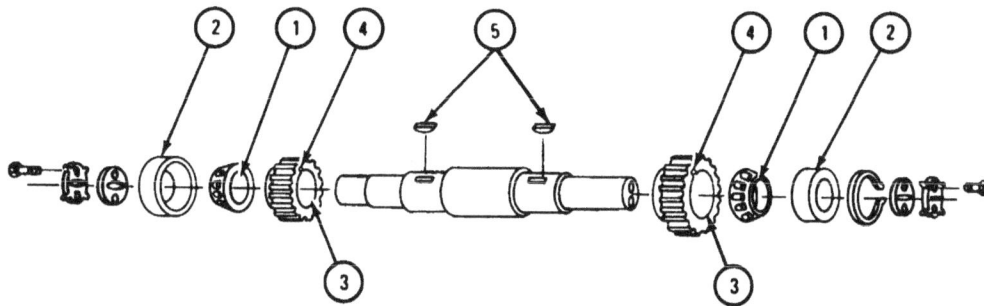

NOTE: CHECK ONLY THOSE PARTS WHICH
ARE CALLED OUT IN THIS FRAME.
PARTS WITHOUT CALLOUTS ARE
SHOWN ONLY FOR REFERENCE
PURPOSES OR ARE CHECKED IN
ANOTHER FRAME.

TA 088041

Table 2-19. Coutershaft Gears and Bearings Wear Limits

Index Number	Item/Point of Measurement	Size and Fit of New Parts (inches)	Wear Limit (inches)
1	Bearing inner race diameter	2.000 to 2.0005	None
2	Bearing outer race diameter	4.000 to 4.0010	None
3	Gear bore	2.375 to 2.376	2.3765
4	Gear keyway width	0.5000 to 0.5025	0.5052
5	Key width	0.5000 to 0.5010	None

FRAME 2

NOTE

Readings must be within limits given in table 2-20. If readings are not within given limits, throw away part and get a new one.

1. Measure width of two countershaft keyways (1).

2. Measure diameter of two countershaft gear surfaces (2).

3. Measure diameter of two countershaft bearing surfaces (3).

GO TO FRAME 3

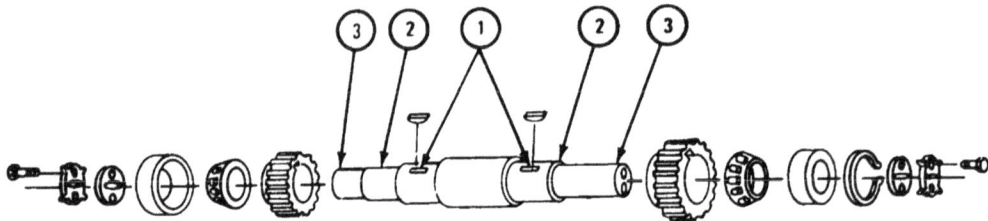

NOTE
CHECK ONLY THOSE PARTS WHICH ARE CALLED OUT IN THIS FRAME. PARTS WITHOUT CALLOUTS ARE SHOWN ONLY FOR REFERENCE PURPOSES OR ARE CHECKED IN ANOTHER FRAME.

TA 088042

Table 2-20. Countershaft Wear Limits

Index Number	Item/Point of Measurement	Size and Fit of New Parts (inches)	Wear Limit (inches)
1	Countershaft keyway width	0.4990 to 0.5010	0. 5028
2	Countershaft gear surface diameter	2.3765 to 2.3775	2.3755
3	Countershaft bearing surface diameter	2.0010 to 2.0015	2.0005

FRAME 3

NOTE

Readings must be within limits given in table 2-21. The
letter L indicates a loose fit and the letter T indicates a
tight fit. If readings are not within given limits, throw
away part and get a new one.

1. Measure fit of two gears (1) on two countershaft gear surfaces (2).

2. Measure fit of two keys (3) in two countershaft keyways (4).

3. Measure fit of two bearings (5) on two countershaft bearing surfaces (6).

END OF TASK

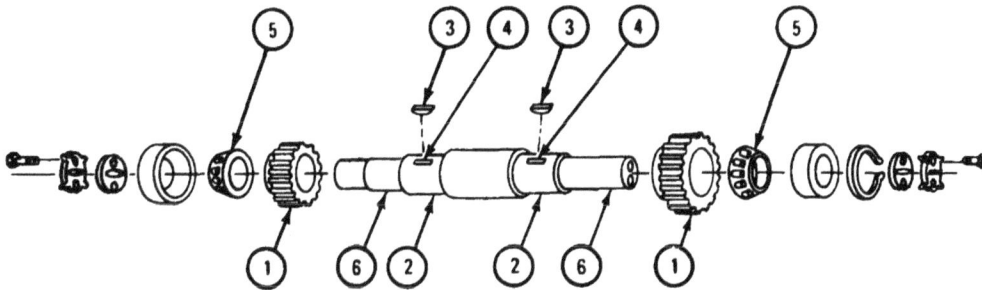

NOTE
CHECK ONLY THOSE PARTS WHICH ARE CALLED OUT IN
THIS FRAME. PARTS WITHOUT CALLOUTS ARE SHOWN
ONLY FOR REFERENCE PURPOSES OR ARE CHECKED IN
ANOTHER FRAME.

TA 088043

Table 2-21. Countershaft Assembly Fits and Tolerances

Index Number	Item/Point of Measurement	Size and Fit of New Parts (inches)	Wear Limit (inches)
1 and 2	Fit of gear on shaft	0.0005T to 0.0025T	0.0010L
3 and 4	Fit of key in keyway	0.0010L to 0.0020T	.0030L
5 and 6	Fit of bearing on shaft	0.0005T to 0.0015T	None

e. Input Shaft Assembly.

| FRAME 1 | |

NOTE

Readings must be within limits given in table 2-22. If readings are not within given limits, throw away part and get a new one.

1. Measure inner bearing race (1) on four gear bearings.

2. Measure outer bearing race (2) on four gear bearings.

3. Measure front inner bearing race (3) and front outer bearing race (4).

4. Measure rear inner bearing race (5) and rear outer bearing race (6).

GO TO FRAME 2

NOTE: CHECK ONLY THOSE PARTS WHICH ARE CALLED OUT IN THIS FRAME. PARTS WITHOUT CALLOUTS ARE SHOWN ONLY FOR REFERENCE PURPOSES OR ARE CHECKED IN ANOTHER FRAME.

TA 088044

Table 2-22. Input Shaft Bearings Wear Limits

Index Number	Item/Point of Measurement	Size and Fit of New Parts (inches)	Wear Limit (inches)
1	Gear bearing inside diameter	1.9680 to 1.9685	None
2	Gear bearing outside diameter	3.5427 to 3.5433	None
3	Front bearing inside diameter	1.9680 to 1.9685	None
4	Front bearing outside diameter	4.3301 to 4.3307	None
5	Rear bearing inner diameter	1.7712 to 1.7717	None
6	Rear bearing outside diameter	3.9364 to 3.9370	None

FRAME 2

NOTE

Readings must be within limits given in table 2-23. If
readings are not within given limits, throw away part
and get a new one.

1. Measure low range gear bore (1) and low range gear outside diameter (2).

2. Measure high range gear bore (3) and high range gear outside diameter (4).

3. Measure inside diameter of two spacer sleeves (5).

4. Measure clutch ring width (6).

5. Measure input shaft diameter (7) in two places and bearing surface diameter (8)

GO TO FRAME 3

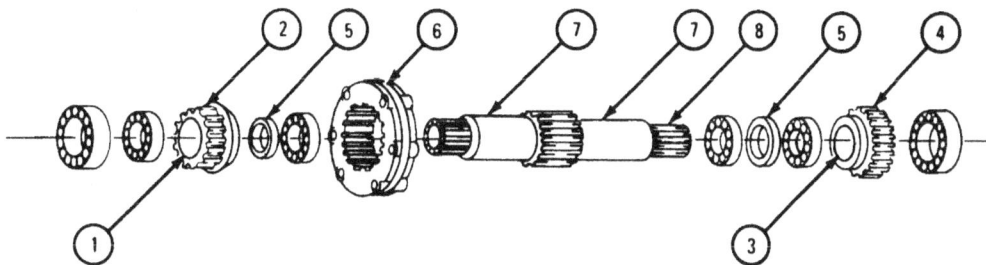

NOTE
CHECK ONLY THOSE PARTS WHICH ARE CALLED OUT IN
THIS FRAME. PARTS WITHOUT CALLOUTS ARE SHOWN
ONLY FOR REFERENCE PURPOSES OR ARE CHECKED IN
ANOTHER FRAME.

TA 088045

Table 2-23. Input Shaft and Gears Wear Limits

Index Number	Item/Point of Measurement	Size and Fit of New Parts (inches)	Wear Limit (inches)
1	Low range input gear bore	3.5425 to 2.5434	3.5439
2	Low range input gear outside diameter	5.3255 to 5.3265	None
3	High range gear bore	3.5425 to 3.5434	3.5439
4	High range gear outside diameter	6.895 to 6.900	None
5	Spacer sleeve inside diameter	1.979 to 1.989	None
6	Clutch ring width	0.682 to 0.687	0.6600
7	Input shaft diameter	1.9682 to 1.9687	None
8	Bearing surface diameter	1.7716 to 1.7721	1.7716

FRAME 3	

NOTE

Readings must be within limits given in table 2-24. The letter L indicates a loose fit and the leter T indicates a tight fit. If readings are not within given limits, throw away part and get a new one.

1. Measure fit of four bearings (1) in bore of two gears (2).

2. Measure fit of fork (3) on ring of clutch (4).

3. Measure fit of four bearings (1) on input shaft bearing surface (5) in two places.

4. Measure fit of front bearing (6) on input shaft bearing surface (5).

5. Measure fit of rear bearing (7) on input shaft bearing surface (8).

END OF TASK

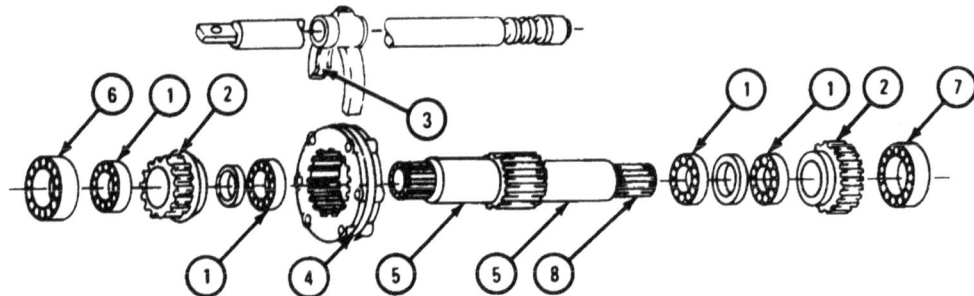

NOTE: CHECK ONLY THOSE PARTS WHICH ARE CALLED OUT. PARTS WITHOUT CALLOUTS ARE SHOWN ONLY FOR REFERENCE PURPOSES.

TA 088046

Table 2-24.   Input Shaft Assembly Fits and Tolerances

Index Number	Item/Point of Measurement	Size and Fit of New Parts (inches)	Wear Limit (inches)
1 and 2	Fit of bearing in gear bore	0.0007T to 0.0009L	0.0012L
3 and 4	Fit of fork on clutch	0.0250L to 0.0380L	0.0800L
1 and 5	Fit of bearing on input shaft	0.0002T to 0.0007T	0. 00 01T
6 and 5	Fit of front bearing on input shaft	0.0002T to 0.0007T	0.000IT
7 and 8	Fit of rear bearing on input shaft	0.0009T to 0.0004T	None

f. Rear Output Shaft Assembly.

FRAME 1

NOTE
Readings must be within limits given in table 2-25. If readings are not within given limits, throw away part and get a new one.

1. Measure clutch jaw taper (1).
2. Measure inner bearing race (2) and outer bearing race (3).
3. Measure inner bearing race (4) and outer bearing race (5).
4. Measure gear bore (6).
5. Measure gear keyway width (7).

GO TO FRAME 2

NOTE: CHECK ONLY THOSE PARTS WHICH ARE CALLED OUT IN THIS FRAME. PARTS WITHOUT CALLOUTS ARE SHOWN ONLY FOR REFERENCE PURPOSES OR ARE CHECKED IN ANOTHER FRAME.

Table 2-25. Rear Output Shaft Clutch and Bearings Wear Limits

Index Number	Item/Point of Measurement	Size and Fit of New Parts (inches)	Wear Limit (inches)
1	Clutch jaw taper	0.006 to 0.012	0.0020
2	Bearing inside diameter	2.6250 to 2.6260	None
3	Bearing race outside diameter	404375 to 4.4385	None
4	Bearing inside diameter	1.7500 to 1.7505	None
5	Bearing race outside diameter	3.8750 to 3.8760	None
6	Driven gear bore	2.375 to 2.376	2.3765
7	Gear keyway width	0.5000 to 0.5025	0.5052

FRAME 2

NOTE

Readings must be within limits given in table 2-26. If readings are not within given limits, throw away part and get a new one.

1. Measure output shaft bushing bore (1).

2. Measure output shaft diameters (2, 3, and 4).

3. Measure output shaft keyway width (5).

4. Measure key width (6).

GO TO FRAME 3

NOTE
CHECK ONLY THOSE PARTS WHICH ARE CALLED OUT IN
THIS FRAME. PARTS WITHOUT CALLOUTS ARE SHOWN
ONLY FOR REFERENCE PURPOSES OR ARE CHECKED IN
ANOTHER FRAME.

TA 088048

Table2-26. Rear Output Shaft Wear Limits

Index Number	Item/Point of Measurement	Size and Fit of New Parts (inches)	Wear Limit (inches)
1	Rear output shaft bushing bore	1.248 to 1.250	1.256
2	Rear output shaft outside diameter	2.6265 to 2.6270	2.6260
3	Rear output shaft outside diameter	2.3765 to 2.3775	2.3755
4	Rear output shaft outside diameter	1.7510 to 1.7515	1.7505
5	Rear output shaft keyway width	0.4985 to 0.5005	0.5028
6	Key width	0.5000 to 0.500	None

FRAME 3

NOTE

Readings must be within limits given in table 2-27.
The letter L indicates a loose fit and the letter T
indicates a tight fit.   If readings are not within
given limits, throw away part and get a new one.

1.   Measure fit of bearing (1) on output shaft bearing surface (2).

2.   Measure fit of bearing (3) on output shaft bearing surface (4).

3.   Measure fit of gear (5) on output shaft gear surface (6).

4.   Measure fit of key (7) in output shaft keyway (8).

5.   Measure fit of front output shaft (9) in rear output shaft bore (10).

END OF TASK

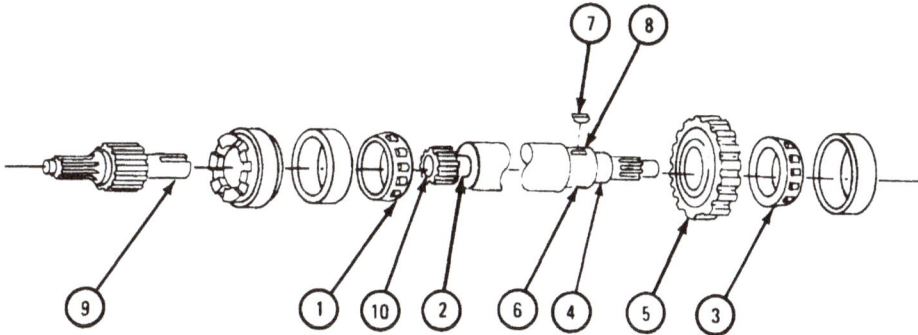

NOTE
CHECK ONLY THOSE PARTS WHICH ARE CALLED OUT IN
THIS FRAME. PARTS WITHOUT CALLOUTS ARE SHOWN
ONLY FOR REFERENCE PURPOSES OR ARE CHECKED IN
ANOTHER FRAME.

TA 088049

Table 2-27.   Rear Output Shaft Assembly Fits and Tolerances

Index Number	Item/Point of Measurement	Size and Fit of New Parts (inches)	Wear Limit (inches)
1 and 2	Fit of bearing on rear output shaft	0.005T to 0.0025T	0.0010L
3 and 4	Fit of bearing on output shaft	0.0015T to 0.0005T	None
5 and 6	Fit of gear on output shaft	0.0005T to 0.0025T	0.0010L
7 and 8	Fit of key in keyway	0.0005L to 0.0025T	0.0028L
9 and 10	Fit of shaft in bore	0.002L to 0.004L	0.0055L

TM 9-2520-246-34-1

g. Transfer Case and Covers.

FRAME 1

NOTE

Readings must be within limits given in table 2-28. If readings are not within given limits, throw away part and get a new one.

1. Measure shifter shaft bore (1).

2. Measure input shaft front bearing bore (2).

3. Measure countershaft front bearing bore (3).

4. Measure rear output shaft front bearing bore (4).

GO TO FRAME 2

TA 088050

Table 2-28.   Transfer Cover Bearing Bores Wear Limits

Index Number	Item/Point of Measurement	Size and Fit of New Parts (inches)	Wear Limit (inches)
1	Shifter shaft housing bore	0.9995 to 1.0015	1.0022
2	Input shaft front bearing bore	4.3305 to 4.3315	4.3320
3	Countershaft front bearing bore	3.998 to 3.999	4.00
4	Rear output shaft front bearing bore	4.4355 to 4.4365	4.4375

| FRAME 2 |

NOTE

Readings must be within limits given in table 2-29. The letter
L indicates a loose fit and the letter T indicates a tight fit.
If readings are not within given limits, throw away part and
get a new one.

1. Measure fit of shifter shaft (1) in bore (2).

2. Measure fit of input shaft front bearing (3) in bearing bore (4).

3. Measure fit of countershaft front bearing outer race (5) in bearing bore (6).

4. Measure fit of front output shaft front bearing outer race (7) in bearing bore (8).

GO TO FRAME 3

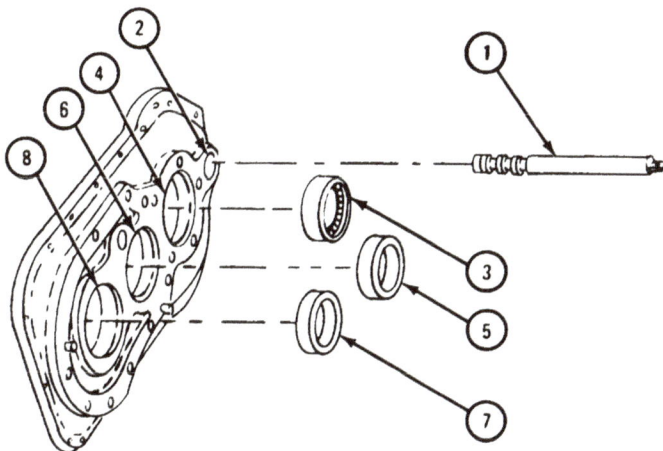

TA 088051

Table 2-29.   Transfer Cover Fits and Tolerances

Index Number	Item/Point of Measurement	Size and Fit of New Parts (inches)	Wear Limit (inches)
1 and 2	Fit of shifter shaft in bore	0.007L to 0.004L	0.0077L
3 and 4	Fit of input shaft bearing in bore	0.0014L to 0.0002T	0.0019L
5 and 6	Fit of counter shaft bearing race in bore	0.001T to 00.003T	None
7 and 8	Fit of output shaft bearing race in bore	0.0010T to 0.0030T	None

FRAME 3	

NOTE

Readings must be within limits given in table 2-30.
If readings are not within given limits, throw away
part and get a new one.

1. Measure shifter shaft bore (1).

2. Measure expansion plug outside diameter (2).

3. Measure input shaft rear bearing bore (3).

4. Measure countershaft rear bearing bore (4).

5. Measure rear output shaft rear bearing bore (5).

GO TO FRAME 4

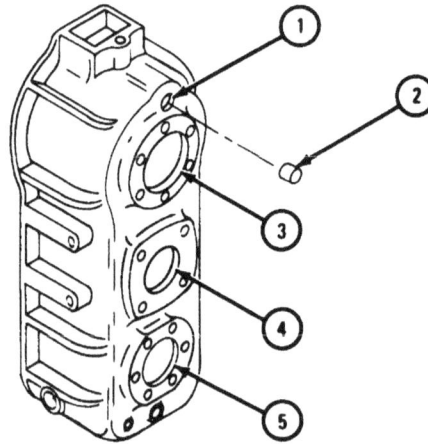

TA 088052

Table 2-30.   Transfer Case Bearing Bores Wear Limits

Index Number	Item/Point of Measurement	Size and Fit of New Parts (inches)	Wear Limit (inches)
1	Shifter shaft bore	1.247 to 1.249	1.2514
2	Expansion plug outside diameter	1.256 to 1.260	None
3	Input shaft rear bearing bore	3.9368 to 3.9378	3.9382
4	Countershaft rear bearing bore	4.000 to 4.001	4.0012
5	Rear output shaft rear bearing bore	3.8750 to 3.8760	3.8765

FRAME 4

NOTE

Readings must be within limits given in table 2-31. The letter
L indicates a loose fit and the letter T indicates a tight fit.
If readings are not within given limits, throw away part and
get a new one.

1. Measure fit of expansion plug (1) in housing bore (2).

2. Measure fit of input shaft rear bearing (3) in bearing bore (4).

3. Measure fit of countershaft rear bearing outer race (5) in bearing bore (6).

4. Measure fit of rear output shaft rear bearing outer race (7) in bearing bore (8).

GO TO FRAME 5

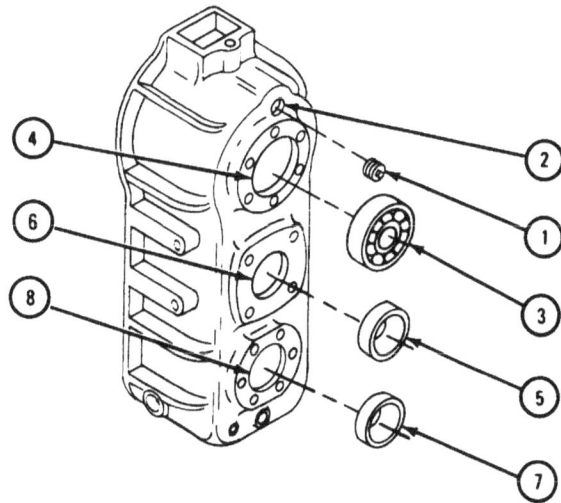

TA 088053

Table 2-31.   Transfer Case Fits and Tolerances

Index Number	Item/Point of Measurement	Size and Fit of New Parts (inches)	Wear Limit (inches)
1 and 2	Fit of expansion plug in bore	0.007T to 0.013T	0.0046T
3 and 4	Fit of input shaft bearing in bore	0.0014L to 0.0002T	0.0018L
5 and 6	Fit of counter shaft bearing race in bore	0.001L to 0.0010T	0.0012L
7 and 8	Fit of output shaft bearing race in bore	0.0010T to 0.0010L	0.0015L

FRAME 5

NOTE

Readings must be within limits given in table 2-32.
The letter T indicates a tight fit. If readings are
not within given limits,   throw away part and get
a new one.

1.   Measure diameter of dowel pin (1).

2.   Measure dowel pin bore (2).

3.   Measure fit of dowel pin (1) in bore (2).

GO TO FRAME 6

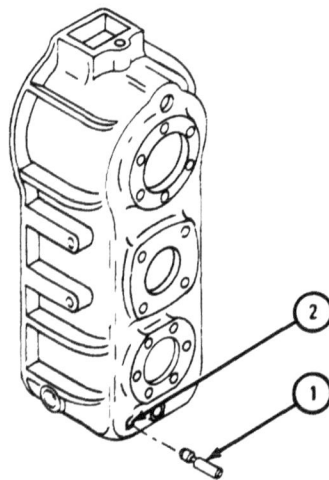

TA 088054

Table 2-32.   Transfer Case Dowel Pin Wear Limits

Index Number	Item/Point of Measurement	Size and Fit of New Parts (inches)	Wear Limits (inches)
1	Dowel pin diameter	0.7501 to 0.7503	None
2	Dowel pin bore	0.748 to 0.749	0.7498
1 and 2	Fit of dowel pin in bore	0.011T to 0.0023T	0.0003T

FRAME 6

NOTE

Readings must be within limits given in table 2-33. The letter L indicates a loose fit and the letter T indicates a tight fit.   If readings are not within given limits, throw away part and get a new one.

1. Measure shifter shaft bore (1).

2. Measure front output shaft bearing bore (2).

3. Measure fit of shifter shaft (3) in shifter shaft bore (1).

4. Measure front output shaft bearing (4) in front output shaft bearing bore (2).

GO TO FRAME 7

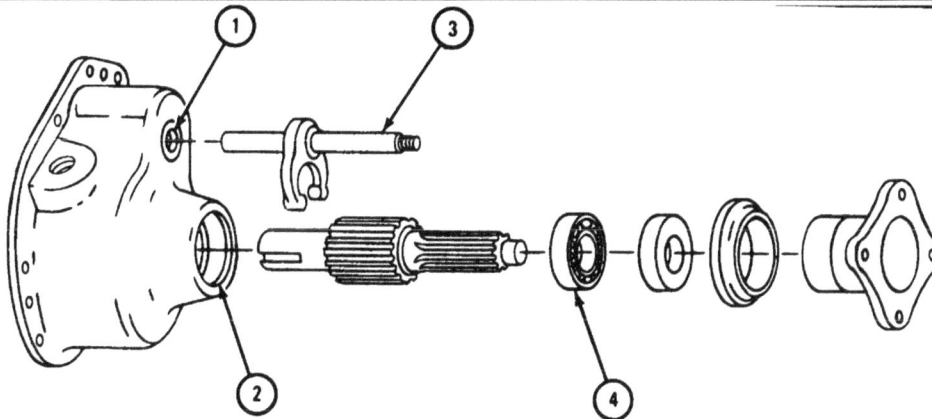

NOTE: CHECK ONLY THOSE PARTS WHICH ARE CALLED OUT IN THIS FRAME. PARTS WITHOUT CALLOUTS ARE SHOWN ONLY FOR REFERENCE PURPOSES OR ARE CHECKED IN ANOTHER FRAME.

TA 088055

Table 2-33. Front Output Shaft Cover Wear Limits

Index Number	Item/Point of Measurement	Size and Fit of New Parts (inches)	Wear Limit (inches)
1	Shifter shaft bore	0.8745 to 0.8765	0.8800
2	Front output shaft bearing bore	3.3464 to 3.3473	3.3483
1 and 3	Fit of shifter shaft in bore	0.0030L to 0.0040L	0.0115L
2 and 4	Fit of output shaft bearing in bore	0.0024L to 0.0002T	0.0024L

FRAME 7

NOTE

Readings must be within limits given in table 2-34. The
letter L indicates a loose fit and the letter T indicates a
tight fit. If readings are not within given limits, throw
away part and get a new one.

1. Measure companion flange inside diameter (1) and outside diameter (2).

2. Measure deflector inside diameter (3).

3. Measure oil seal outside diameter (4).

4. Measure fit of companion flange inside diameter (1) on front output shaft (5).

5. Measure fit of companion flange outside diameter (2) in deflector (3).

6. Measure fit of oil seal outside diameter (4) in output shaft bore (6).

GO TO FRAME 8

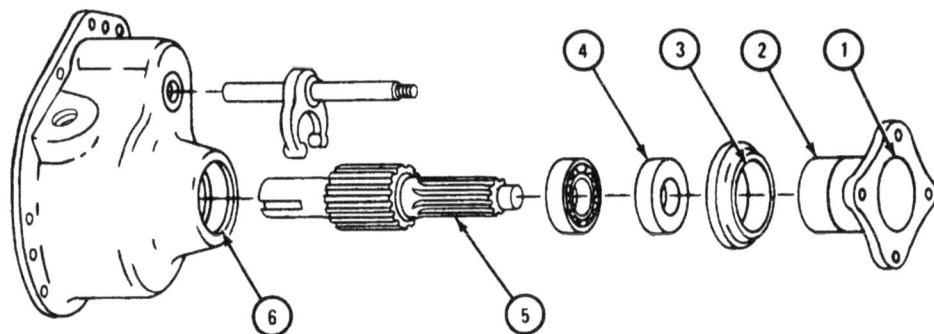

NOTE
CHECK ONLY THOSE PARTS WHICH ARE CALLED OUT IN
THIS FRAME. PARTS WITHOUT CALLOUTS ARE SHOWN
ONLY FOR REFERENCE PURPOSES OR ARE CHECKED IN
ANOTHER FRAME.

TA 088056

Table 2-34.   Front Output Shaft and Companion Flange Wear Limits

Index Number	Item/Point of Measurement	Size and Fit of New Parts (inches)	Wear Limit (inches)
1	Companion flange inside diameter	1.621 to 1.624	1.627
2	Companion flange outside diameter	2.522 to 2.530	2.5150
3	Deflector inside diameter	2.506 to 2.511	None
4	Oil seal outside diameter	3.353 to 3.357	None
1 and 5	Fit of flange on shaft	0.0020T to 0.0030L	0.0073L
2 and 3	Fit of deflector on flange	0.011T to 0.024T	0.0008T
4 and 6	Fit of oil seal in output shaft bore	0.003T to 0.009T	0.0020T

FRAME 8

NOTE

Readings must be within limits given in table 2-35. If readings are not within given limits, throw away part and get a new one.

1. Measure flange outside diameter (1).

2. Measure deflector inside diameter (2).

3. Measure oil seal outside diameter (3).

4. Measure input shaft cover bore (4).

5. Measure oil seal outside diameter (5).

GO TO FRAME 9

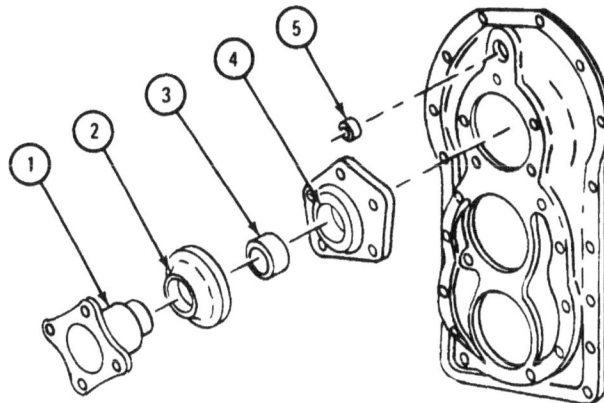

NOTE
CHECK ONLY THOSE PARTS WHICH ARE CALLED OUT IN
THIS FRAME. PARTS WITHOUT CALLOUTS ARE SHOWN
ONLY FOR REFERENCE PURPOSES OR ARE CHECKED IN
ANOTHER FRAME.

TA 088057

Table 2-35.   Input Shaft Cover Wear Limits

Index Number	Item/Point of Measurement	Size and Fit of New Parts (inches)	Wear Limit (inches)
1	Input shaft flange outside diameter	2.522 and 2.530	2.5150
2	Deflector inside diameter	2.506 and 2.511	None
3	Oil seal outside diameter	3.353 and 3.357	None
4	Input shaft cover bore	3.348 and 3.350	3.3495
5	Oil seal outside diameter	1.501 and 1.505	None

FRAME 9

NOTE

Readings must be within limits given in table 2-36.
The letter T indicates a tight fit. If readings are
not within given limits,   throw away part and get
a new one.

1. Measure fit of oil seal (1) in bore (2).

2. Measure fit of oil seal (3) in input shaft cover bore (4).

3. Measure fit of deflector (5) on flange (6).

GO TO FRAME 10

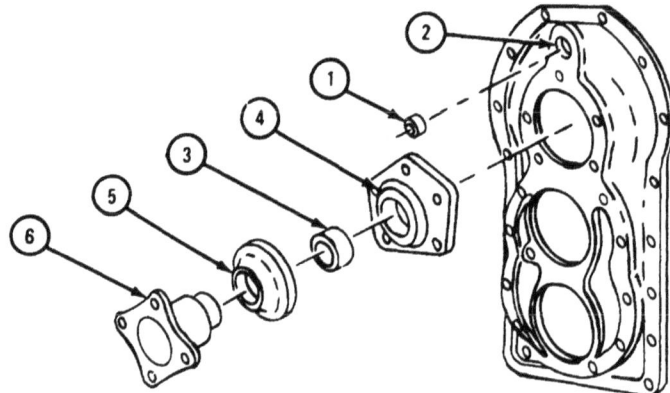

TA 088058

Table 2-36.   Input Shaft Cover Fits and Tolerances

Index Number	Item/Point of Measurement	Size and Fit of New Parts (inches)	Wear Limit (inches)
1 and 2	Fit of oil seal in bore	0.001T to 0.007T	0.0065T
3 and 4	Fit of oil seal in input shaft cover bore	0.003T to 0.009T	0.0020T
5 and 6	Fit of deflector on flange	0.011T to 0.024T	0.0009T

FRAME 10

NOTE

Readings must be within limits given in table 2-37. The letter L indicates a loose fit and the letter T indicates a tight fit. If readings are not within given limits, throw away part and get a new one.

1. Measure outside diameter of oil seal (1).

2. Measure rear bearing retainer bore (2).

3. Measure dowel pin bore (3).

4. Measure fit of oil seal (1) in rear bearing retainer bore (2).

5. Measure fit of dowel pin (4) in dowel pin bore (3).

END OF TASK

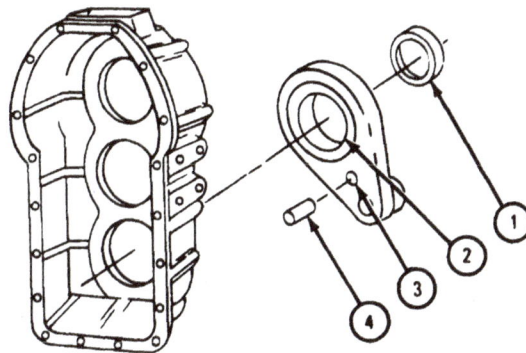

NOTE
CHECK ONLY THOSE PARTS WHICH ARE CALLED OUT IN
THIS FRAME. PARTS WITHOUT CALLOUTS ARE SHOWN
ONLY FOR REFERENCE PURPOSES OR ARE CHECKED IN
ANOTHER FRAME.

TA 088059

Table 2-37.   Rear Bearing Retainer Wear Limits

Index Number	Item/Point of Measurement	Size and Fit of New Parts (inches)	Wear Limit (inches)
1	Oil seal outside diameter	3.353 to 3.357	None
2	Rear bearing retainer bore	3.348 to 3.350	3.3510
3	Dowel pin bore	0.7500 to 0.7510	0.7525
1 and 2	Fit of oil seal in retainer bore	0. 003T to 0.009T	0.0020T
3 and 4	Fit of dowel pin in bore	0.0009L to 0.0003T	0.0024L

2-19. WEAR LIMIT INSPECTION (MODEL T-136-21). The following   paragraphs give
the minimum and maximum wear limits for each subassembly to which   a part or parts
may be worn before a new part is needed.

    a.   Front Output Shaft Assembly.

FRAME 1

NOTE

Readings must be within limits given in table 2-38. If readings are not within given limits, throw away part and get a
new one.

1.   Measure front output shaft diameter (1) and diameter (2).

2.   Measure width of key (3).

3.   Measure width of front output shaft keyway (4).

4.   Measure inner race outside diameter (5) and inside diameter (6).

5.   Measure width of inner race keyway (7).

6.   Measure rear output shaft bore (8).

GO TO FRAME 2

NOTE
CHECK ONLY THOSE PARTS WHICH ARE CALLED OUT IN THIS
FRAME. PARTS WITHOUT CALLOUTS ARE SHOWN ONLY FOR
REFERENCE PURPOSES OR ARE CHECKED IN ANOTHER FRAME.   TA 088060

Table 2-38. Front Output Shaft Wear Limits

Index Number	Item/Point of Measurement	Size and Fit of New Parts (inches)	Wear Limit (inches)
1	Output shaft inside diameter	1.8096 to 1.8106	1.8087
2	Output shaft outside diameter	2.0028 to 2.0036	2.0023
3	Key width	0.3750 to 0.3760	None
4	Output shaft keyway width	0.3736 to 0.3756	None
5	Inner race outside diameter	3.0875 to 3.0885	3.0865
6	Inner race bore	2.0028 to 2.0040	2.0045
7	Inner race keyway width	0.3740 to 0.3760	0.3780
8	Rear output shaft bore	1.8126 to 1.8136	1.8142

FRAME 2

NOTE

Readings must be within limits given in table 2-39. If readings are not within given limits, throw away part and get a new one.

1. Measure reverse shift collar groove width (1).

2. Measure reverse shift shaft fork thickness (2).

3. Measure rear outer race inside diameter (3).

4. Measure rear sprag unit width (4).

5. Measure front sprag unit width (5).

6. Measure front outer race inside diameter (6).

7. Measure free length of spring (7) and spring length at pressures given in table 2-39.

GO TO FRAME 3

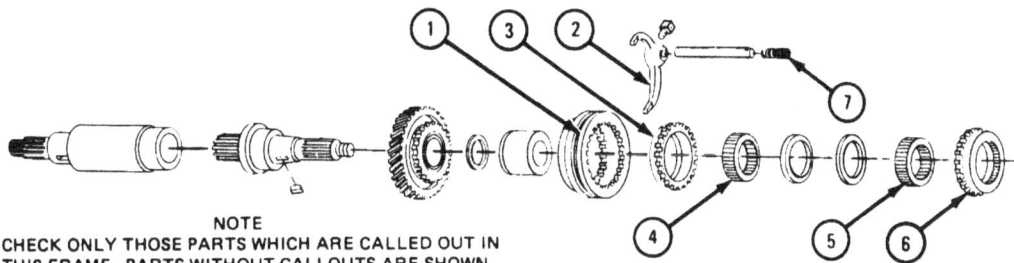

NOTE
CHECK ONLY THOSE PARTS WHICH ARE CALLED OUT IN THIS FRAME. PARTS WITHOUT CALLOUTS ARE SHOWN ONLY FOR REFERENCE PURPOSES OR ARE CHECKED IN ANOTHER FRAME.

TA 088061

Table 2-39.   Front Output Shaft Sprag Unit Wear Limits

Index Number	Item/Point of Measurement	Size and Fit of New Parts (inches)	Wear Limit (inches)
1	Shift collar groove width	0.6060 to 0.6100	0.6180
2	Shift shaft fork thickness	0.5705 to 0.5665	0.5590
3	Rear race inside diameter	3.8338 to 3.8348	3.8378
4	Rear sprag unit width	0.3770 to 0.3780	0.3750
5	Front sprag unit width	0.3770 to 0.3780	0.3750
6	Front race inside diameter	3.8338 to 3.8348	3.8378
7	Spring free length	2.290 to 2.440	None
7	Spring length at pressure	2 3/32 at 12 to 18 pounds	None
		1 1/2 at 41 to 44 pounds	None

FRAME 3

NOTE

Readings must be within limits given in table 2-40. The letter L indicates a loose fit and the letter T indicates a tight fit. If readings are not within given limits, throw away part and get a new one.

1. Measure fit of front output shaft (1) in rear output shaft bore (2).

2. Measure fit of key (3) in front output shaft keyway (4).

3. Measure fit of key (3) in inner race keyway (5).

4. Measure fit of front output shaft (6) in inner race bore (7).

5. Measure fit of reverse shift shaft fork (8) in reverse shift collar groove (9).

GO TO FRAME 4

NOTE
CHECK ONLY THOSE PARTS WHICH ARE CALLED OUT IN
THIS FRAME. PARTS WITHOUT CALLOUTS ARE SHOWN
ONLY FOR REFERENCE PURPOSES OR ARE CHECKED IN
ANOTHER FRAME.

TA 088062

Table 2-40. Front Output Shaft Assembly Fits and Tolerances

Index Number	Item/Point of Measurement	Size and Fit (inches)	Wear Limit (inches)
1 and 2	Fit of shaft in bore	0.0020L to 0.0040L	0.0055L
3 and 4	Fit of key in keyway	0.0006L to 0.0024L	0.0006L
3 and 5	Fit of key in keyway	0.0020T to 0.0010L	0.0030L
6 and 7	Fit of shaft in bore	0.0008T to 0.0012L	0.0015L
8 and 9	Fit of fork in groove	0.0355L to 0.0435L	0.0590L

FRAME 4

NOTE

When checking sprags, anvil and spindle ends of micro-meter and flat back of sprag must all rest on a flat surface as shown in view A.

Since wear on all sprags in any one sprag unit will be the same, it is only necessary to check 5 sprags in each assembly.

1. Measure five sprags (1) as shown in view A.   If three or more sprags are worn to 0.375 inch or smaller, put new sprags in sprag unit.

2. Measure five sprags (1) as shown in view B.   If three or more sprags are worn more than 1/16 inch on the polished edge, put new sprags in sprag unit.

END OF TASK

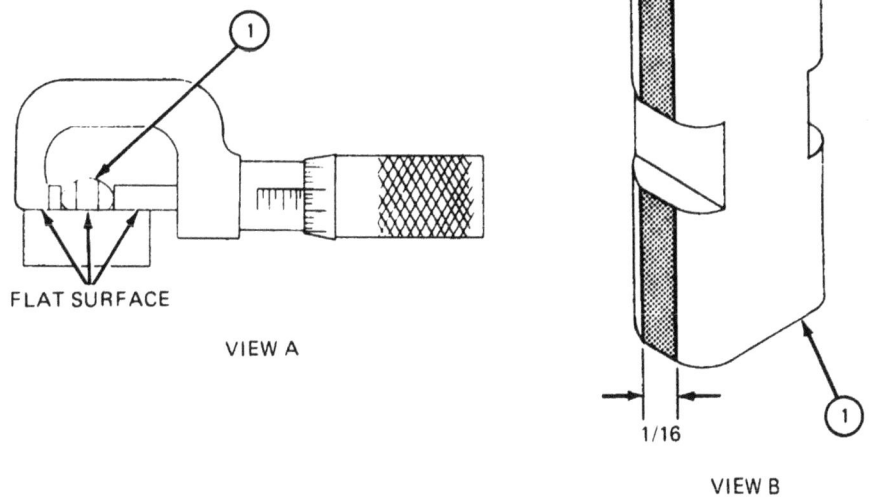

FLAT SURFACE

VIEW A

1/16

VIEW B

TA 088063

b. Countershaft Assembly.

FRAME 1

NOTE

Readings must be within limits given in table 2-41. If
readings are not within given limits, throw away part
and get a new one.

1. Measure high range gear bore (1) and high range gear keyway width (2).

2. Measure countershaft diameter (3) in two places and two countershaft keyway
widths (4).

3. Measure two key widths (5).

4. Measure low range gear bore (6) and low range gear keyway width (7).

GO TO FRAME 2

NOTE: CHECK ONLY THOSE PARTS WHICH
ARE CALLED OUT IN THIS FRAME.
PARTS WITHOUT CALLOUTS ARE
SHOWN ONLY FOR REFERENCE
PURPOSES OR ARE CHECKED IN
ANOTHER FRAME.

TA 088064

Table 2-41. Countershaft Assembly Wear Limits

Index Number	Item/Point of Measurement	Size and Fit of New Parts (inches)	Wear Limits (inches)
1	High range gear bore	2.3750 to 2.3760	2.3765
2	High range gear keyway width	0.5000 to 0.5032	0.5052
3	Countershaft diameter	2.3765 to 2.3775	2.3755
4	Countershaft keyway width	0.4984 to 0.5000	0.5024
5	Key width	0.5000 to 0.5012	None
6	Low range gear bore	2.3750 to 2.3760	2.3765
7	Low range gear keyway width	0.5000 to 0.5032	2.3765

FRAME 2

NOTE

Readings must be within limits given in table 2-42. The letter L indicates a loose fit and the letter T indicates a tight fit. If readings are not within given limits, throw away part and get a new one.

1. Measure fit of two gears (1) on countershaft (2) in two places.

2. Measure fit of two keys (3) in two countershaft keyways (4).

3. Measure fit of two keys (3) in two gear keyways (5).

END OF TASK

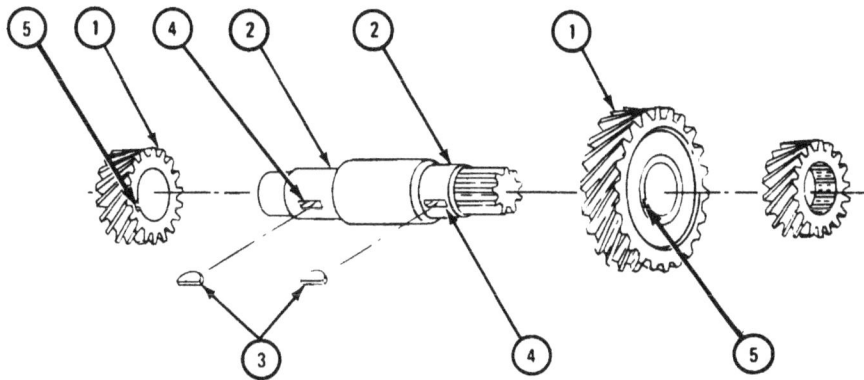

NOTE
CHECK ONLY THOSE PARTS WHICH ARE CALLED OUT IN THIS FRAME. PARTS WITHOUT CALLOUTS ARE SHOWN ONLY FOR REFERENCE PURPOSES OR ARE CHECKED IN ANOTHER FRAME.

TA 088065

Table 2-42. Countershaft Assembly Fits and Tolerances

Index Number	Item/Point of Measurement	Size and Fit of New Parts (inches)	Wear Limit (inches)
1 and 2	Fit of gear on countershaft	0.0005 to 0.0025T	0.0010L
3 and 4	Fit of key in shaft keyway	0.000T to 0.0032L	0.0052L
3 and 5	Fit of key in gear keyway	0.0028 to 0.0004L	0.0024L

c. Input Shaft Assembly.

FRAME 1	

NOTE

Readings must be within limits given in table 2-43. If readings are not within given limits, throw away part and get a new one.

1. Measure inner bearing race (1).

2. Measure high speed gear bore (2).

3. Measure inner bearing race (3) and outer bearing race (4).

4. Measure synchronizer ring width (5).

5. Measure width of groove in shifter fork (6).

6. Measure shifter shaft outside diameter (7).

GO TO FRAME 2

NOTE: CHECK ONLY THOSE PARTS WHICH ARE CALLED OUT IN THIS FRAME. PARTS WITHOUT CALLOUTS ARE SHOWN ONLY FOR REFERENCE PURPOSES OR ARE CHECKED IN ANOTHER FRAME.

Table 2-43. Input Shaft Gear and Shifter Shaft Wear Limits

Index Number	Item/Point of Measurement	Size and Fit of New Parts (inches)	Wear Limit (inches
1	Bearing inside diameter	1.9680 to 1.9685	None
2	High speed gear bore	3.5426 to 3.5435	3.5439
3	Bearing inside diameter	1.9680 to 1.9685	None
4	Bearing outside diameter	3.5427 to 3, 5433	None
5	Synchronizer ring width	0.6890 to 0.6930	0.6650
6	Shifter fork groove width	0.7090 to 0.7170	0.7320
7	Shifter shaft outside diameter	0.9949 to 0.0061	0.9941

FRAME 2

NOTE

Readings must be within limits given in table 2-44. If readings are not within given limits, throw away part and get a new one.

1. Measure input shaft diameter (1) in two places.

2. Measure inner bearing race (2).

3. Measure low speed gear bore (3).

4. Measure inner bearing race (4) and outer bearing race (5).

GO TO FRAME 3

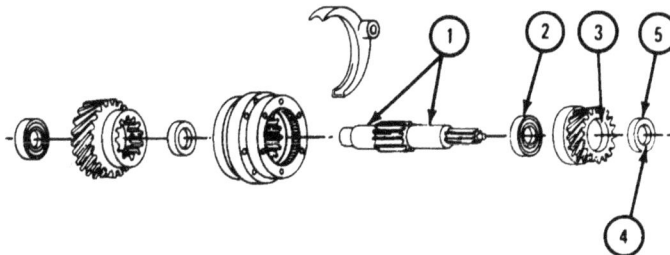

NOTE
CHECK ONLY THOSE PARTS WHICH ARE CALLED OUT IN
THIS FRAME. PARTS WITHOUT CALLOUTS ARE SHOWN
ONLY FOR REFERENCE PURPOSES OR ARE CHECKED IN
ANOTHER FRAME.

TA 088067

Table 2-44. Input Shaft and Bearing Wear Limits

Index Number	Item/Point of Measurement	Size and Fit of New Parts (inches)	Wear Limit (inches)
1	Input shaft diameter	1.9683 to 1.9689	None
2	Bearing inside diameter	1.9680 to 1.9685	None
3	Low speed gear bore	3.5426 to 3.5435	3.5439
4	Bearing inside diameter	1.9680 to 1.9685	None
5	Bearing outside diameter	3.5427 to 3.5433	None

FRAME 3

NOTE

Readings must be within limits given in table 2-45.
The letter L indicates a loose fit and the letter T
indicates a tight fit.   If readings are not within
given limits, throw away part and get a new one.

1.   Measure fit of two bearings (1) in two gear bores (2).

2.   Measure fit of four bearings (1 and 3) on input shafts bearing surface (4) in
     two places.

3.   Measure fit of fork (5) on ring of synchronizer (6).

END OF TASK

TA 088068

Table 2-45.   Input Shaft Assembly Fits and Tolerances

Index Number	Item/Point of Measurement	Size and Fit of New Parts (inches)	Wear Limit (inches)
1 and 2	Fit of bearing gear bore	0.0007T to 0.008L	0.0012L
3 and 4	Fit of bearing on input shaft	0.0002L to 0.0009T	0.0005L
5 and 6	Fit of fork on synchronizer	0.0160L to 0.0280L	0.06706

d. Rear Output Shaft Assembly.

FRAME 1

NOTE

Readings must be within limits given in table 2-46. If readings are not within given limits, throw away part and get a new one.

1. Measure inner race of two   bearings (1 and 2).

2. Measure rear output shaft   outside diameter (3 and 4).

3. Measure rear output shaft   bore (5).

GO TO FRAME 2

NOTE: CHECK ONLY THOSE PARTS WHICH
ARE CALLED OUT IN THIS FRAME.
PARTS WITHOUT CALLOUTS ARE
SHOWN ONLY FOR REFERENCE
PURPOSES OR ARE CHECKED IN
ANOTHER FRAME.

TA 088069

Table 2-46. Rear Output Shaft and Bearings Wear Limits

Index Number	Item/Point of Measurement	Size and Fit of New Parts (inches)	Wear Limit (inches)
1	Bearing inside diameter	1.7500 to 1.7505	None
2	Bearing inside diameter	2.6250 to 2.6260	None
3	Shaft outside diameter	1.7510 to 1.7515	2.7505
4	Shaft outside diameter	2.6265 to 2.6275	2.6260
5	Output shaft bore	1.8126 to 1.8136	1.8142

FRAME 2

NOTE

Readings must be within limits given in table 2-47.
The letter T indicates a tight fit. If readings are
not within given limits, throw away part and get
a new one.

1. Measure fit of bearing cone (1) on rear output shaft (2).

2. Measure fit of bearing cone (3) on rear output shaft (4).

END OF TASK

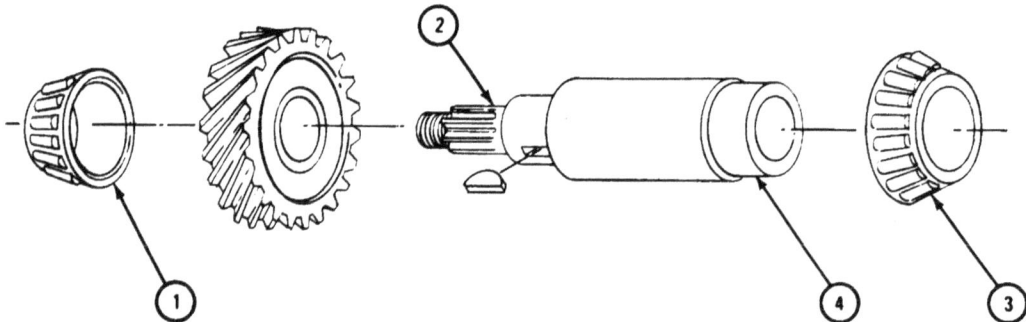

NOTE
CHECK ONLY THOSE PARTS WHICH ARE CALLED OUT IN
THIS FRAME. PARTS WITHOUT CALLOUTS ARE SHOWN
ONLY FOR REFERENCE PURPOSES OR ARE CHECKED IN
ANOTHER FRAME.

TA 088070

Table 2-47.   Rear Output Shaft Assembly Fits and Tolerances

Index Number	Item/Point of Measurement	Size and Fit of New Parts (inches)	Wear Limit (inches)
1 and 2	Fit of bearing cone on shaft	0.0005T to 0.0015T	0.000T
3 and 4	Fit of bearing cone on shaft	0.0005T to 0.0025T	0.000T

e. Transfer Case and Covers.

FRAME 1

NOTE

Readings must be within limits given in table 2-48. If readings
are not within given limits, throw away part and get a new one.

1. Measure shifter shaft bore (1).

2. Measure input shaft front bearing bore (2).

3. Measure countershaft front bearing bore (3).

4. Measure rear output shaft front bearing bore (4).

GO TO FRAME 2

TA 088050

Table 2-48. Transfer Cover Bearing Bores Wear Limits

Index Number	Item/Point of Measurement	Size and Fit of New Parts (inches)	Wear Limit (inches)
1	Shifter shaft housing bore	1.000 to 1.0012	1.0020
2	Input shaft front bearing bore	4.3300 to 4.3309	4.3316
3	Countershaft front bearing bore	3.9980 to 3.9990	4.000
4	Rear output shaft front bearing bore	4.4355 to 4.4365	4.4375

FRAME 2

NOTE

Readings must be within limits given in table 2-49.
The letter L indicates a loose fit and the letter T
indicates a tight fit.   If readings are not within
given limits, throw away part and get a new one.

1.   Measure fit of shifter shaft (1) in bore (2).

2.   Measure fit of input shaft front bearing (3) in bearing bore   (4).

3.   Measure fit of countershaft front bearing outer race (5) in bearing bore (6).

4.   Measure fit of front output shaft front bearing outer race (7) in bearing
       bore (8)

GO TO FRAME 3

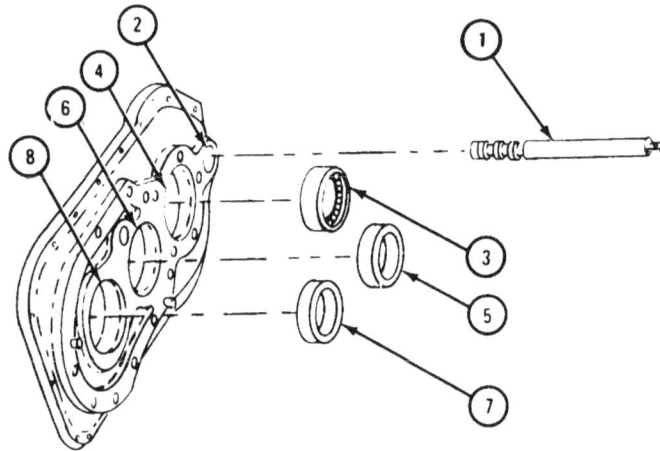

TA 088051

Table 2-49.   Transfer Cover Fits and Tolerances

Index Number	Item/Point of Measurement	Size and Fit of New Parts (inches)	Wear Limit (inches)
1 and 2	Fit of shifter shaft in bore	0.0039L to 0.0063L	0.070L
3 and 4	Fit of input shaft bearing in bore	0.008L to 0.007L	0.001L
5 and 6	Fit of countershaft bearing race in bore	0.003T to 0.001T	0.00T
7 and 8	Fit of output shaft bearing in bore	0.0010T to 0.0030T	0.000T

FRAME 3

NOTE

Readings must be within limits given in table 2-50. If readings
are not within given limits,   throw away part and get a new one.

1.   Measure shifter shaft bore (1).

2.   Measure expansion plug outside diameter (2).

3.   Measure input shaft rear bearing bore (3).

4.   Measure countershaft rear bearing bore (4).

5.   Measure rear output shaft rear bearing bore (5).

GO TO FRAME 4

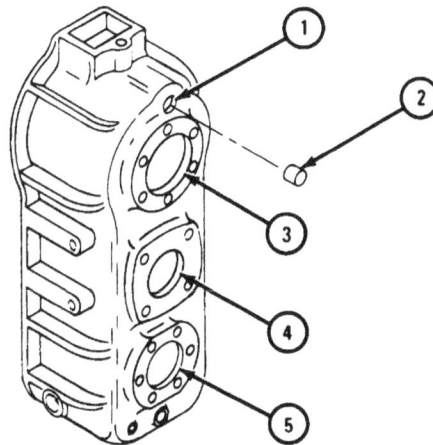

TA 088052

Table 2-50.   Transfer Case Bearing Bores Wear Limits

Index Number	Item/Point of Measurement	Size and Fit of New Parts (inches)	Wear Limit (inches)
1	Shifter shaft bore	1.2480 to 1.2496	1.2520
2	Expansion plug outside diameter	1.2540 to 1.2580	None
3	Input shaft rear bearing bore	3.9363 to 3.9372	3.9379
4	Countershaft rear bearing bore	2.000 to 2.0005	None
5	Rear output shaft rear bearing bore	1.7500 to 1.7505	None

FRAME 4

NOTE

Readings must be within limits given in table 2-51.
The letter L indicates a loose fit and the letter T
indicates a tight fit.   If readings are not within
given limits, throw away part and get a new one.

1.   Measure fit of   expansion plug (1) in housing bore (2).

2.   Measure fit of   input shaft rear bearing (3) in bearing bore (4).

3.   Measure fit of   countershaft rear bearing outer race (5) in bearing bore (6).

4.   Measure fit of   rear output shaft rear bearing outer race (7) in bearing
      bore (8).

GO TO FRAME 5

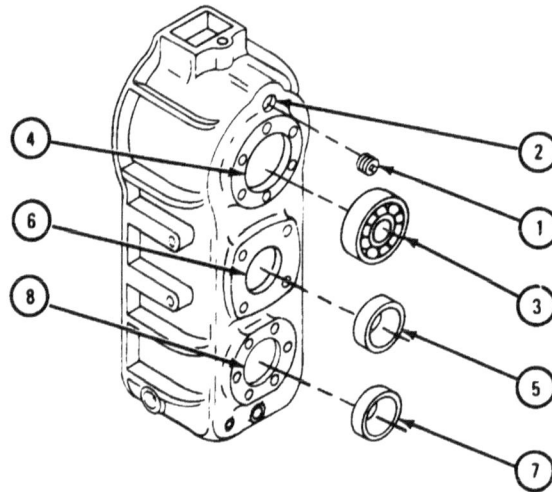

TA 088053

Table 2-51.   Transfer Case Fits and Tolerances

Index Number	Item/Point of Measurement	Size and Fit of New Parts (inches)	Near Limit (inches)
1 and 2	Fit of expansion plug in bore	0.0024T to 0.0100T	0.0020T
3 and 4	Fit of input shaft bearing in bore	0.000TT to 0.0008L	0.0015L
5 and 6	Fit of countershaft bearing race in bore	0.005L to 0.0015T	0.0010L
7 and 8	Fit of output shaft bearing race in bore	0.0010T to 0.0010L	0.0015L

FRAME 5

NOTE

Readings must be within limits given in table 2-52.
The letter T indicates a tight fit. If readings are
not within given limits,   throw away part and get
a new one.

1.   Measure diameter of dowel pin (1).

2.   Measure dowel pin bore (2).

3.   Measure fit of dowel pin (1) in bore (2).

GO TO FRAME 6

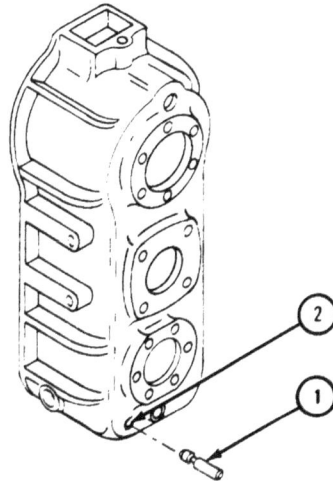

TA 088054

Table 2-52.   Transfer Case Dowel Pin Wear Limits

Index Number	Item/Point of Measurement	Size and Fit of New Parts (inches)	Wear Limit (inches)
1	Dowel pin diameter	0.7500 to 0.7510	None
2	Dowel pin bore	0.7461 to 0.7472	0.7480
1 and 2	Fit of dowel pin in bore	0.0028T to 0.0049T	0.0020T

FRAME 6

NOTE

Readings must be within limits given in table 2-53. The letter L indicates a loose fit and the letter T indicates a tight fit. If readings are not within given limits, throw away part and get a new one.

1. Measure shifter shaft bore (1).

2. Measure front output shaft bearing bore (2).

3. Measure fit of shifter shaft (3) in shifter shaft bore (1).

4. Measure front output shaft bearing (4) in front output shaft bearing bore (2).

GO TO FRAME 7

NOTE: CHECK ONLY THOSE PARTS WHICH ARE CALLED OUT IN THIS FRAME. PARTS WITHOUT CALLOUTS ARE SHOWN ONLY FOR REFERENCE PURPOSES OR ARE CHECKED IN ANOTHER FRAME.

TA 088055

Table 2-53.   Front Output Shaft Cover Wear Limits

Index Number	Item/Point of Measurement	Size and Fit of New Parts (inches)	Wear Limit (inches)
1	Shifter shaft bore	0.8760 to 0.8780	0.8800
2	Front output shaft bearing bore	3.3465 to 3.3485	3.3495
1 and 3	Fit of shifter shaft in bore	0.0020L to 0.0052L	0.0100L
2 and 4	Fit of output shaft bearing in bore	0.007T to 0.00081L	0.0015L

FRAME 7

NOTE

Readings must be within limits given in table 2-54.
The letter L indicates a loose fit and the letter T
indicates a tight fit.   If readings are not within
given limits,   throw away part and get a new one.

1. Measure companion flange inside diameter (1) and outside diameter (2).

2. Measure deflector inside diameter (3).

3. Measure oil seal outside diameter (4).

4. Measure fit of companion flange inside diameter (1) on front output shaft (5).

5. Measure fit of companion flange outside diameter (2) in deflector (3).

6. Measure fit of oil seal outside (4) in output shaft bore (6).

GO TO FRAME 8

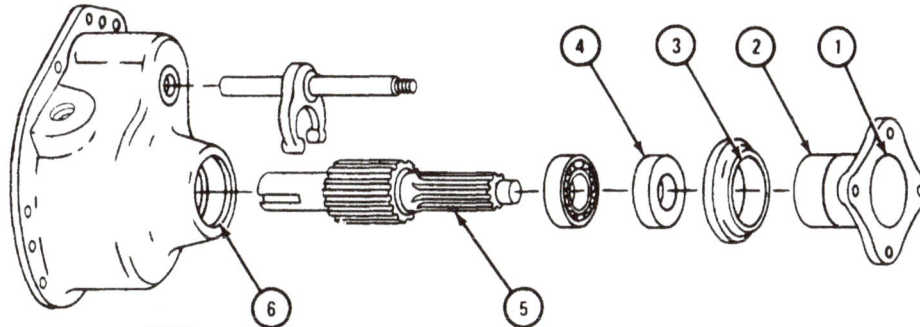

NOTE
CHECK ONLY THOSE PARTS WHICH ARE CALLED OUT IN
THIS FRAME. PARTS WITHOUT CALLOUTS ARE SHOWN
ONLY FOR REFERENCE PURPOSES OR ARE CHECKED IN
ANOTHER FRAME.

TA 088056

Table 2-54.   Front Output Shaft and Companion Flange Wear Limits

Index Number	Item/Point of Measurement	Size and Fit of New Parts (inches)	Wear Limit (inches)
1	Companion flange inside diameter	1.621 to 1.624	1.627
2	Companion flange outside diameter	2.4016 to 2.4056	2.4010
3	Deflector inside diameter	2.3956 to 2.3996	None
4	Oil seal outside diameter	3.3524 to 3.3563	None
1 and 5	Fit of flange on shaft	0.0020T to 0.0030L	0.0073L
2 and 3	Fit of deflector on flange	0.0020T to 0.00100T	0.0010T
4 and 6	Fit of oil seal in output shaft bore	0.0038T to 0.0098T	0.0025T

FRAME 8

NOTE

Readings must be within limits given in table 2-55. If readings are not within given limits, throw away part and get a new one.

1. Measure flange outside diameter (1).

2. Measure deflector inside diameter (2).

3. Measure oil seal outside diameter (3).

4. Measure input shaft cover bore (4).

5. Measure oil seal outside diameter (5).

GO TO FRAME 9

NOTE
CHECK ONLY THOSE PARTS WHICH ARE CALLED OUT IN
THIS FRAME. PARTS WITHOUT CALLOUTS ARE SHOWN
ONLY FOR REFERENCE PURPOSES OR ARE CHECKED IN
ANOTHER FRAME.

TA 088057

Table 2-55. Input Shaft Cover Wear Limits

Index Number	Item/Point of Measurement	Size and Fit of New Parts (inches)	Wear Limit (inches)
1	Input shaft flange outside diameter	2.4016 to 2.4056	2.4010
2	Deflector inside diameter	2.3956 to 2.3996	None
3	Oil seal outside diameter	3.3524 to 3.3563	None
4	Input shaft cover bore	3.3465 to 3.3485	3.3495
5	Oil seal outside diameter	1.5000 to 1.5039	None

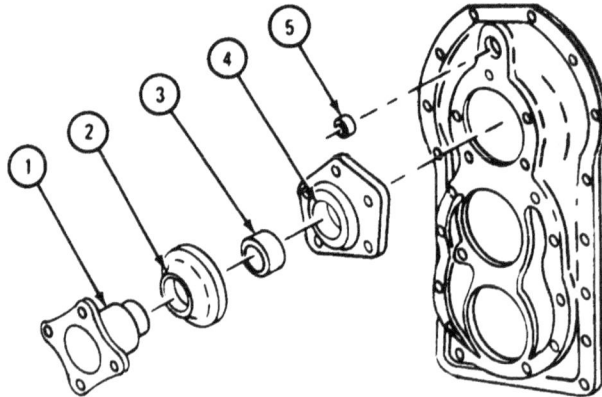

2-216

FRAME 9

NOTE

Readings must be within limits given in table 2-56.
The letter T indicates a tight fit. If readings are
not within given limits,   throw away part and get
a new one.

1.   Measure fit of oil seal (1) in bore (2).

2.   Measure fit of oil seal (3) in input shaft cover bore (4).

3.   Measure fit of deflector (5) on flange (6).

GO TO FRAME 10

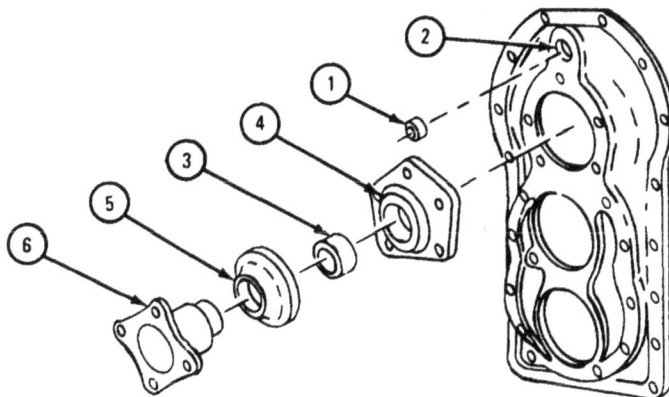

TA 088058

Table 2-56.   Input Shaft Cover Fits and Tolerances

Index Number	Item/Point of Measurement	Size and Fit of New Parts (inches)	Wear Limit (inches)
1 and 2	Fit of oil seal in bore	0.0024T to 0.0078T	0.0015T
3 and 4	Fit of seal in input shaft cover	0.0038T to 0.0098T	0.0025T
5 and 6	Fit of deflector on flange	0.0020T to 0.0100T	0.0010T

FRAME 10

NOTE

Readings must be within limits given in table 2-57.
The letter L indicates a loose fit and the letter T
indicates a tight fit.   If readings are not within
given limits,   throw away part and get a new one.

1.   Measure outside diameter of oil seal (1).

2.   Measure rear bearing retainer bore (2).

3.   Measure dowel pin bore (3).

4.   Measure fit of oil seal (1) in rear bearing retainer bore (2).

5.   Measure fit of dowel pin (4) in dowel pin bore (3).

END OF TASK

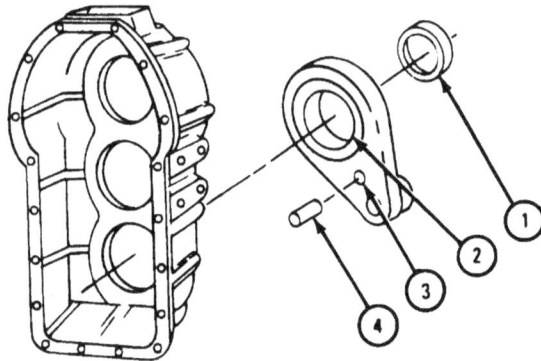

NOTE
CHECK ONLY THOSE PARTS WHICH ARE CALLED OUT IN
THIS FRAME. PARTS WITHOUT CALLOUTS ARE SHOWN
ONLY FOR REFERENCE PURPOSES OR ARE CHECKED IN
ANOTHER FRAME.   TA 088059

Table 2-57.   Rear Bearing Retainer Wear Limits

Item Number	Item/Point of Measurement	Size and Fit of New Parts (inches)	Wear Limit (inches)
1	Oil seal outside diameter	3.3524 to 3.3563	None
2	Rear bearing retainer bore	3.3465 to 3.3485	3.3495
3	Dowel pin bore	0.7510 to 0.7530	0.7530
1 and 2	Fit of oil seal in retainer bore	0.0038T to 0.0098T	0.0025T
3 and 4	Fit of dowel pin in bore	0.0000T to 0.0020L	0.0025L

2-20. REPAIR. This paragraph gives instructions to repair the transmission transfer case, covers, gear shafts, and gears.

a.   Smooth out any chips, scratches or burrs on gear shafts and gears with a honing stone.

b.   Weld cracks and small holes in housing and cover castings. Refer to TM 9-237.

b.   Drill out any bolts or studs broken off in tapped holes.

d.   Drill out threaded holes that are stripped or out-of-round to the next larger size and retap them.   When putting together transmission transfer, use a bolt or stud the size of the newly tapped hole.

2-21. ASSEMBLY OF SUBASSEMBLIES.   The following paragraphs give instructions to assemble the transmission transfer subassemblies.

NOTE

Keep all parts clean and protected from dust and dirt.   Coat all bearings with multipurpose lubricant during assembly.   Coat shafts and bores of gears with white lead pigment during assembly. Use new seals and snaprings during assembly.

a.   Rear Output Shaft Assembly.

FRAME 1

1.   Put in key (1).

2.   Aline key (1) with keyway in driven gear (2). Set up hydraulic press as shown.

3.   Press in rear output shaft (3).

GO TO FRAME 2

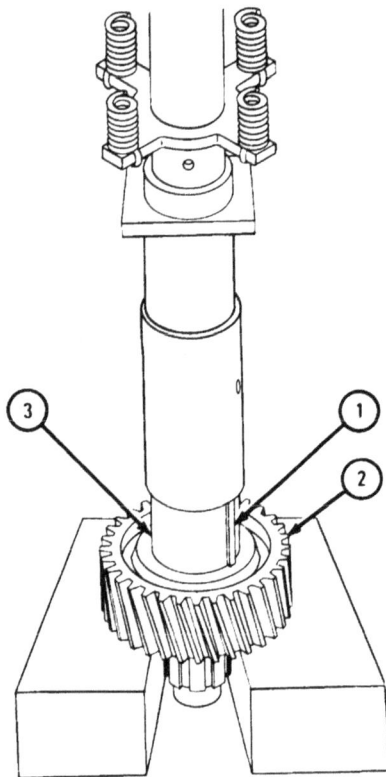

MODEL T 136-21

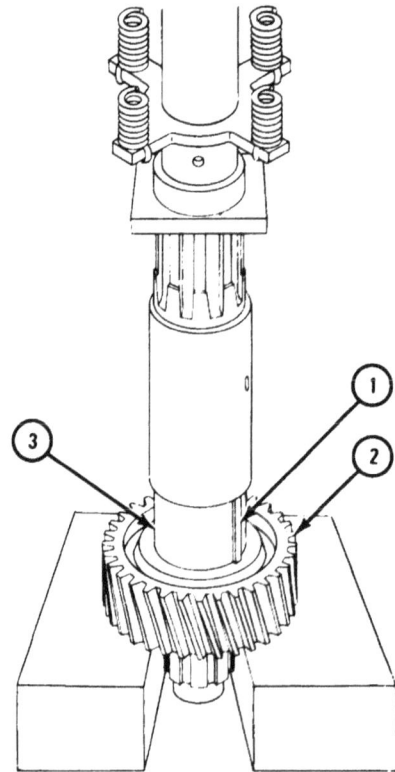

MODEL T 136-27

TA 088071

FRAME 2

1. Set up hydraulic press as shown.
2. Press rear bearing (1) on rear output shaft assembly (2).

GO TO FRAME 3

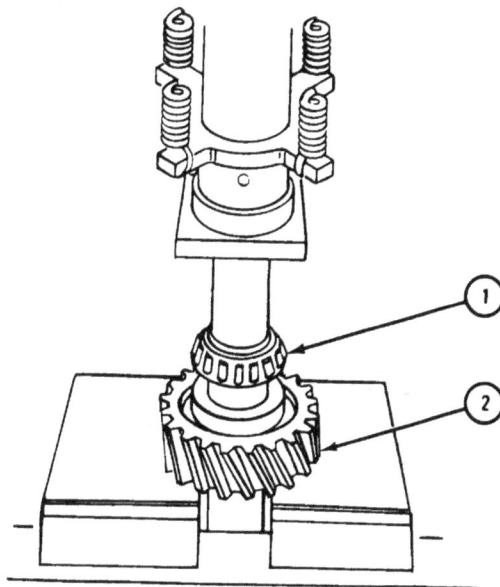

TA 088072

FRAME 3

1. Set up hydraulic press as shown.
2. Press front tapered roller bearing (1) onto rear output shaft assembly (2).

END OF TASK

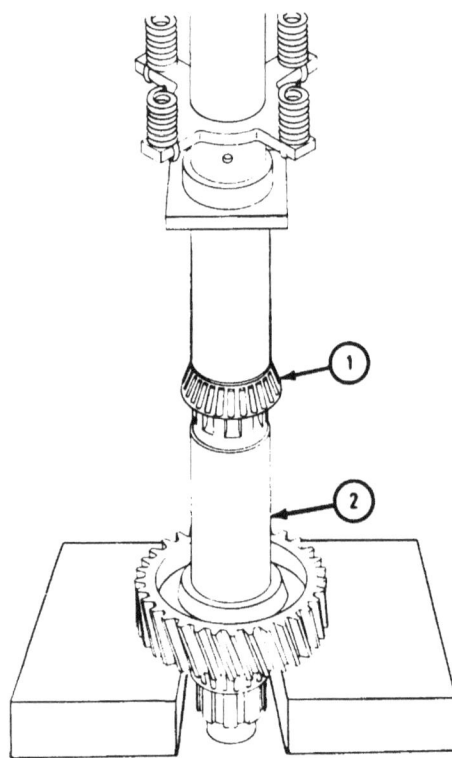

MODEL T-136-27

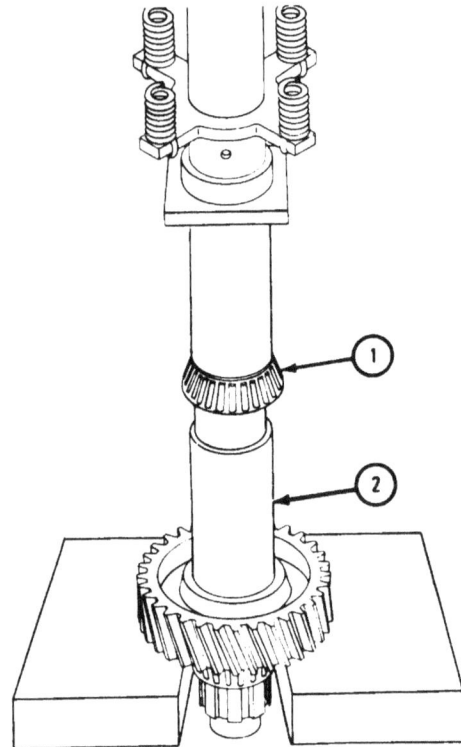

MODEL T-136-21

TA 088073

b.   <u>Countershaft Assembly.</u>

FRAME 1

1.   Put in two keys (1).
2.   Aline key with keyway in low range gear (2) and set up hydraulic press as shown.
3.   Press in countershaft (3).

GO TO FRAME 2

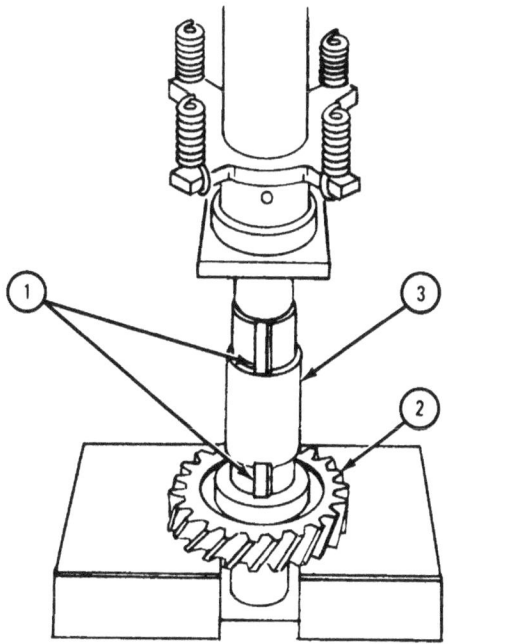

MODEL T-136-27

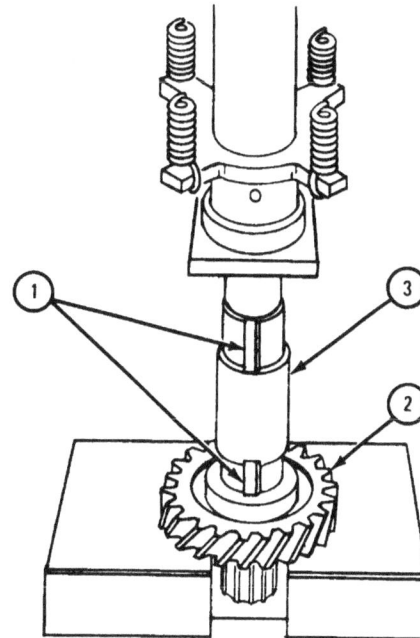

MODEL T-136-21

TA 088074

FRAME 2

1. Set up hydraulic press as shown.
2. Press front bearing (1) onto countershaft (2) as shown.
GO TO FRAME 3

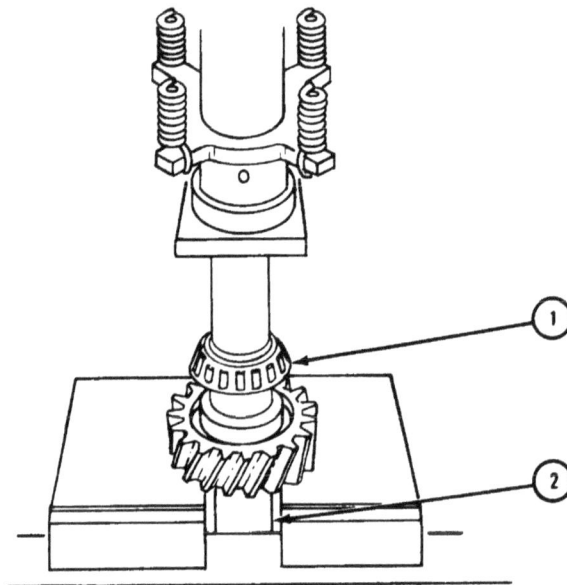

TA 088075

FRAME 3

1. Set up hydraulic press as shown.
2. Aline key in countershaft (1) with keyway in high range gear (2).
3. Press on high range gear (2).

GO TO FRAME 4

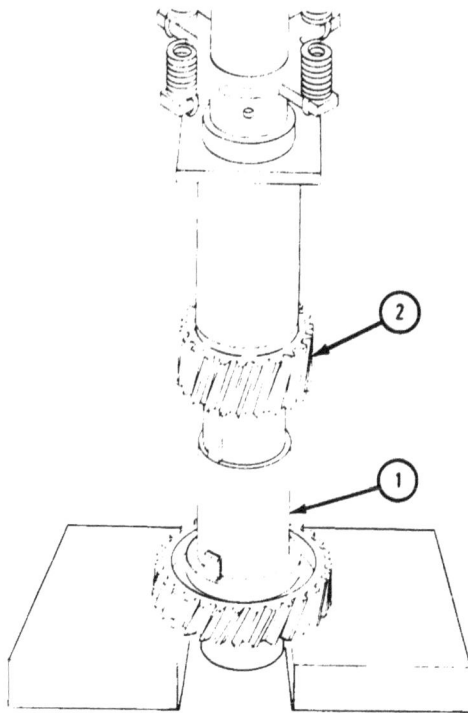

MODEL T-136-27

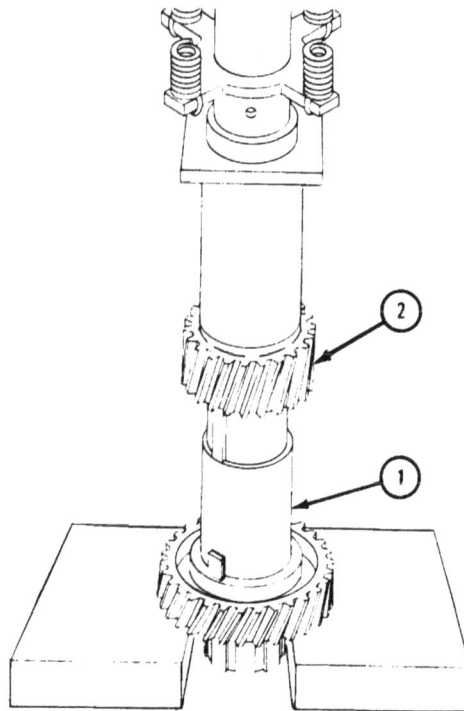

MODEL T-136-21

TA 088076

FRAME 4

1. Set up hydraulic press as shown.
2. Press rear bearing (1) onto countershaft (2).

GO TO FRAME 5

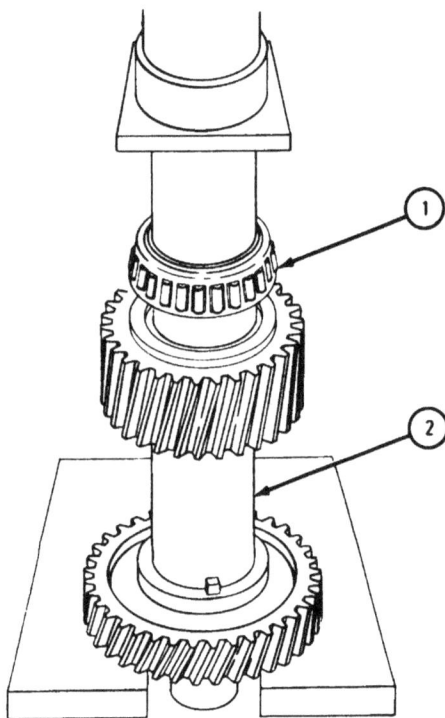

MODEL T-136-27

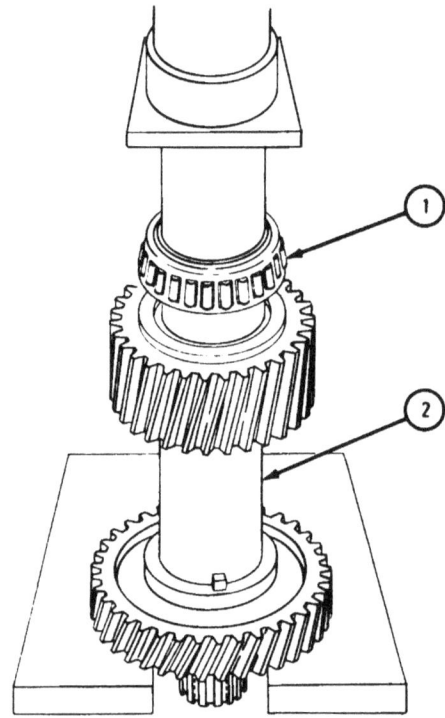

MODEL T-136-21

TA 088077

FRAME 5

1. Put retainer plate (1) and retainer lock assembly (2) into place and aline screw holes.

2. Put in two capscrews (3).   Bend ears on retainer lock assembly (2) against capscrews.

IF WORKING ON MODEL T-136-27, GO TO FRAME 6.
IF WORKING ON MODEL T-136-21, END OF TASK

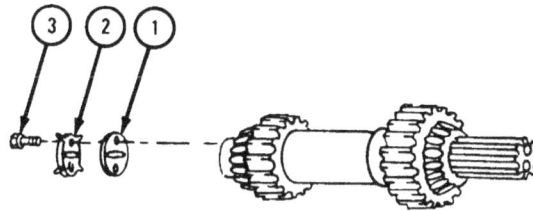

MODEL T-136-21

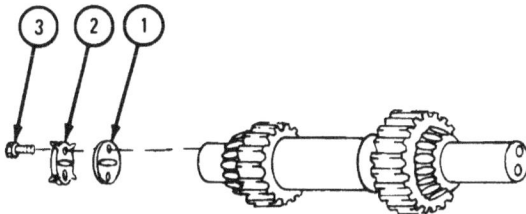

MODEL T-136-27

TA 088078

FRAME 6

1. Put on sleeve (1).

2. Put retainer plate (2) and retainer lock assembly (3) into place and aline screw holes.

3. Put in two capscrews (4).   Bend ears on retainer lock assembly (3) against capscrews.

END OF TASK

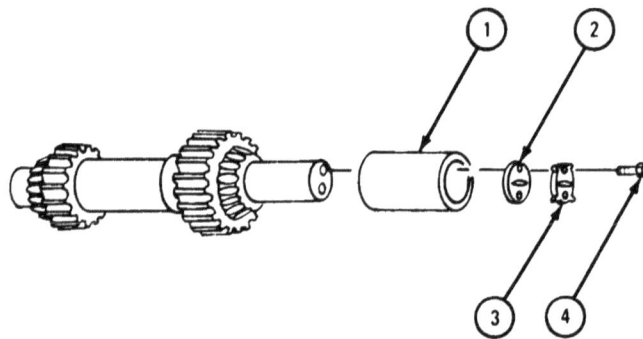

TA088079

c. Input Shaft Assembly.

FRAME 1

1. Using hammer and brass drift, tap in inner bearing (1).
2. Turn over gear (2), put in spacer (3), and tap in outer bearing (4) .
GO TO FRAME 2

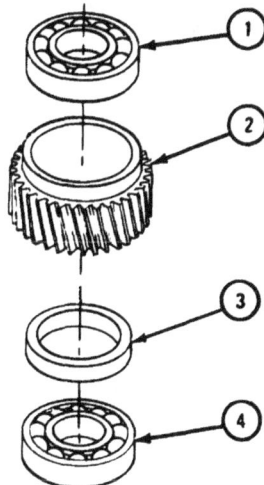

TA 088080

FRAME 2

1. Using hammer and brass drift, tap in inner bearing (1).
2. Turn over gear (2), put in spacer (3), and tap in outer bearing (4).

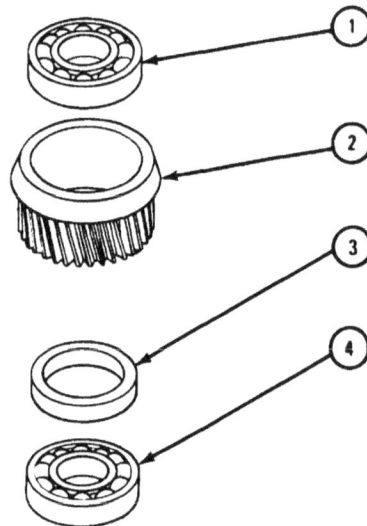

GO TO FRAME 3

TA 088081

FRAME 3

1. Slide on high range gear (1) and thrust washer (2).
2. Set up input shaft assembly (3) in hydraulic press as shown.
3. Press on rear bearing (4).

GO TO FRAME 4

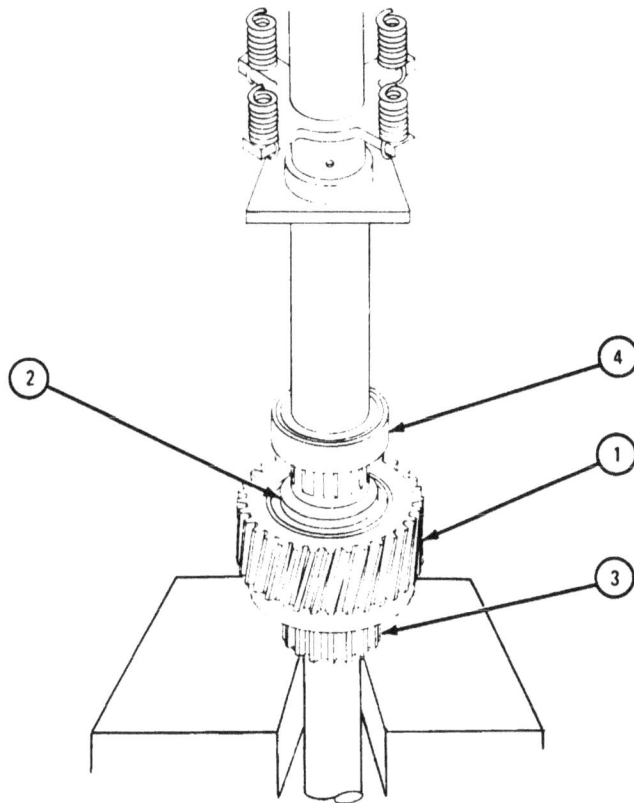

TA 088082

FRAME 4

1. Slide on synchronizer (1), low range gear (2), and thrust washer (3).
2. Put on snapring (4).

END OF TASK

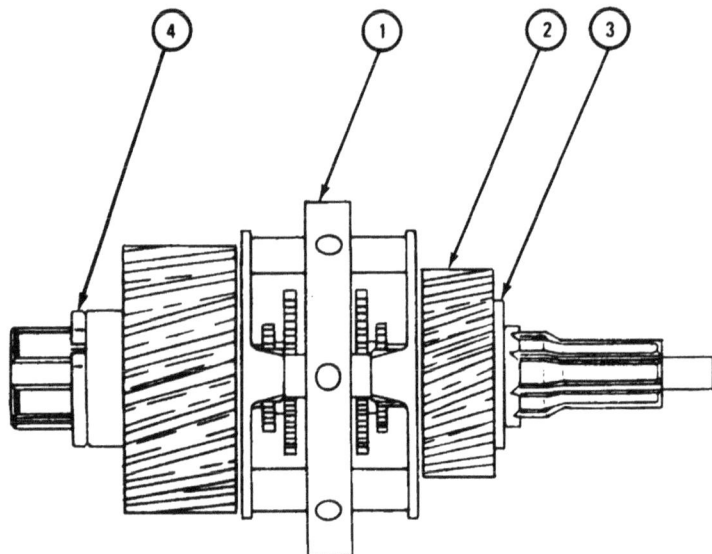

TA 088083

d. Shifter Shaft Assembly (Model T-136-27).

FRAME 1

1. Slide shifter fork (1) onto shifter shaft (2). Aline setscrew hole in shifter fork with setscrew hole in shifter shaft.

2. Put in setscrew (3) and tighten setscrew to 90 to 118 pound-feet.

3. Put on safety wire (4).

END OF TASK

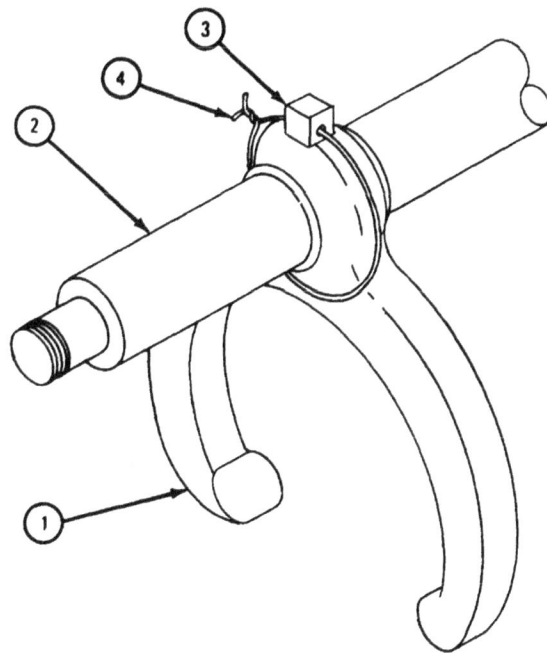

TA 088084

e. <u>Front Output Case Assembly (Model T-136-27).</u>

FRAME 1

1. Put in filler plug (1).
2. Working at rear side of front output shaft cover (2), put in snapring (3).
3. Using bearing remover and replacer, tap in output shaft bearing (4).
4. Working at front side of front output shaft cover (2), put in snapring (5).

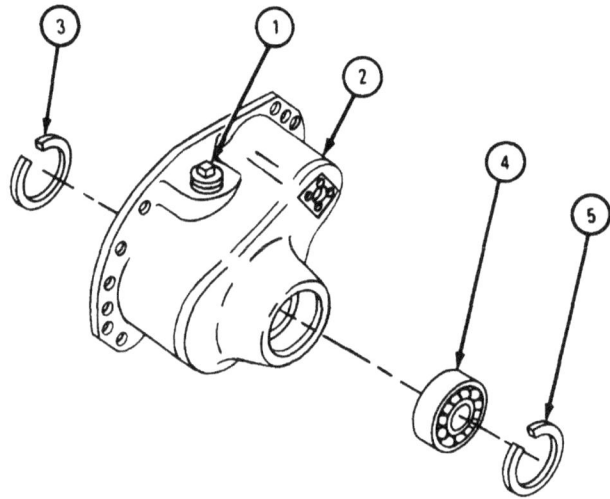

GO TO FRAME 2

TA 088085

FRAME 2

1. Put in thrust washer (1) and snapring (2).
2. Working at front side of output shaft case (3), tap in seal (4).

GO TO FRAME 3

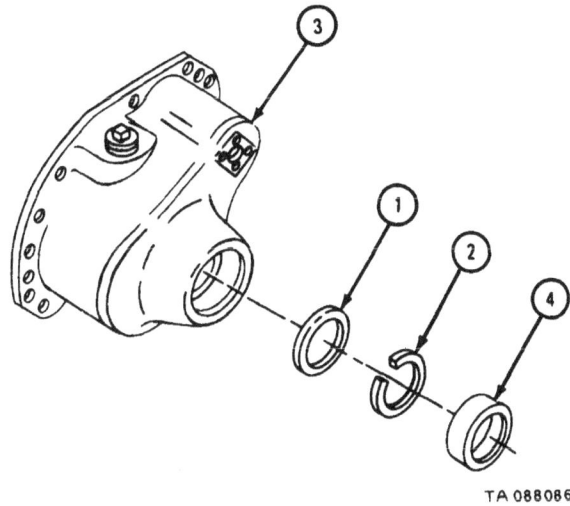

TA 088086

FRAME 3

1. Set up output shaft cover assembly (1) on hydraulic press as shown.
2. Press in front output shaft (2).

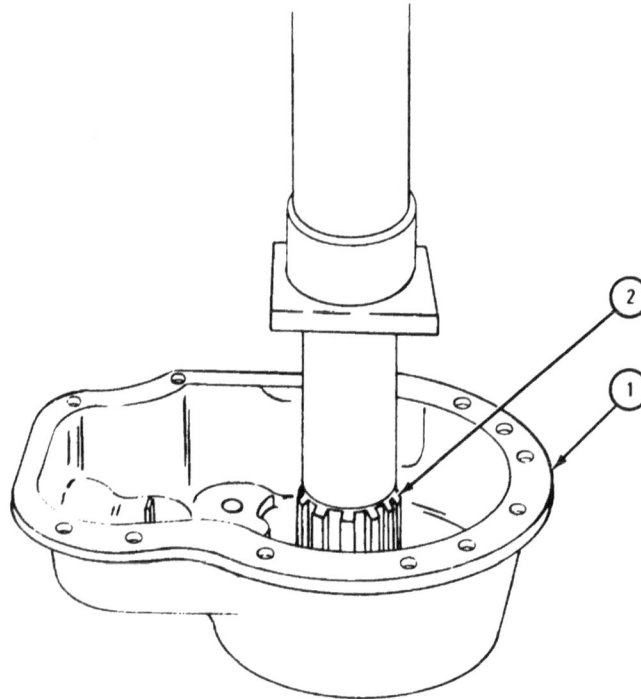

GO TO FRAME 4

TA 088087

FRAME 4

1. Put shifter fork (1) into groove in sliding clutch (2) as shown.

2. Put shifter shaft (3) through hole in front output case cover (4) and at the same time, put sliding clutch (2) on output shaft (5).

GO TO FRAME 5

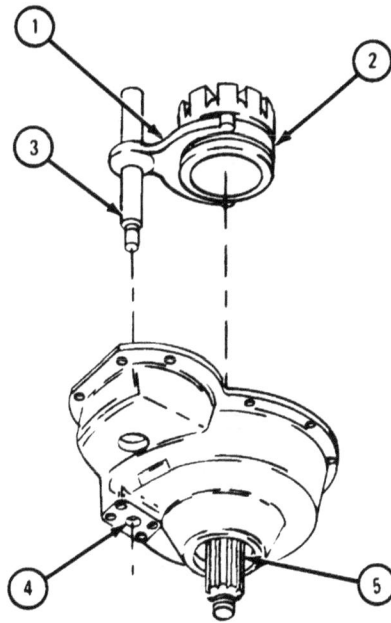

TA 088088

FRAME 5

1. Put on brass gasket (1).
2. Put on washer (2) and piston (3). Put on second washer (2) and nut (4).
3. Slide on air tube (5).
4. Put on brass gasket (6) and air tube cover (7) and aline screwholes.
5. Put in four capscrews (8) with four key washers (9).
6. Bend ears on four key washers (9) against four capscrews (8).

END OF TASK

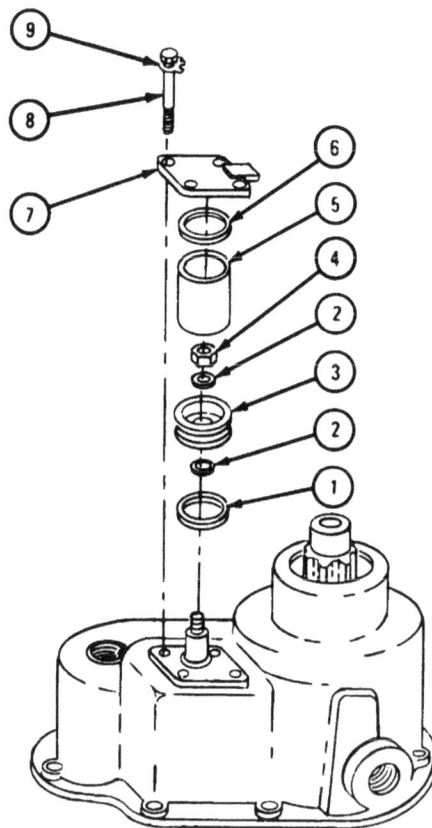

TA 088089

f. Front Output Shaft Case Assembly (Model T-136-21).

FRAME 1

1. Put in filler plug (1).
2. Put in snapring (2).
3. Using bearing remover and replacer, tap in output shaft bearing (3).
4. Put in snapring (4).
GO TO FRAME 2

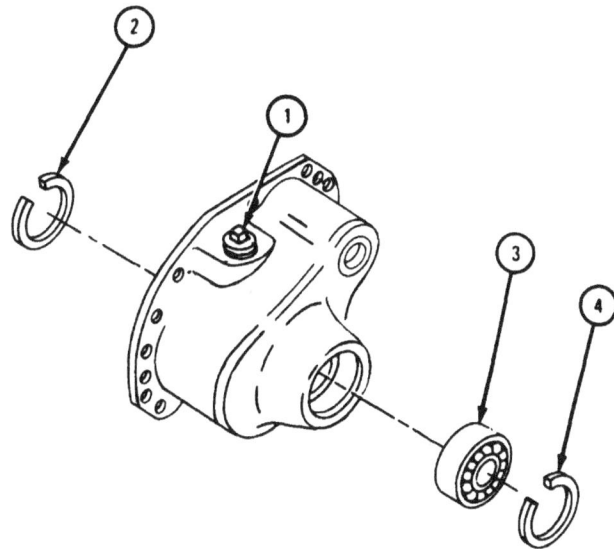

TA 088090

FRAME 2

1. Put in thrust washer (1) and snap ring (2).
2. Working at front side of output shaft case (3), tap in oil seals (4 and 5).
GO TO FRAME 3

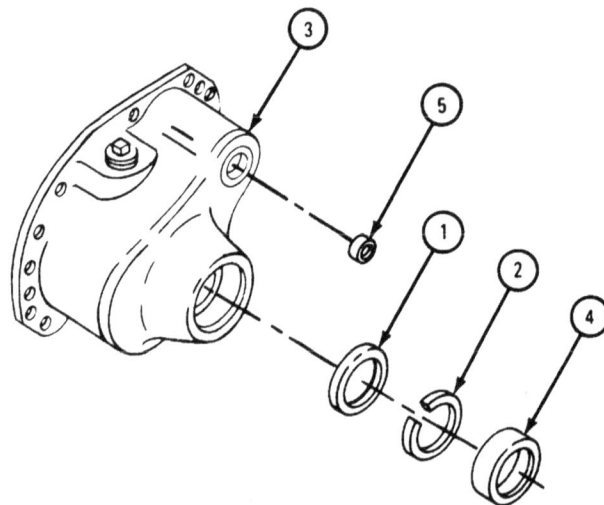

TA 088091

FRAME 3

1. Slide shifter fork (1) on shifter shaft (2). Aline setscrew hole in shifter fork with setscrew hole in shifter shaft.

2. Put in setscrew (3) and safety wire (4).

GO TO FRAME 4

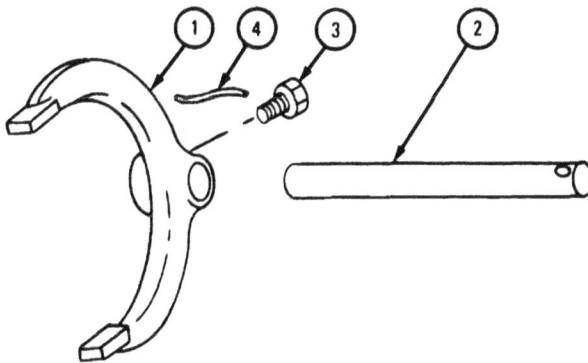

TA 088092

FRAME 4

1.   Put shifter fork (1) in groove in clutch collar (2) as shown.
2.   Tap shifter shaft (3) with clutch collar (2) into hole in output shaft case (4).
END OF TASK

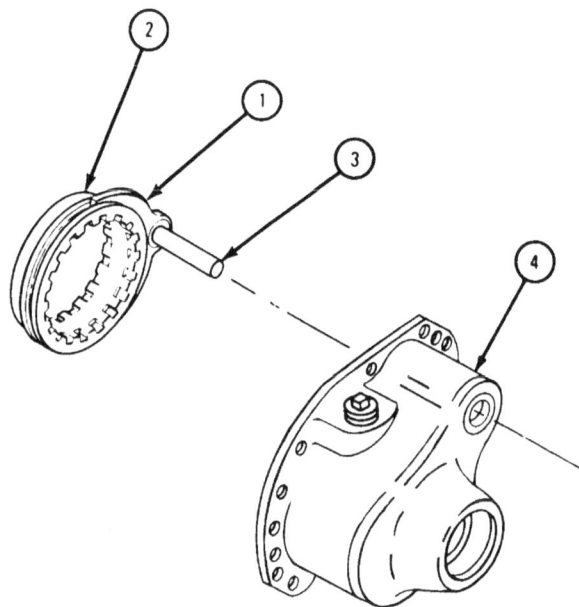

TA 088093

g. <u>Front Output Shaft (Model T-136-21).</u>

NOTE

Sprags (1) have two notches, one large and one
small notch.   Ring (2) fits in small notch and
spring (3) fits in large notch. Tips of all sprags
(1) must face the same way.

FRAME 1

1.   Put four sprags (1) into place on ring (2) as shown,

2.   Put spring (3) into slots in sprags (1) as shown.

3.   Do steps 1 and 2 again for rear sprag unit.

GO TO FRAME 2

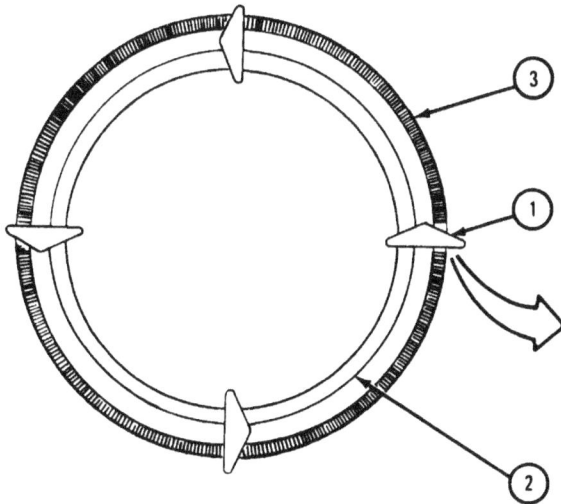
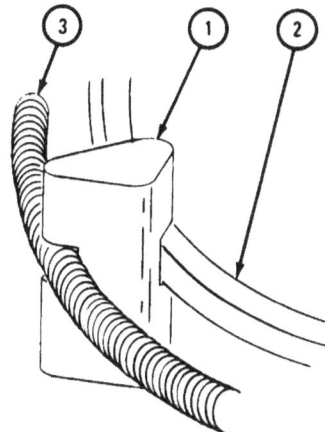

TA 088094

FRAME 2

1. Put one sprag (1) between ring (2) and spring (3) as shown in view A.
2. Turn sprag (1) to lock as shown in view B.
3. Do steps 1 and 2 again for 37 sprags (1).
4. Do steps 1 through 3 again for rear sprag unit.

GO TO FRAME 3

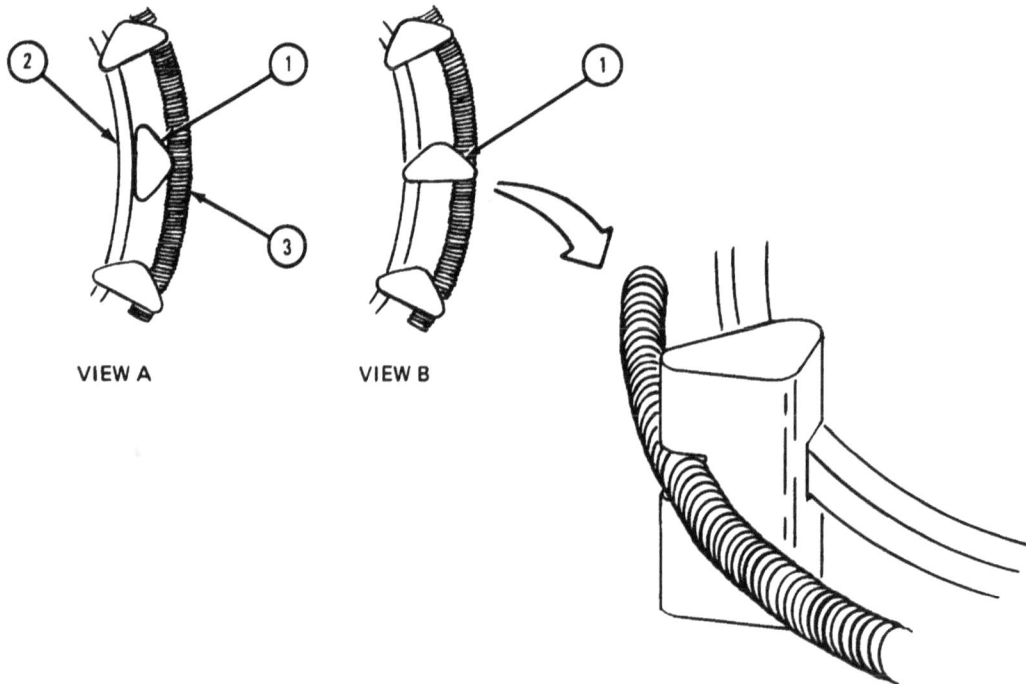

VIEW A   VIEW B

TA 088095

FRAME 3

NOTE

Before putting front sprag unit (1) into front outer race (2),
be sure that all tips of sprags point to the left as in view A.

1.   Put front sprag unit (1) into front outer race (2).

NOTE

Before putting rear sprag unit (3), into rear outer race (4),
be sure that all tips of sprags point to the right as in view B.

2.   Put rear sprag unit (3) into rear outer race (4).

GO TO FRAME 4

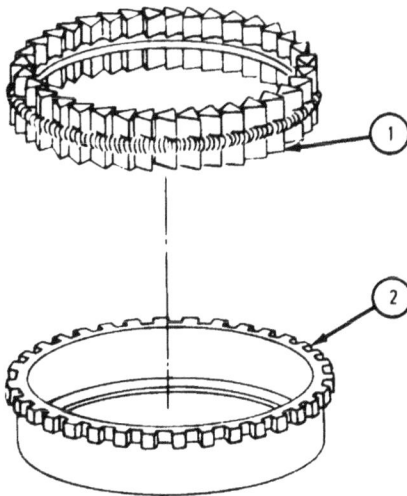

VIEW A

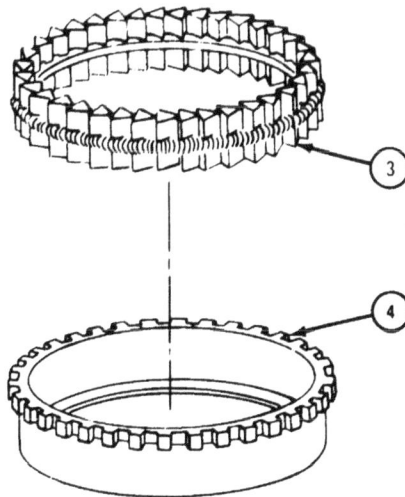

VIEW B

TA 088096

FRAME 4

1. Put in two keys (1 and 2).

2. Set up hydraulic press as shown.

3. Aline key (1) with keyway in transmission gear (3).

4. Press front output shaft (4) into transmission gear (3).

GO TO FRAME 5

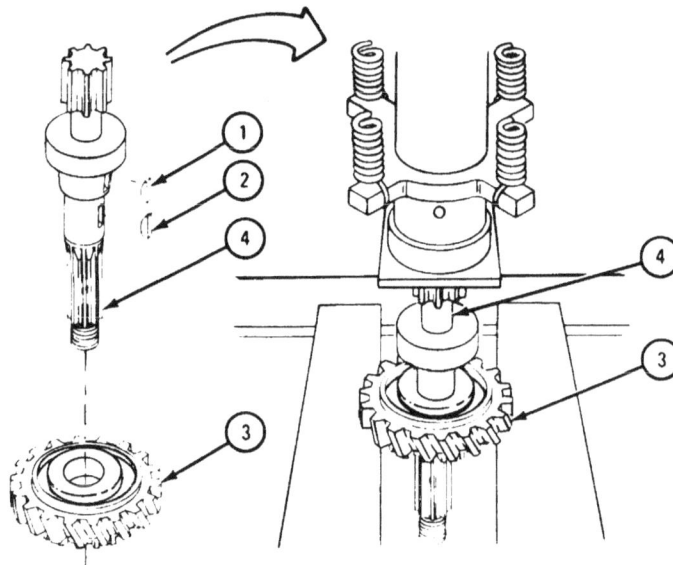

TA 088097

FRAME 5

1. Aline key (1) with keyway in inner race (2).
2. Set up hydraulic press as shown.
3. Press front output shaft (3) into inner race (2).

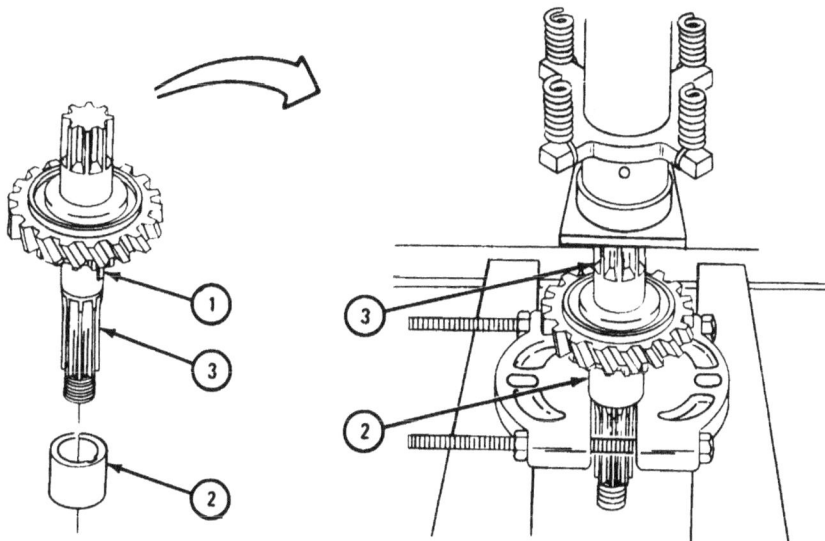

GO TO FRAME 6

TA 088098

FRAME 6

1. Put on snap ring (1).

2. Put rear outer race with rear sprag unit (2) on front output shaft (3) as shown.

3. Put on two spacers (4).

4. Put on front outer race with front sprag unit (5).

5. Put on thrust washer (6) and snap ring (7).

END OF TASK

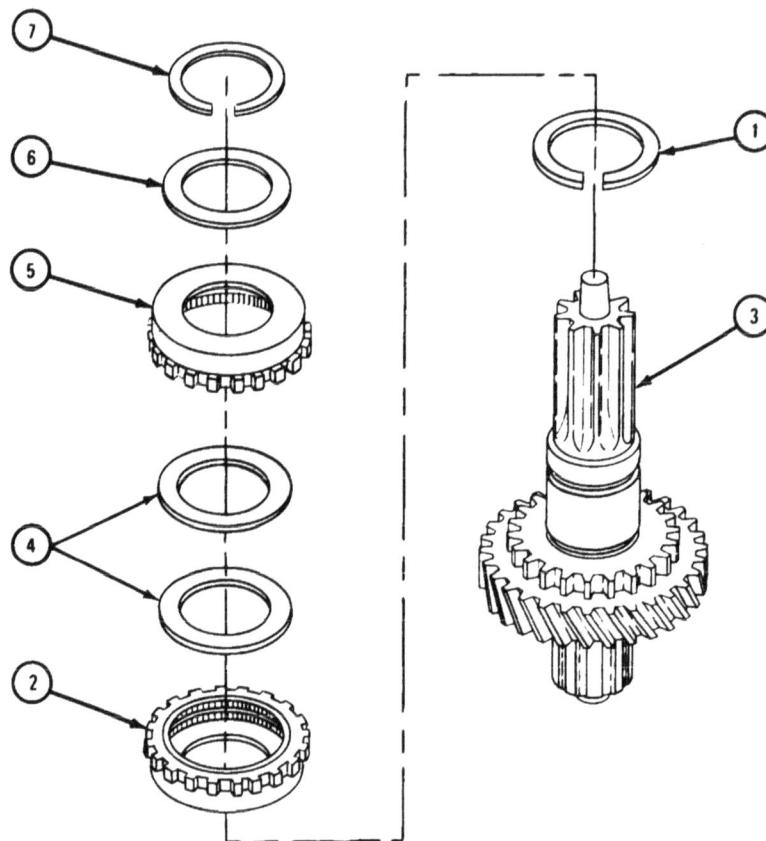

TA 088099

1

h.   Case Cover.

FRAME 1

1.   Working at front side of cover (1), put in two snaprings (2).

2.   Using bearing remover and replacer and working from rear side of cover (1), tap in countershaft bearing cup (3) and rear output shaft bearing cup (4).

3.   Tap in seal (5).

GO TO FRAME 2

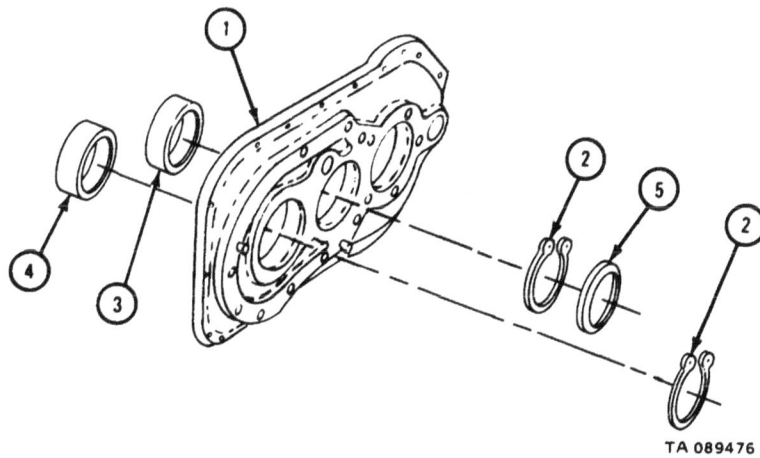

TA 089476

FRAME 2

1. Using hammer and 12-inch long 1/2-inch round stock and working from front side of cover (1), tap in shifter shaft oil seal (2).

2. Working from front side of cover (1), tap in input shaft front bearing (3).

3. Put in snap ring (4).

END OF TASK

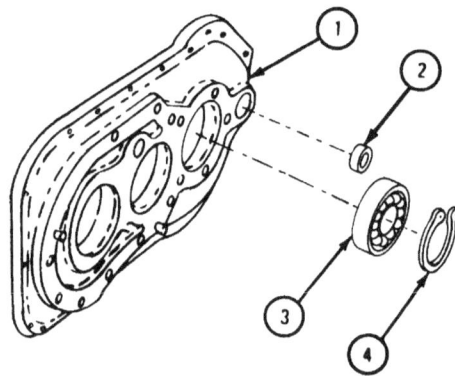

TA 089477

i. Transfer Case Housing.

FRAME 1

1. Using hammer and punch and working from outside transfer case (1), tap in dowel pin (2).

2. Put in magnetic drain plug (3) and drain plug (4).

3. Using bearing remover and replacer and working from outside transfer case (1), tap in rear output shaft bearing cup (5) and countershaft bearing cup (6).

END OF TASK

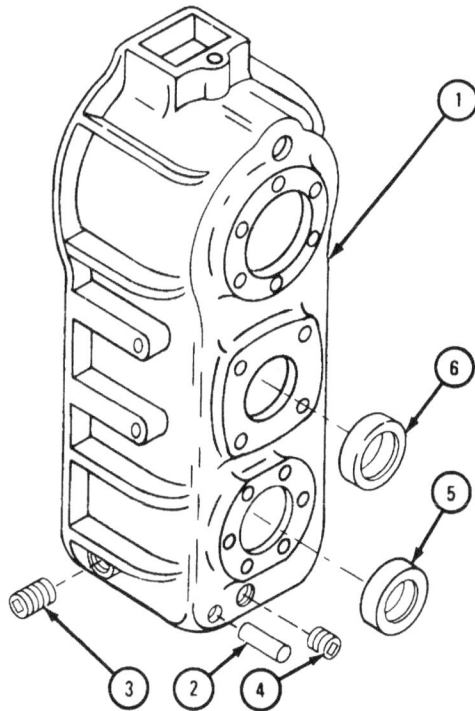

TA 089478

j.   Handbrake Brakeshoe Assembly.

FRAME 1

1.   Put inner brakeshoe (1) on lever pin (2) as shown.
2.   Put outer brakeshoe (3) on lever pin (4) as shown.
3.   Put on two retaining rings (5 and 6).
GO TO FRAME 2

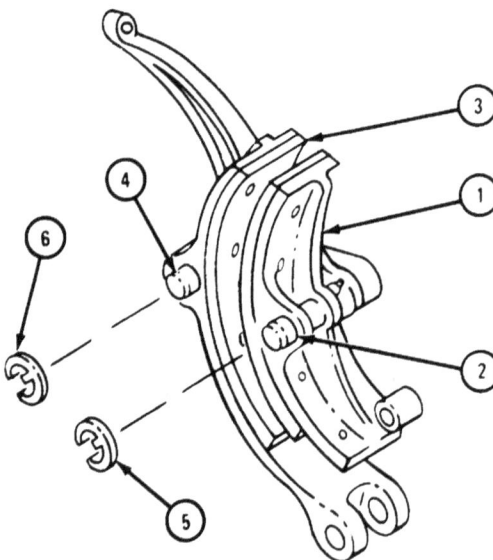

TA 089479

FRAME 2

1. Put on four washers (1).
2. Put on stabilizer spring (2).
3. Put on two snaprings (3).
END OF TASK

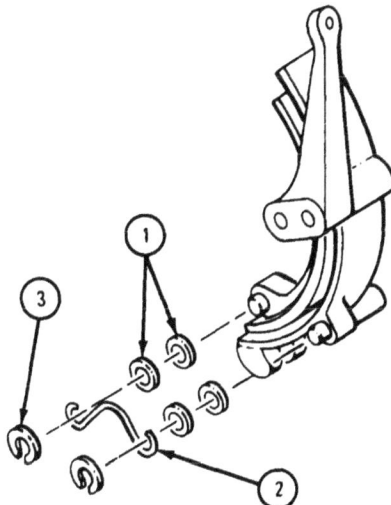

TA 089480

k. Handbrake Brake Drum Assembly.

FRAME 1

1. Tap four capscrews (1) into handbrake brake drum (2).
2. Put brake drum shield (3) and companion flange (4) into place.
3. Put on two 5/16-inch nuts (5).
4. Tap on shield (6).
5. Take off two 5/16-inch nuts (5).

END OF TASK

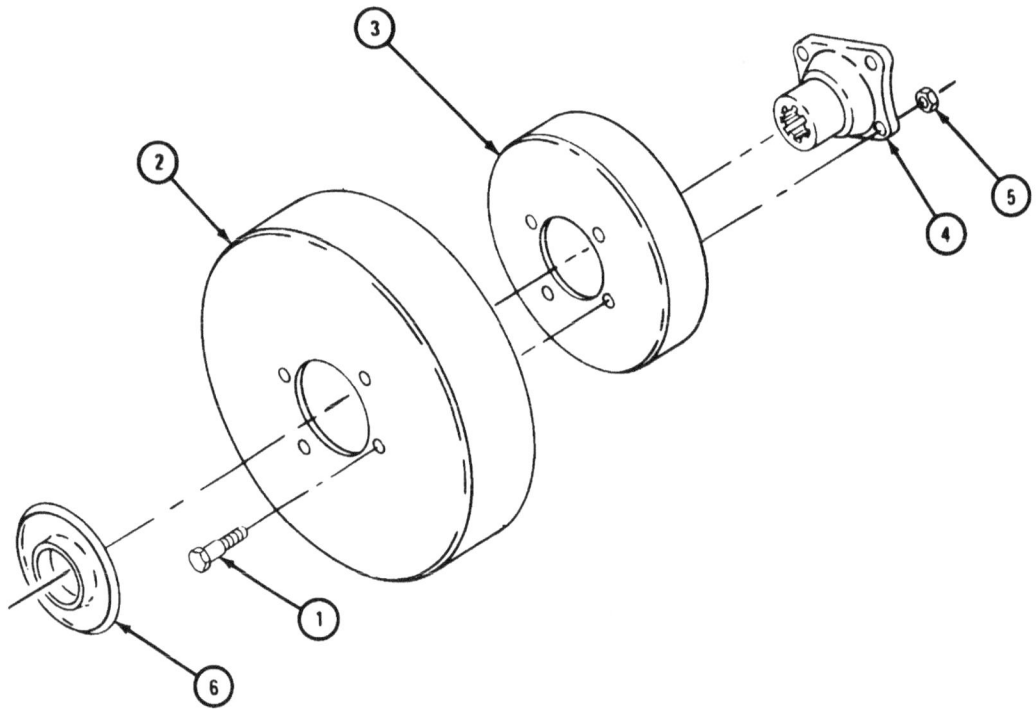

TA 089481

1. <u>Companion Flanges and Deflectors.</u>

FRAME 1

1.   Using brass hammer, tap two deflectors (1) onto two companion flanges (2).
END OF TASK

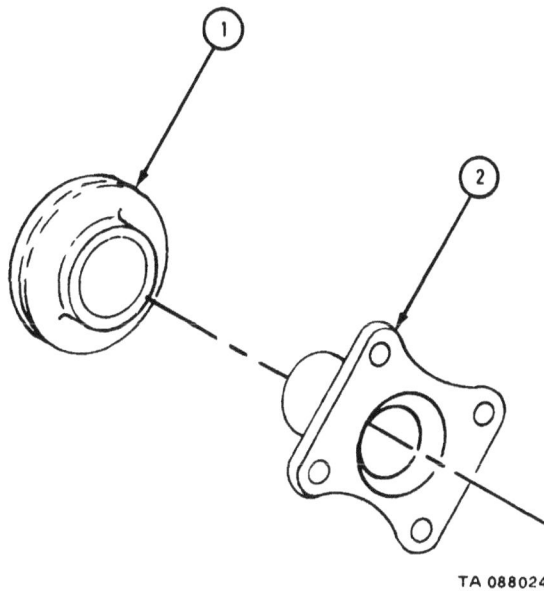

TA 088024

m. Rear Output Shaft Rear Bearing Retainer.

FRAME 1

1. Using hammer and brass drift, tap seal (1) into retainer (2).

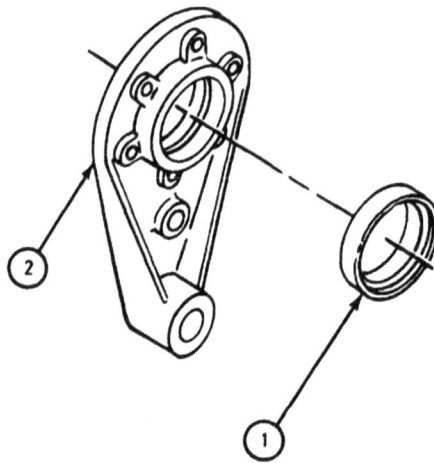

END OF TASK

TA 088025

2-22. FINAL ASSEMBLY. The following paragraphs give instructions to assemble the transmission transfer subassemblies into a final assembly.

NOTE

Keep all parts clean and protected from dust and dirt. Coat all oil seals and all gears and shafts with engine lubricating oil during assembly. Use new snaprings and gaskets during assembly.

a.   Input Shaft Assembly.

FRAME 1

1.   Put shifter fork (1) on synchronizer sleeve (2).

2.   Put input shaft assembly (3) in transfer case (4) so that bearing (5) goes into bearing bore (6).

END OF TASK

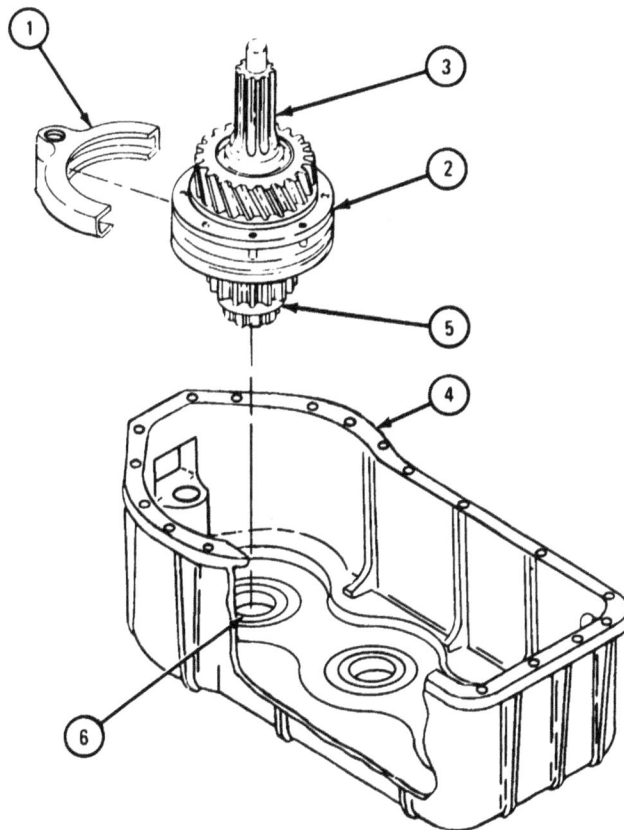

TA 089482

b. Countershaft Assembly.

FRAME 1

NOTE

Input shaft assembly (1) will have to be moved back and forth
to seat countershaft assembly (2) in transfer case (3).

1. Put countershaft assembly (2) in transfer case (3) so that bearing cone (4) sits in
bearing cup (5).

END OF TASK

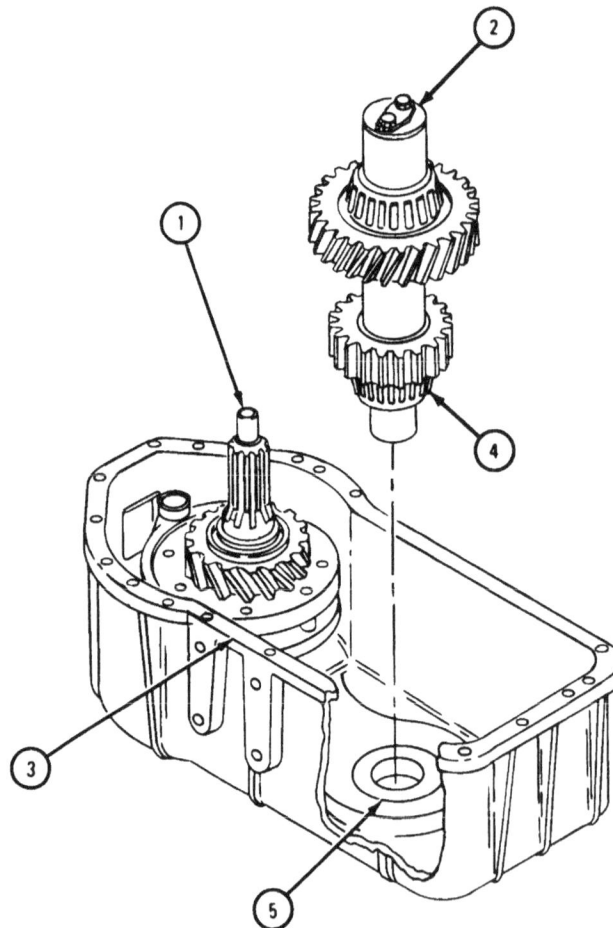

TA 089483

c. Rear Output Shaft Assembly.

FRAME 1

1. Put rear output shaft assembly (1) into transfer case (2) so that bearing cone (3) sits in bearing cup (4).

END OF TASK

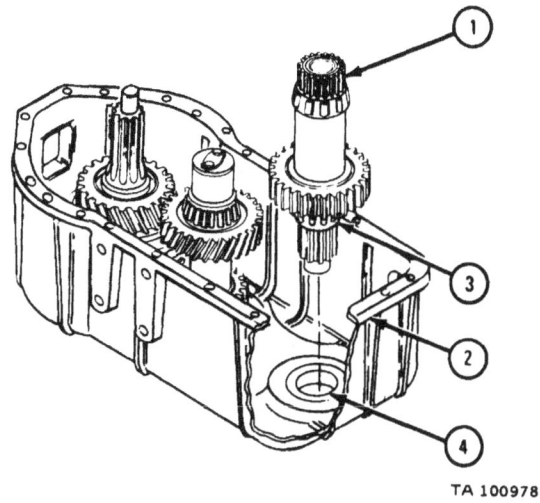

TA 100978

d. Shifter Shaft and Top Cover.

FRAME 1

1. Slide shifter shaft (1) into place in shifter fork (2). Aline setscrew hole in shifter fork with setscrew hole in shifter shaft.

2. Put in setscrew (3) and tighten setscrew to 90 to 118 pound-feet.

3. Put on safety wire (4).

GO TO FRAME 2

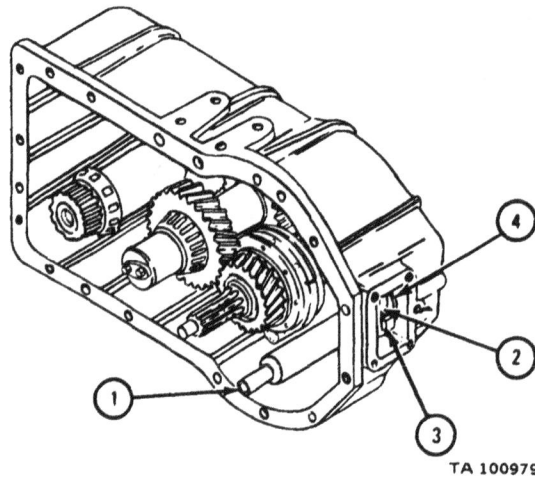

TA 100979

FRAME 2

1. Put in ball (1), plunger (2), and spring (3).

2. Put gasket (4) and cover (5) into place and aline screw holes.

3. Put in four capscrews (6) and four lockwashers (7) and tighten capscrews to 60 to 77 pound-feet.

END OF TASK

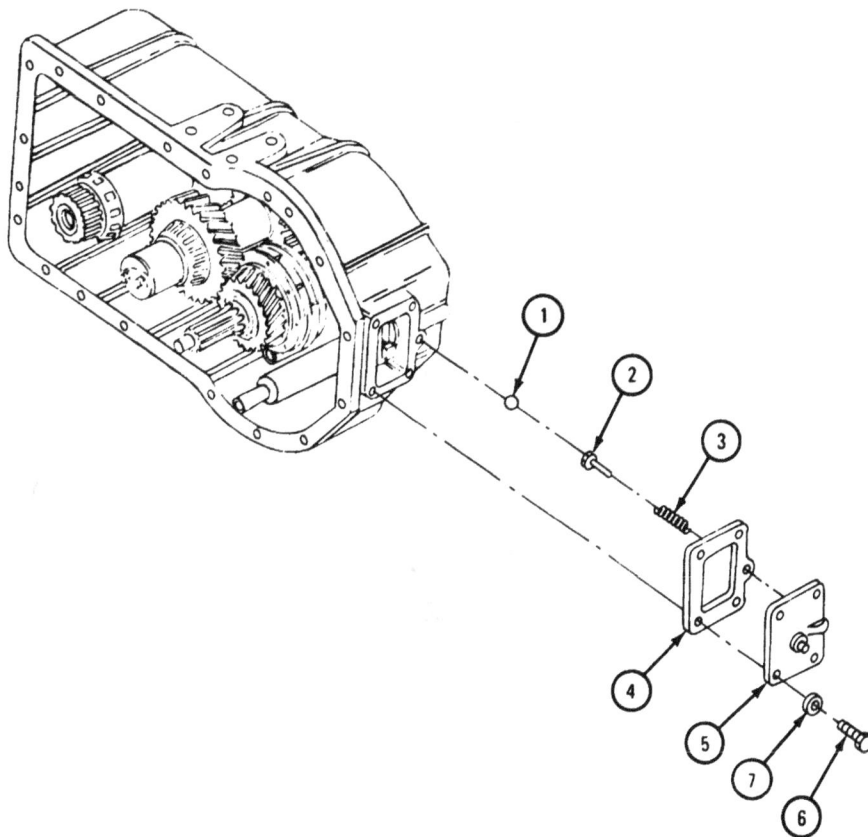

TA 100980

e. Case Cover.

FRAME 1

Soldiers 1. Aline three taper pinholes in gasket (1) and case cover (2) with three
A and B   taper pins (3) in case (4).

Soldier A 2. Put in six capscrews (5) and six lockwashers (6) and evenly tighten
         capscrews to 60 to 77 pound-feet.

GO TO FRAME 2

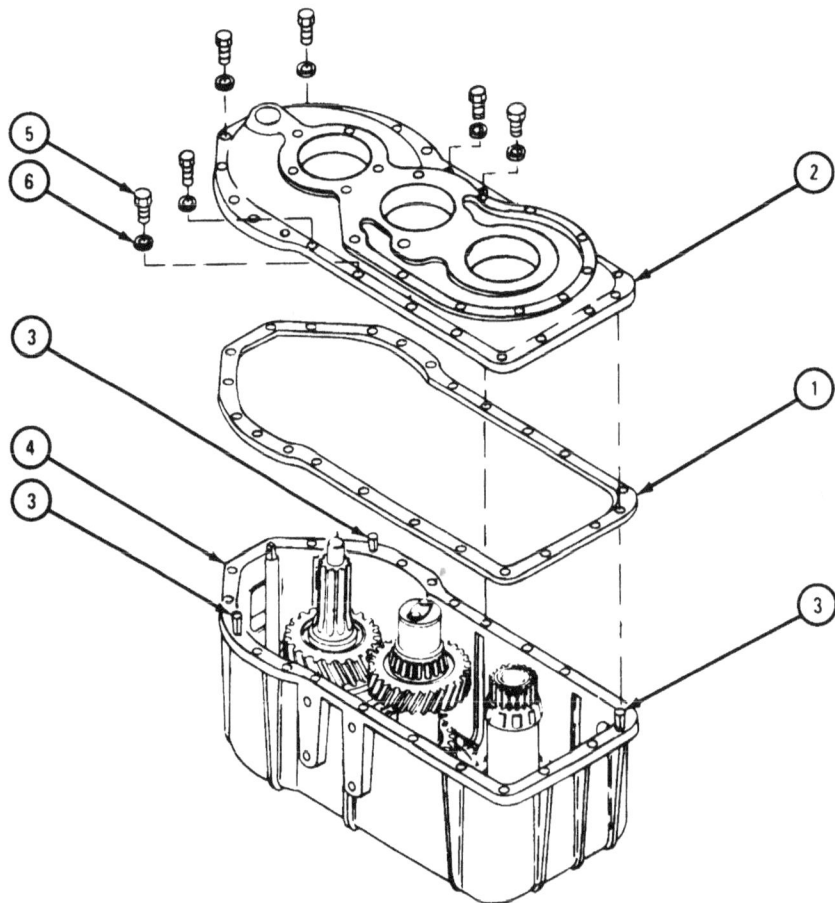

TA 100981

FRAME 2

1. Put in 12 capscrews (1), 12 lockwashers (2), and 12 nuts (3) and evenly tighten capscrews to 67 to 87 pound-feet. Switch tightening from one side to the other.

END OF TASK

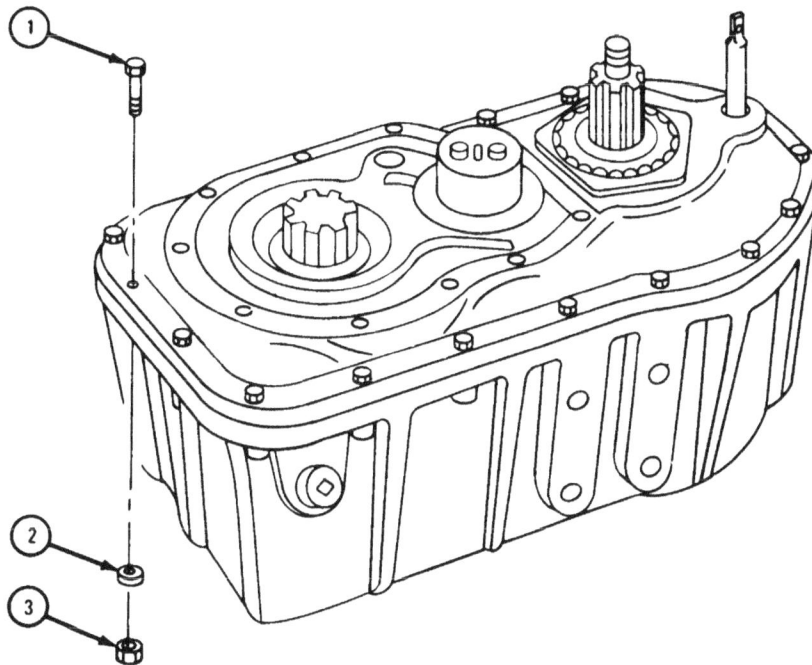

TA 100982

f. Input Shaft Front Bearing Cover.

FRAME 1

1. Put oil seal (1) in bearing cover (2).

2. Put outer thrust washer (3) into place on input shaft (4).

3. Put gasket (5) and bearing cover (2) into place on transfer case cover (6) and aline screw holes.

4. Put in five capscrews (7) and five lockwashers (8) and tighten capscrews to 60 to 77 pound-feet.

END OF TASK

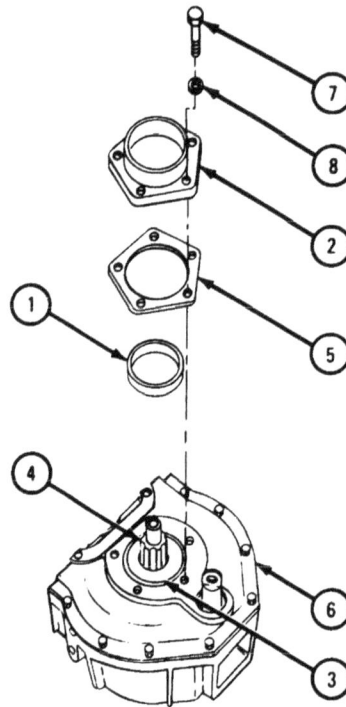

TA 100983

g. <u>Countershaft Rear Bearing Cover.</u>

FRAME 1

1. Put gasket (1) and countershaft rear bearing cover   (2) into place on case (3) and aline screw holes.

2. Put in four capscrews (4) and four lockwashers (5)   and evenly tighten capscrews to 60 to 77 pound-feet.

GO TO FRAME 2

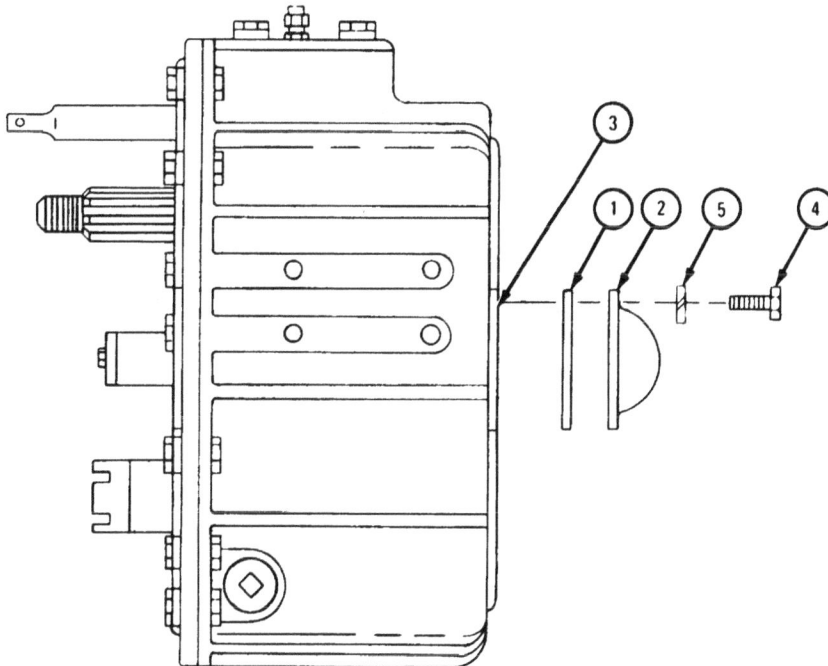

TA 100984

FRAME 2

1. Set up dial indicator (1) as shown.

2. Push countershaft (2) toward rear of transfer case (3).

3. Set dial indicator (1) to zero.

4. Pull countershaft (2) toward dial indicator (1) and note dial indicator reading.

IF READING IN STEP 4　IS MORE THAN 0.005-INCH, GO TO FRAME　3.
IF READING IN STEP 4　IS 0.003 TO 0.005-INCH, END OF TASK

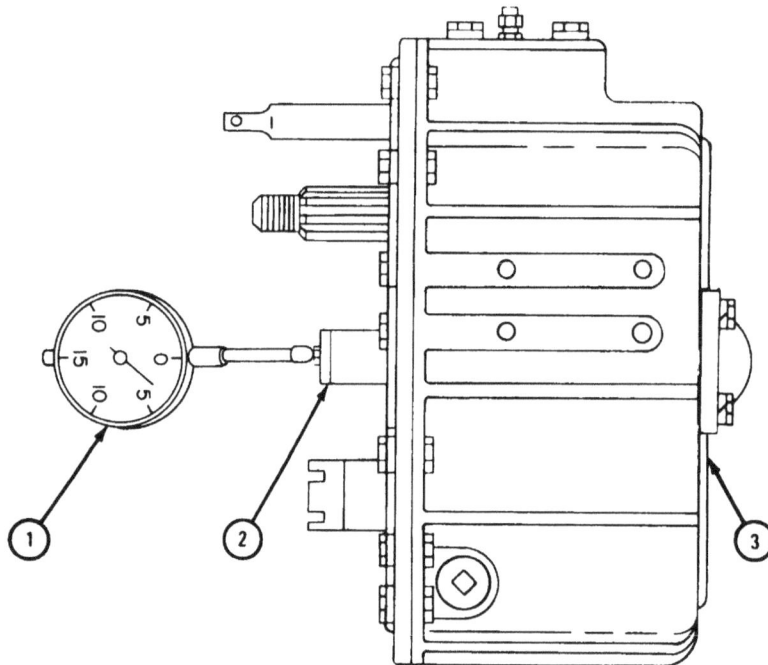

TA 100985

FRAME 3

1. Take out four capscrews (1) and four lockwashers (2). Take off countershaft rear bearing cover (3) and gasket (4).

2. Subtract 0.005-inch from dial indicator reading in step 4 of frame 2. The difference is the thickness of shims (5) needed.

3. Put on shims (5), gasket (4), and countershaft rear bearing cover (3). Aline all screw holes and put in four capscrews (1) and four lockwashers (2).

4. Evenly tighten four capscrews (1) to 60 to 77 pound-feet.

GO BACK TO FRAME 2

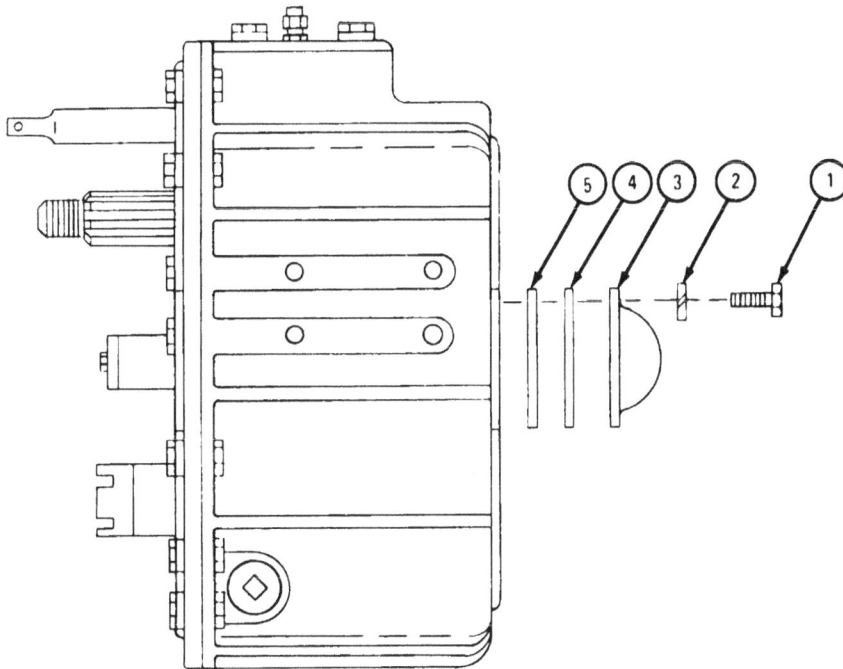

TA 100986

TM 9-2520-246-34-1

h. <u>Rear Output Shaft Rear Bearing Retainer.</u>

FRAME 1

1. Put rear output shaft rear bearing retainer (1) into place on case (2) and aline screw holes.
2. Put in six capscrews (3) and six lockwashers (4) and evenly tighten capscrews to 60 to 77 pound-feet.

GO TO FRAME 2

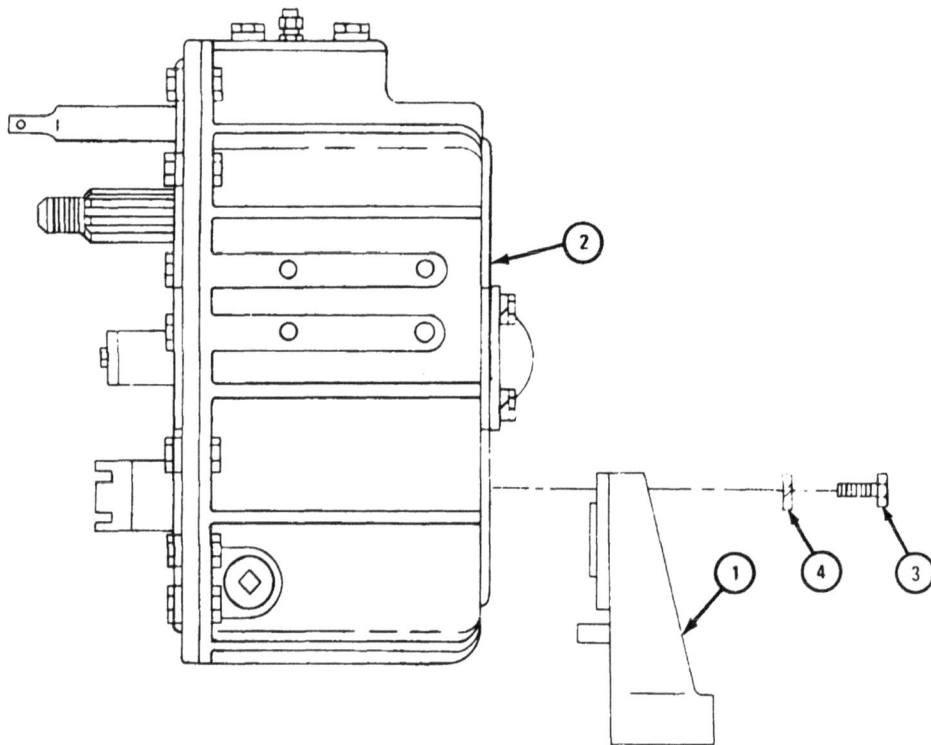

TA 100987

2-268

FRAME 2

1. Set up dial indicator (1) as shown.

2. Push rear output shaft (2) toward rear of transfer case (3).

3. Set dial indicator (1) to zero.

4. Pull rear output shaft (2) toward dial indicator (1) and note dial indicator reading.

5. Take off dial indicator (1).

IF READING IN STEP 4 IS MORE THAN 0.005-INCH, GO TO FRAME 3.
IF READING IN STEP 4 IS 0.003 TO 0.00-INCH, END OF TASK

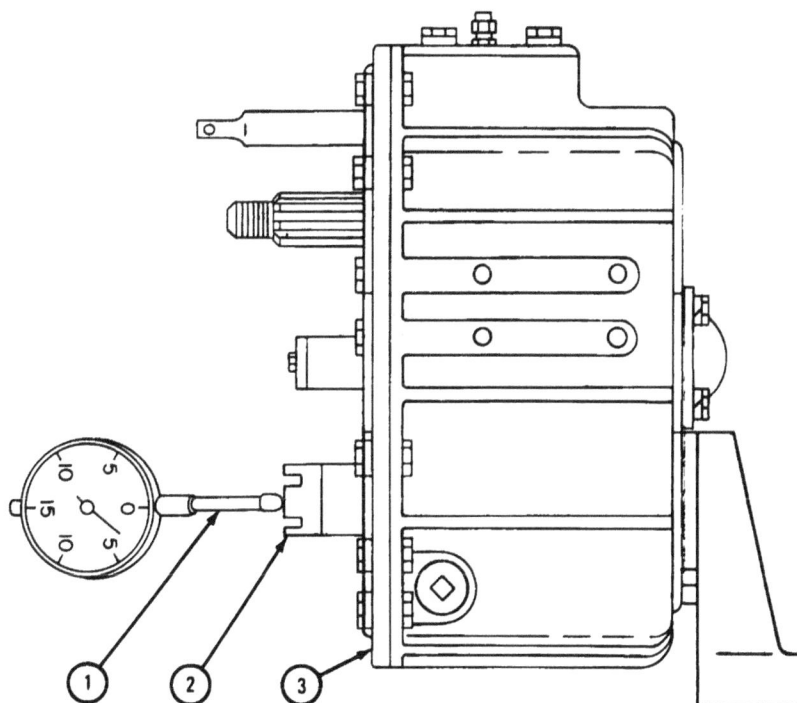

TA 100988

FRAME 3

1. Take out six capscrews (1) and six lockwashers (2). Take off rear output shaft rear bearing retainer (3).

2. Subtract 0.005-inch from dial indicator reading in step 4 of frame 2. The difference is the thickness of shims (4) needed.

3. Put on shims (4) and rear bearing retainer (3). Aline all screw holes and put   in six capscrews (1) and six lockwashers (2).

4. Evenly tighten six capscrews (1) to 60 to 77 pound-feet.

GO BACK TO FRAME 2

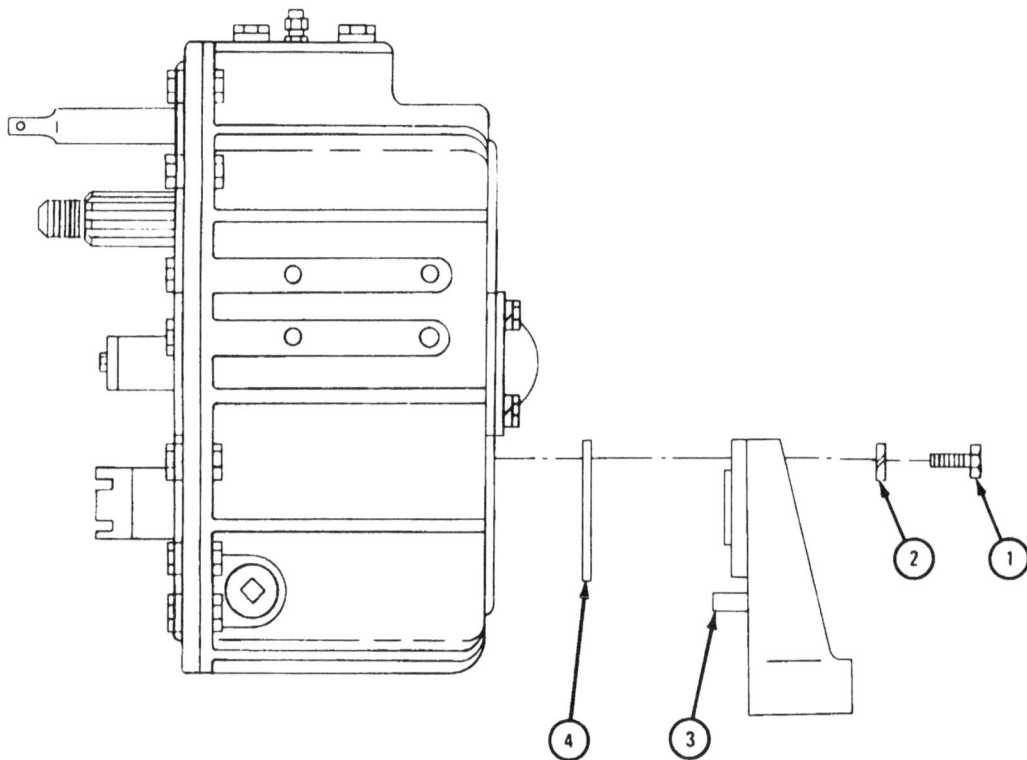

TA 100989

i. Input Shaft Rear Bearing   Cover (Trucks Without Transfer Power Takeoff
    Installed).

FRAME 1

1. Put gasket (1) and bearing cover (2) into place and aline screw holes.

2. Put in six capscrews (3) and six lockwashers (4) and evenly tighten cap-
    screws to 60 to 77 pound-feet.

END OF TASK

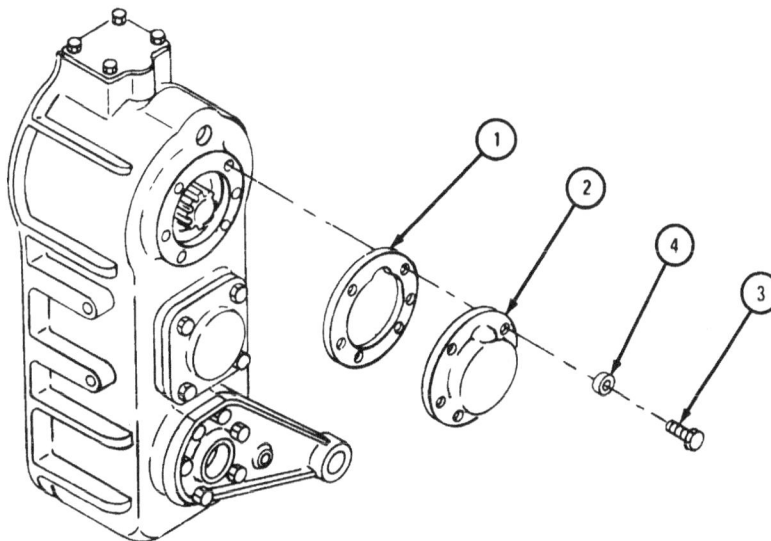

TA 100990

j. <u>Front Output Shaft and Drive Gear (Model T-136-21).</u>

FRAME 1

1. Put output shaft assembly (1) in place as shown.

2. Put helical gear (2) into place as shown.

3. Put on retainer plate (3) and retainer lock assembly (4).

4. Put in two capscrews (5) . Bend tabs on retainer lock assembly (4) toward capscrews.

END OF TASK

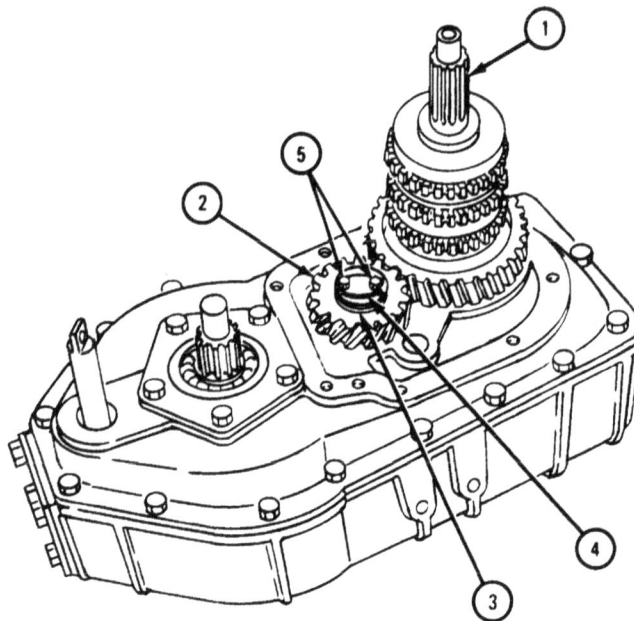

TA 100991

k. Front Output Case Assembly (Model T-136-1).

FRAME 1

1. Put shifter shaft spring (1) on hole (2).

2. Aline three taper pin holes in gasket (3) and front output case (4) with three taper pins (5) in transfer case cover (6).

3. Put gasket (3) and front output case (4) in place. Make sure that shifter shaft (7) goes in shifter shaft spring (1) and hole (2).

4. Put in ten capscrews (8) and ten lockwashers (9) and evenly tighten capscrews to 45 to 55 pound-feet.

END OF TASK

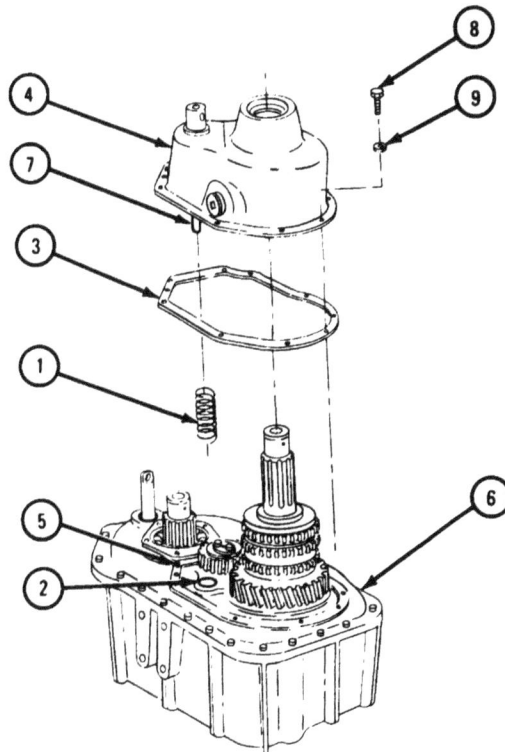

TA 100992

1. <u>Front Output Clutch (Model T-136- 27).</u>

FRAME 1

1. Put on clutch (1) as shown.
2. Put in setscrew (2) and tighten setscrew to 45 to 57 pound-feet.
3. Put on safety wire (3).

END OF TASK

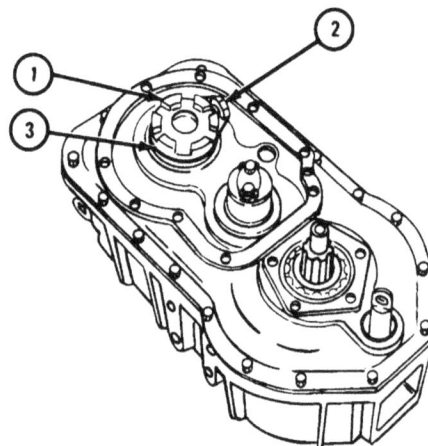

TA 100993

m. <u>Front Output Case Assembly (Model T- 136-27).</u>

FRAME 1

1. Put shifter shaft spring (1) with two spring caps (2) on hole (3).

2. Aline three taper pin holes in gasket (4) and front output case (5) with three taper pins (6) in transfer case cover (7).

3. Put gasket (4) and front output case (5) in place. Make sure that   shifter shaft (8) goes in shifter shaft spring (1) and hole (3).

4. Put in ten capscrews (9) and ten lockwashers (10) and evenly tighten capscrews to 45 to 55 pound-feet.

END OF TASK

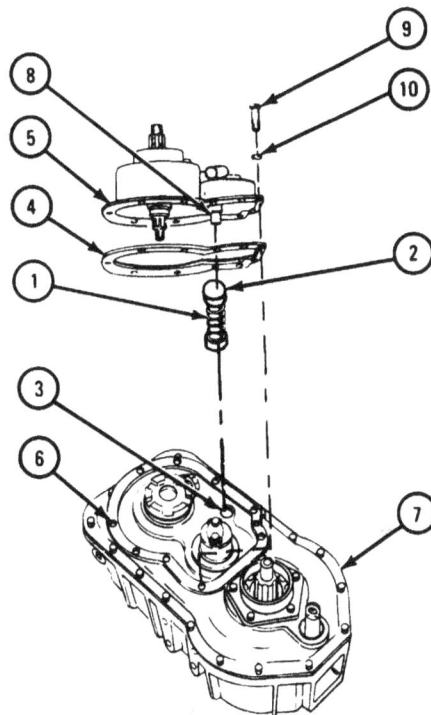

TA 100994

n.   Handbrake Brake Drum and Brakeshoes.

FRAME 1

1.   Take out three capscrews (1) and three lockwashers (2).
2.   Put brake cable bracket (3) into place and aline screw holes.
3.   Put two capscrews (1) and two lockwashers (2) into cable bracket (3).
4.   Put shoe stop bracket (4) into place and aline screw holes.
5.   Put two capscrews (1) and two lockwashers (2) into shoe stop bracket (4).
6.   Put on brake drum assembly (5).
GO TO FRAME 2

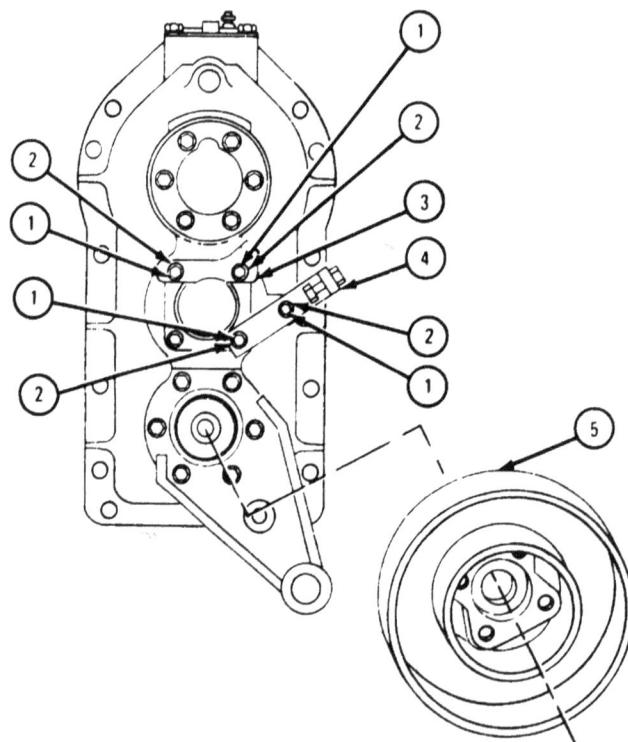

TA 100995

FRAME 2

1. Put lever assembly (1) with brakeshoe (2) into place on handbrake brake drum (3).

2. Put in anchor pin (4).

3. Check that brakeshoe (2) is flush with handbrake brake drum (3) or not more than 1/16-inch below handbrake brake drum. If brakeshoe is more than 1/16-inch below handbrake brake drum,   add spacer (5).

4. Put on jamnut (6).

5. Screw in anchor pin (4).

6. Tighten jamnut (6).

GO TO FRAME 3

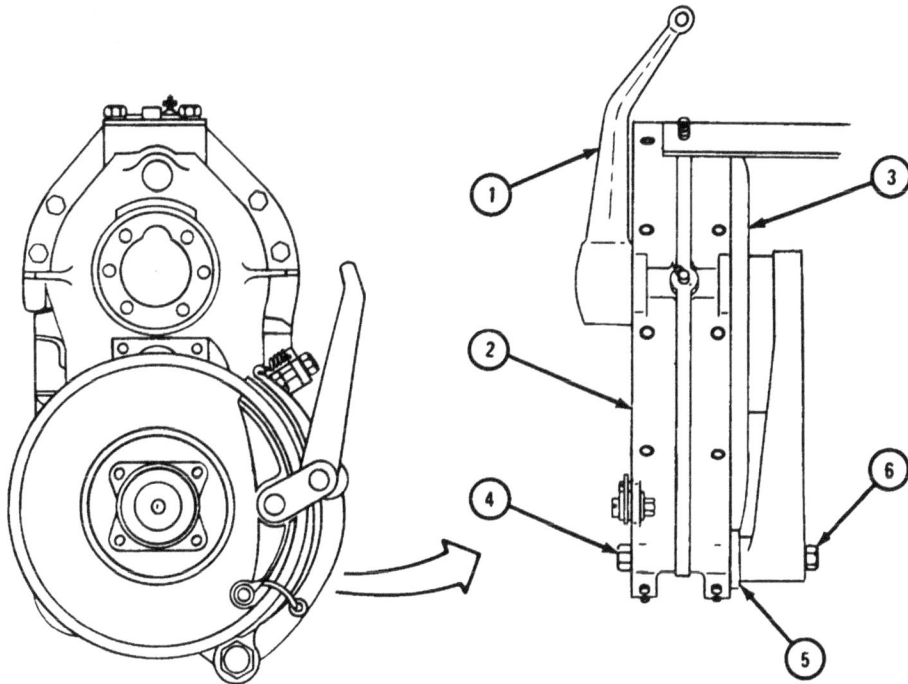

TA 100996

FRAME 3

1.   Put on spring (1).   Put in adjusting screw (2) and tighten jamnut (3).
END OF TASK

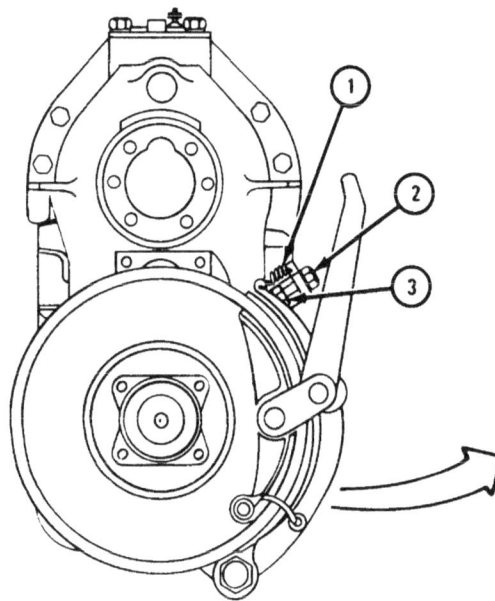

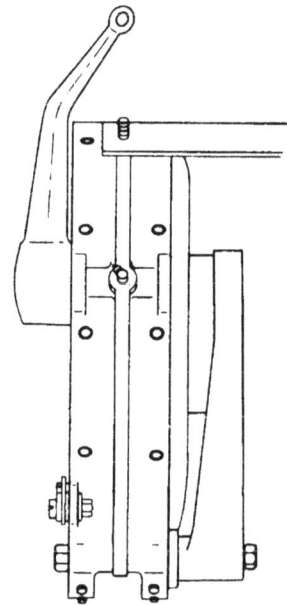

TA 100997

o. Companion Flanges.

FRAME 1

Soldier A 1. Push in shifter shaft (1).

2. Put deflector (2) and companion flange (3) on rear output shaft (4).

3. Using adjustable wrench as shown, hold companion flange (2).

Soldier B 4. Put on washer (5) and slotted nut (6) and tighten nut to 300 to 400 pound-feet.

5. Tighten slotted nut (6)   until slot is alined with cotter pin hole in rear output shaft (4) .

6. Put in cotter pin (7).

GO TO FRAME 2

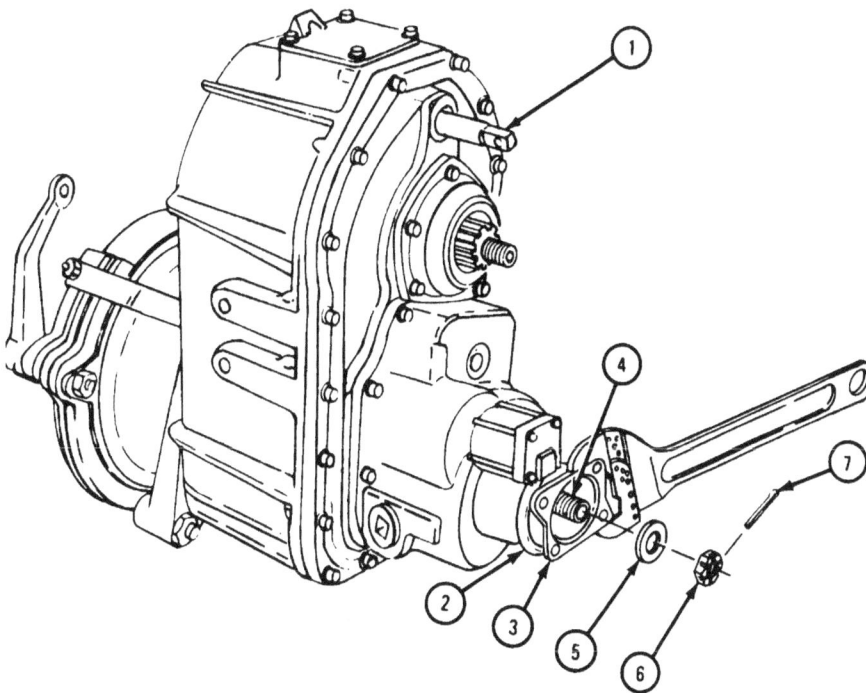

TA 100998

FRAME 2

1. Aline tab on sliding drive shaft (1) with slot in retainer plate on countershaft (2).
    Put in sliding drive shaft.

2. Put drive adapter (3) on sliding drive shaft (l).

GO TO FRAME 3

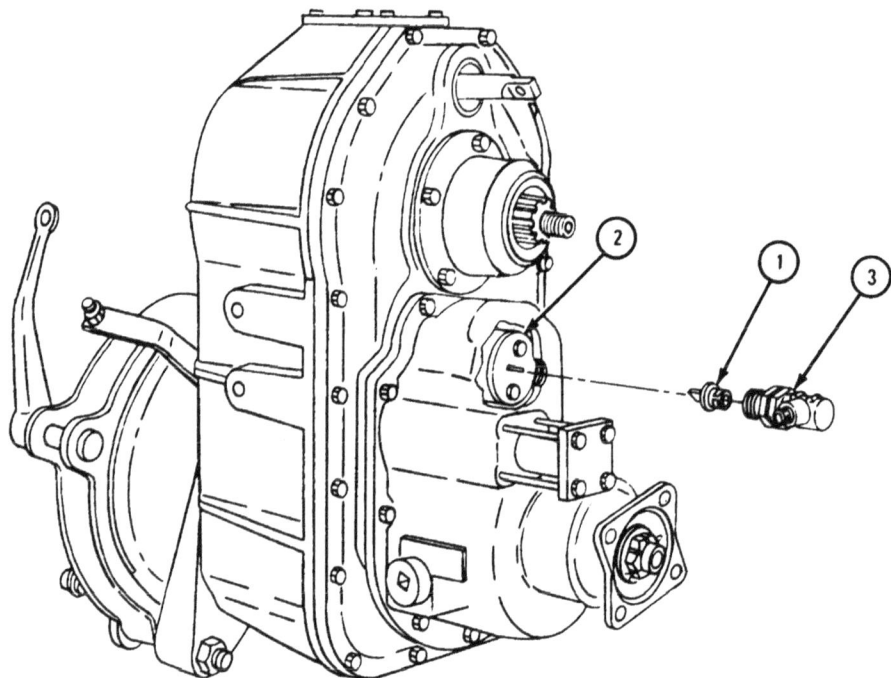

TA 102091

FRAME 3

Soldier A 1. Put deflector (1) and companion flange (2) on input shaft (3).

2. Using adjustable wrench as shown, hold companion flange (2).

Soldier B 3. Working at front of transfer (4), put washer (5) and slotted nut (6) on input shaft (3) and tighten nut to 300 to 400 pound–feet.

4. Tighten slotted nut (6) until slot is alined with cotter pin hole in shaft.

5. Put in cotter pin (7).

6. Working at rear of transfer (4), do steps 3 through 5 again for rear output shaft flange (8).

END OF TASK

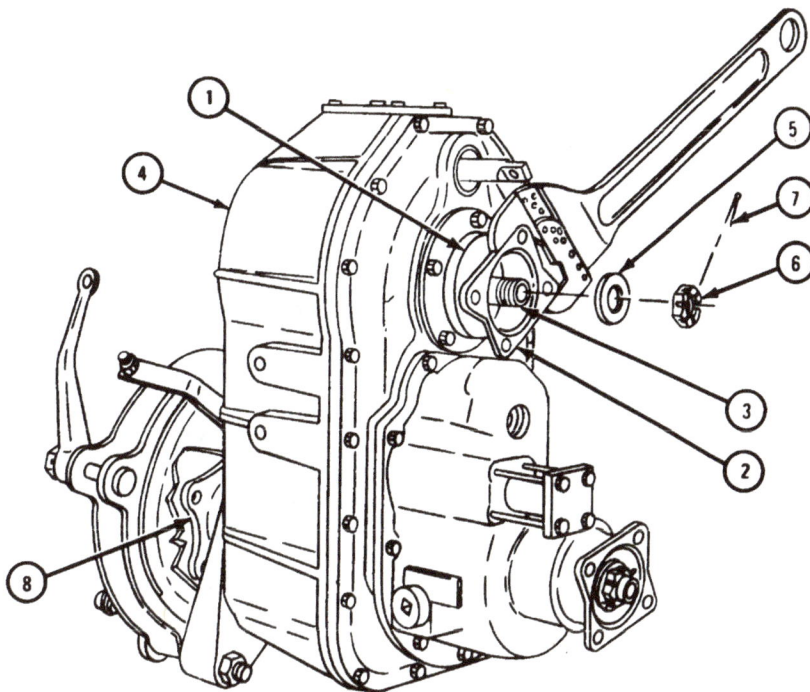

TA 100999

p. Handbrake Brakeshoe Adjustment.

FRAME 1

1. Loosen jamnut (1).

2. Check that clearance between outer brakeshoe (2) and brakedrum (3) is at least 0.015-inch all along.

3. Screw in adjusting screw (4) until clearance in step 2 is 0.015-inch at the highest point.

4. Tighten jamnut (1).

END OF TASK

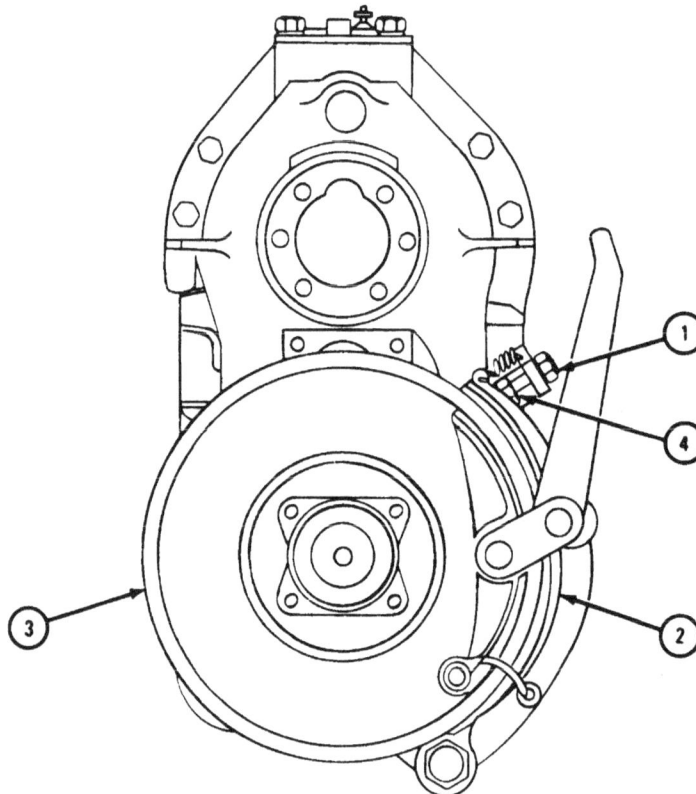

TA 101000

2-23. SHIFT TEST. The following paragraphs give instructions for testing the transmission transfer after final assembly and includes removing the transmission transfer assembly from stand.

NOTE

Before making tests, fill transmission assembly with 1/2-pint of gear oil. Refer to LO 9-2320-209-12/1.

Attach a tag to filler plug saying transmission transfer must be filled after it is put back in truck.

a.   Neutral Position.

FRAME 1

1.   Push shifter shaft (1) all the way in.

2.   Pull shifter shaft (1) out until you feel it go into the first detent.

3.   Turn input shaft (2) by hand.   Front output shaft (3) and rear output shaft (4) should not turn.   If rear output shaft (4) turns, do the following:

    a.   Remove top cover and shifter shaft. Refer to para 2-14k, frame 2.

    b.   Replace top cover and shifter shaft. Refer to para 2-22d, frame 3.

4.   If front output shaft (3) turns, do the following:

    a.   Remove front output case assembly. For model T-136-27, refer to para 2-14d.   For model T-136-21, refer to para 2-14f.

    b.   Make sure that all parts are free to turn in front output case assembly.

    c.   Replace front output case assembly.   For model T-136-27, refer to para 2-22m.   For model T-136-21, refer to para 2-22k.

END OF TASK

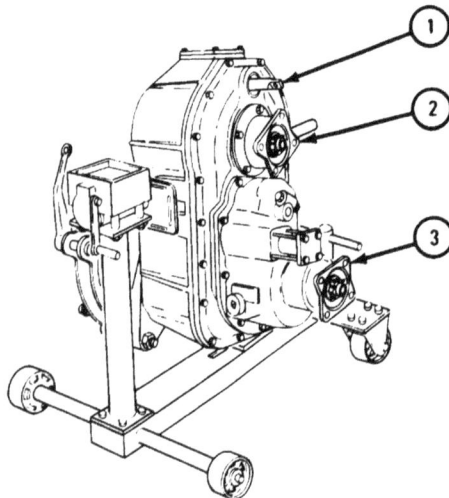 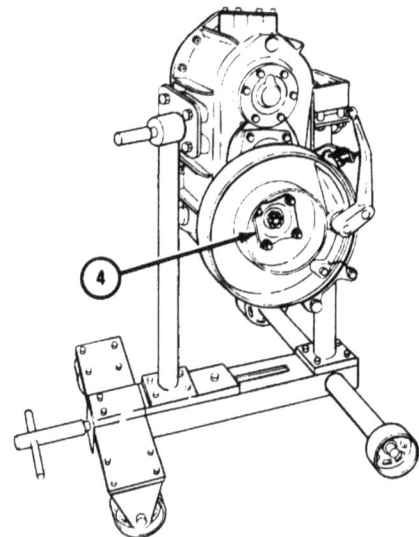

TA 102067

b. <u>High Range Position.</u>

FRAME 1

1. Push shifter shaft (1) all the way in.

2. Turn input shaft (2) to the right by hand.

3. Rear output shaft (3) should turn and f rent output shaft (4) should not turn.

4. If rear output shaft (3) does not turn, do the following:

    a. Remove top cover and shifter shaft.   Refer to para 2-14k, frame 1.

    b. Replace top cover and shifter shaft.   Refer to para 2-22d, frame 2.

5. If front output shaft (4) turns, do the following:

    a. Remove front output case assembly. For model T-136-27, refer to para 2-14d.   For model T-136-21, refer to para 2-14f.

    b. Make sure that all parts are free to turn in front output case assembly.

    c. Replace front output case assembly. For model T-136-27, refer to para 2-22m.   For model T-136-21, refer to para 2-22k.

END OF TASK

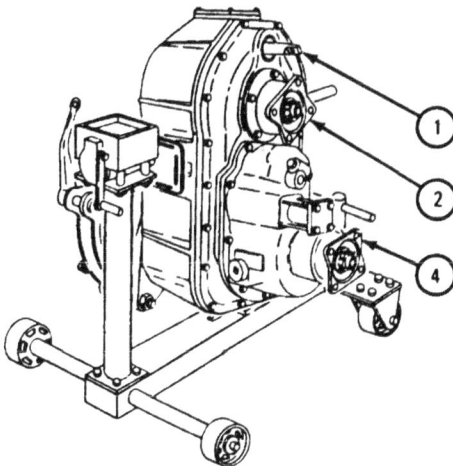 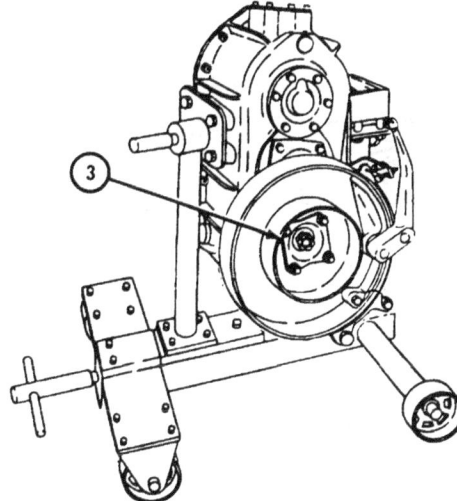

TA 102068

c. <u>Low Range Position.</u>

FRAME 1

1. Pull shifter shaft (1) all the way out.

2. Turn input shaft (2) by hand. Rear output shaft (3) should turn. Front output shaft (4) should not turn.

3. If rear output (3) does not turn, do the following:

   a. Remove top cover and shifter shaft. Refer to para 2-14k, frame 1.

   b. Replace top cover and shifter shaft. Refer to para 2-22d, frame 1.

4. If front output shaft (4) turns, do the following:

   a. Remove front output case assembly. For model T-136-27, refer to para 2-14d. For model T-136-21, refer to para 2-14f.

   b. Make sure that all parts are free to turn in front output case assembly.

   c. Replace front output case assembly. For model T-136-27, refer to para 2-22m. For model T-136-21, refer to para 2-22k.

END OF TASK

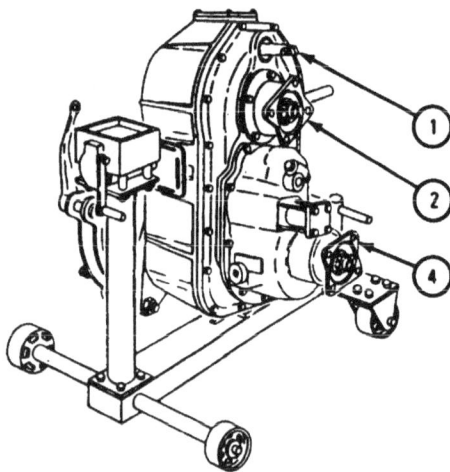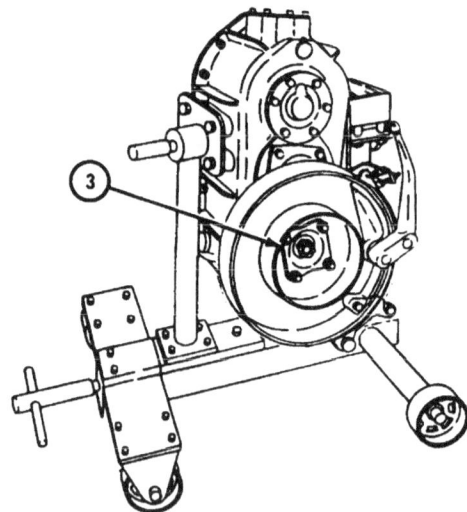

TA 102068

---

d.   Front Output Shaft Air Lockup (Model T-136-27).

FRAME 1

1.   Pull shifter shaft (1) all the way out.

2.   Using adapters, connect shop air hose (2) to air cylinder (3) and turn on air supply.

3.   Turn input shaft (4) by hand. Front output shaft (5) should be engaged and turn.

4.   Turn off air supply and disconnect shop air hose (2).

5.   Turn input shaft (4) by hand. Front output shaft (5) should be disengaged and not turn.

6.   If front output shaft (5) did not engage or disengage, do the following:

    a.   Disassemble air cylinder.   Refer to para 2-15d, frame 1.

    b.   Make sure that all parts move freely.

    c.   Assemble air cylinder.   Refer to para 2-21e, frame 5.

END OF TASK

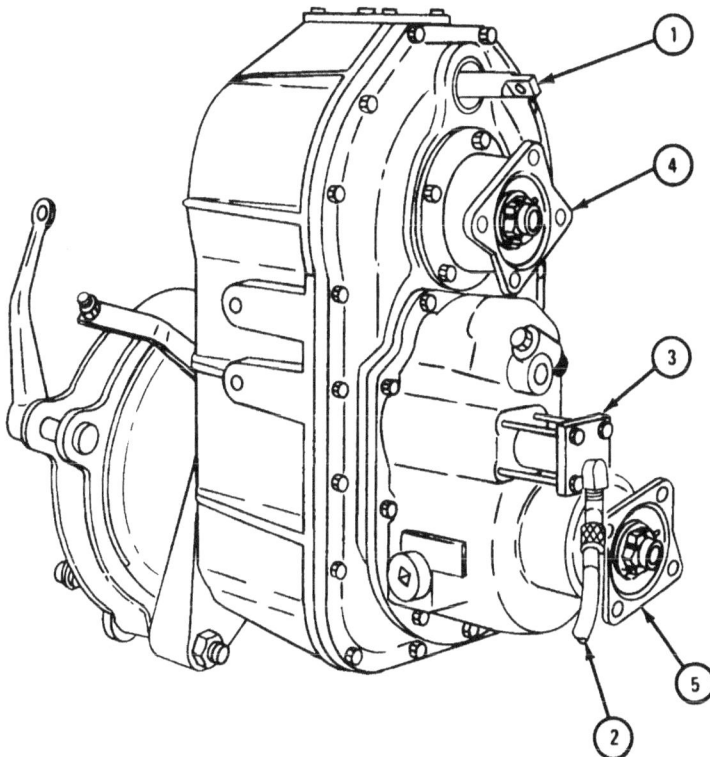

TA 102069

e. Front Output Shaft Sprag Unit (Model T-136-21).

FRAME 1		
Soldier A	1.	Pull shifter shaft (1) all the way out.
	2.	Hold input shaft (2). Turn front output shaft (3) to the right and check that it turns freely. Turn front output shaft to the left and check that it does not turn.
Soldier B	3.	Push in and hold shifter shaft (4).
Soldier A	4.	Hold input shaft (2). Turn front output shaft (3) to the right and check that it does not turn. Turn front output shaft to the left and check that it turns freely.
	5.	If tests made in steps 2 and 4 do not check out, do the following:

        a.   Remove front output case assembly. Refer to para 2-14f.

        b.   Disassemble sprag units.   Refer to para 2-15g, frames 1 and 4.

        c.   Inspect sprag unit.   Refer to para 2-17a, frame 1.

        d.   Assemble sprag units. Refer to para 2-21g.

        e. Replace front output case assembly. Refer to para 2-22k.

END OF TASK

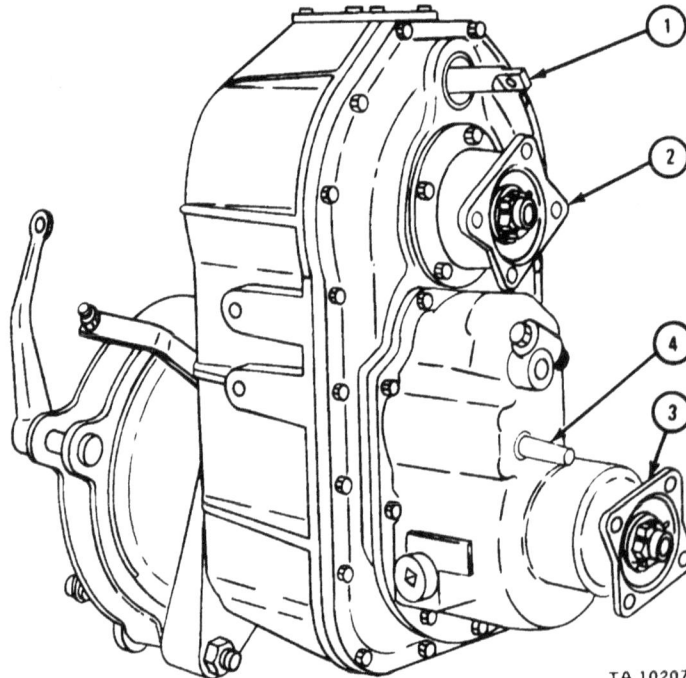

TA 102070

f. Removing the Transmission Transfer from Stand.

FRAME 1

1. Using chain hoist, lift transmission transfer assembly (1) from stand (2) just enough to take weight of transmission transfer assembly off stand.

2. Take out four bolts (3), two on each side of stand (2).

3. Using chain hoist, lift transmission transfer assembly (1) out of stand (2).

END OF TASK

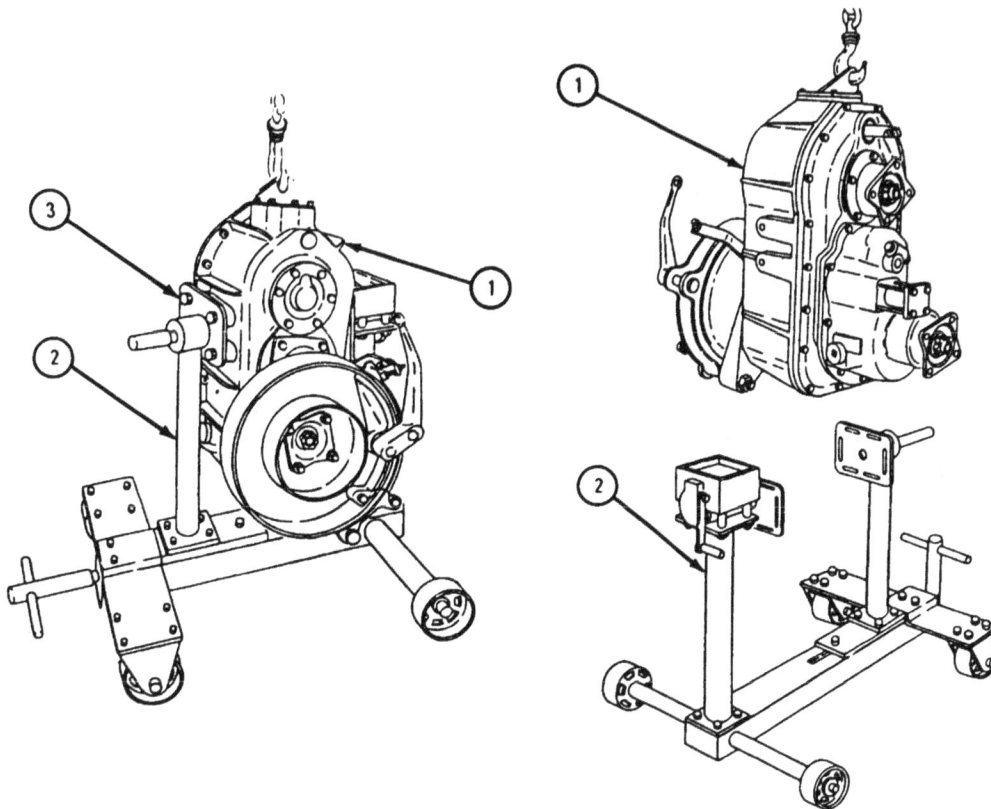

TA 102071

Section IV.   MAINTENANCE OF TRANSMISSION POWER TAKEOFFS

NOTE

Procedures given are for model WND-7-28 transmission
power takeoff.   Procedures are the same for model
WN-7-28 except where noted.

TOOLS: Power train rebuild tool kit, pn 7950356

SUPPLIES: Dry cleaning solvent, type II (SD-2), Fed. Spec P-D-680
White lead pigment, NSN 8010-00-290-6643
Multipurpose gear lubricating oil, GO 85/140, MIL-L-2105
Lubricating oil, ICE, OE/HDO 10, MIL-L-2104
Artillery and automotive grease, type GAA, MIL-G-10924
Safety wire, MS20995F47
Case and cover gasket
Accessory drive rear output cover gasket
Output shaft front bearing retainer gasket
Reverse shaft gear thrust washer (2)
Input shaft gear thrust washer (2)
Fork rod shifter seal (2) (model WN-7-28)
Fork rod shifter seal (3) (model WND-7-28)
Accessory drive drive shaft seal (model WND-7-28)
Output shaft oil seal
Compressed air source, 30   psi max

PERSONNEL: One

EQUIPMENT CONDITION: Transmission   power takeoff on workbench.

2-24. CLEANING BEFORE DISASSEMBLY.   This paragraph gives instructions for
cleaning the transmission power takeoff before disassembly.   Note and scribe
transmission power takeoff case in places that have oil soaked road mud. It is not
necessary to mark oil spots around gaskets or seals since the gaskets and seals will
be replaced.   Scrape, brush, and steam clean all dirt and road mud from the
transmission power takeoff assembly.

2-25. DISASSEMBLY. This paragraph gives instructions to completely disassemble the transmission power takeoff assembly.

FRAME 1

1. Take out six capscrews (1) and six starwashers (2).

NOTE

If power takeoff cover (3) does not come off easily, push a scraper between power takeoff cover and gasket (4) to work it loose.

2. Take off power takeoff cover (3) and gasket (4). Throw away gasket.

3. Take out poppet retainer (5).

4. Turn power takeoff assembly (6) on its side and take out poppet spring (7) and ball (8).

GO TO FRAME 2

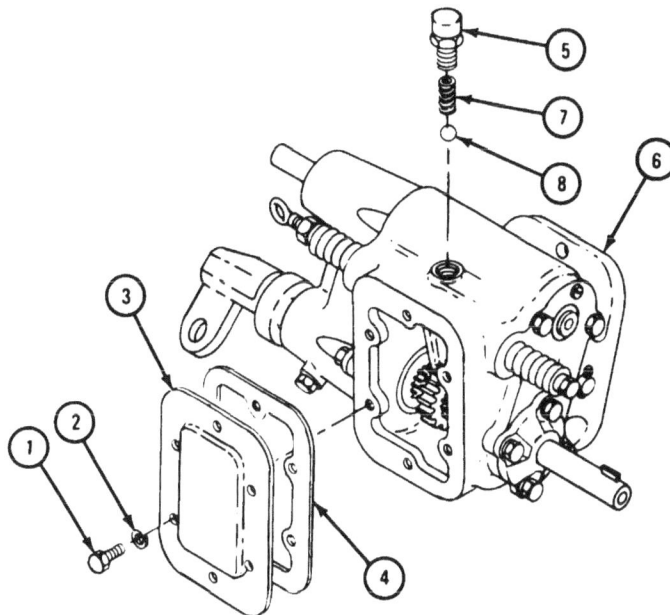

TA 087930

FRAME 2

1. Take out capscrew (1) and washer (2).

2. Take out eyebolt (3) and locknut with flat washer (4).

3. Working from inside power takeoff assembly (5), take off safety wire (6).

4. Take out setscrew (7).

5. Hold shifter fork (8) and pull out shifter shaft (9). Take out shifter fork.

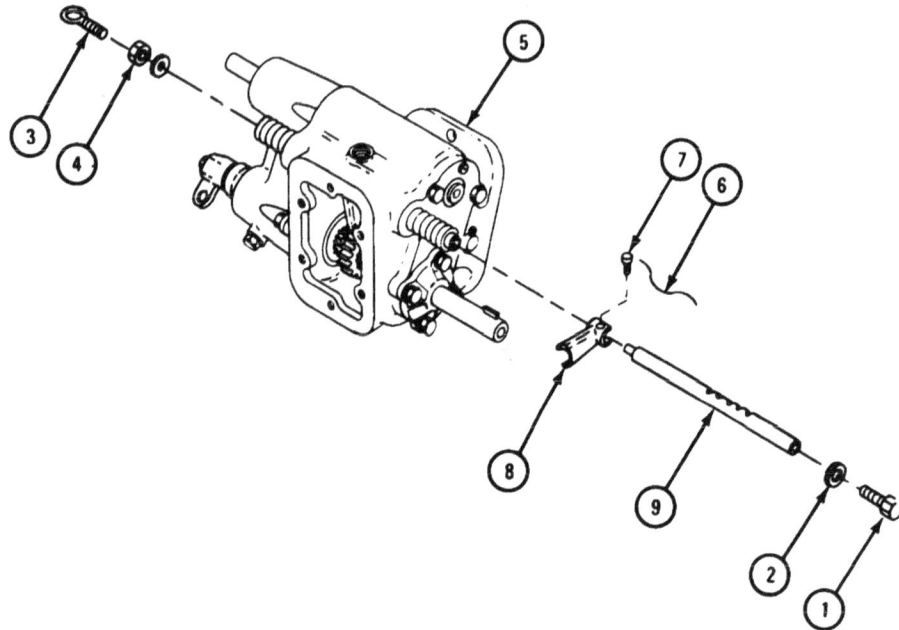

GO TO FRAME 3

TA 087931

FRAME 3

1. Take out cotter pin (1).

2. Take out input gear shaft (2).

3. Take input gear (3) out of power takeoff housing (4).

4. Take out two sets of roller bearings (5) from input gear (3).

5. Take out and throw away two thrust washers (6).

IF WORKING ON DOUBLE OUTPUT POWER TAKEOFF (MODEL WND-7-28),
GO TO FRAME 4.
IF WORKING ON SINGLE OUTPUT POWER TAKEOFF (MODEL WN-7-28),
GO TO FRAME 8

TA 087932

FRAME 4

1. Take out five capscrews (1) and five starwashers (2).

NOTE

If accessory drive housing (3) is stuck to power takeoff housing case (4), tap sides of accessory drive housing with soft faced hammer.

2. Take off accessory drive housing (3) and gasket (5). Throw away gasket.

3. Take out plug (6), spring (7), and ball (8).

GO TO FRAME 5

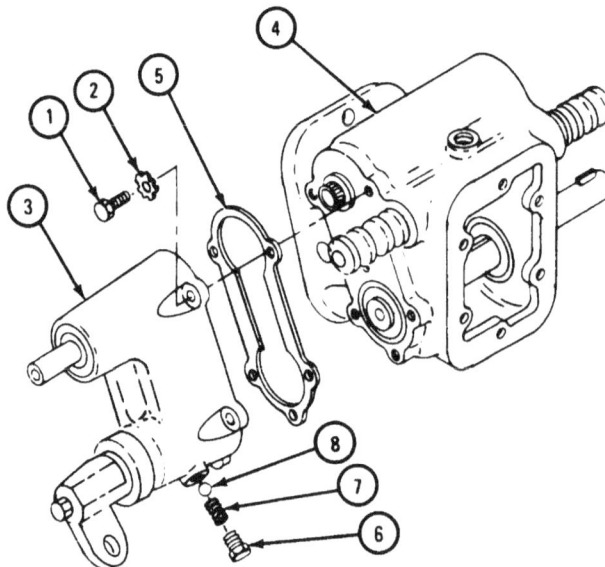

TA 087934

FRAME 5

1. Take out capscrew (1) and connector (2).
2. Take off safety wire (3).
3. Take out setscrew (4).
4. Hold shifter fork (5) and slide out shifter shaft (6). Take out shifter fork.
5. Tap out and throw away boot (7) and seal (8).

GO TO FRAME 6

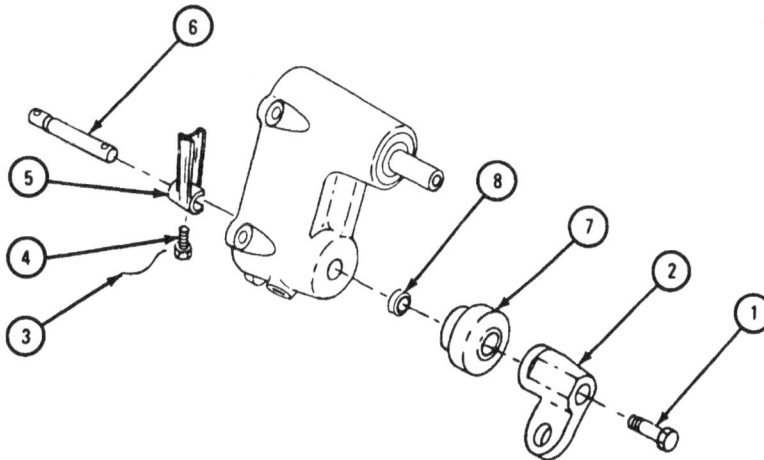

TA 087935

FRAME 6

1. Take key (1) out of output shaft (2).
2. Pry out seal (3).
3. Take off snapring (4) from accessory drive housing (5).
4. Working from inside accessory drive housing (5), tap out output shaft (2).

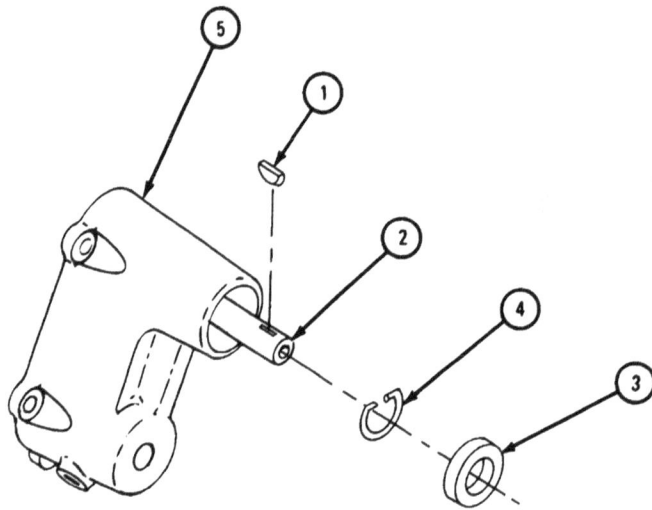

GO TO FRAME 7

TA 087936

FRAME 7

1. Take clutch sliding sleeve (1) out of accessory drive housing (2).

2. Tap out output shaft (3).

3. Take snapring (4) out of accessory drive housing (2).

4. Take off other snapring (4) and pull off bearing (5).

GO TO FRAME 9

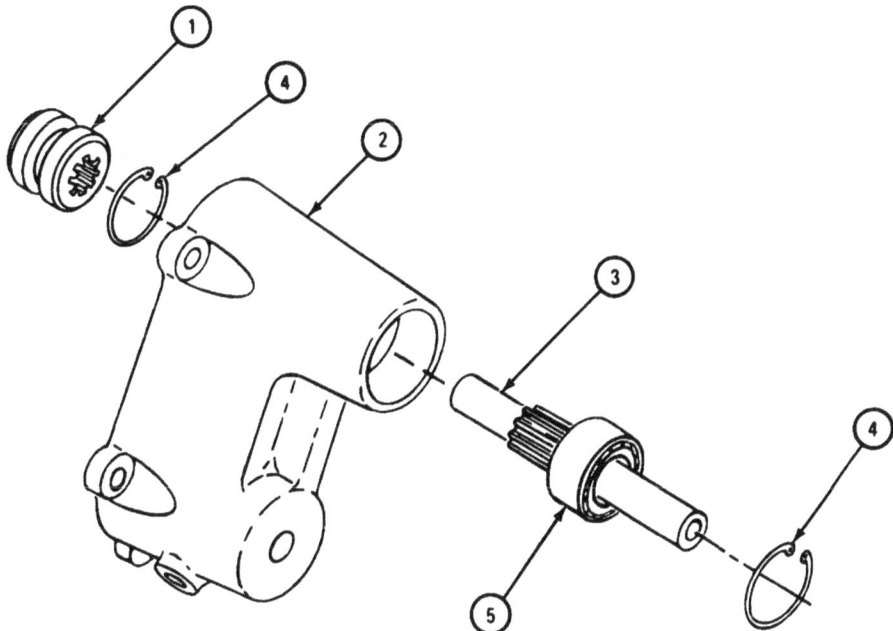

TA 087937

FRAME 8

1. Take off five capscrews (1) and five starwashers (2).

NOTE

If rear bearing cap (3) is stuck to power takeoff housing
case (4), tap sides of bearing cap with a soft faced hammer.

2. Take off rear bearing cap (3) and gasket (5). Throw away gasket.

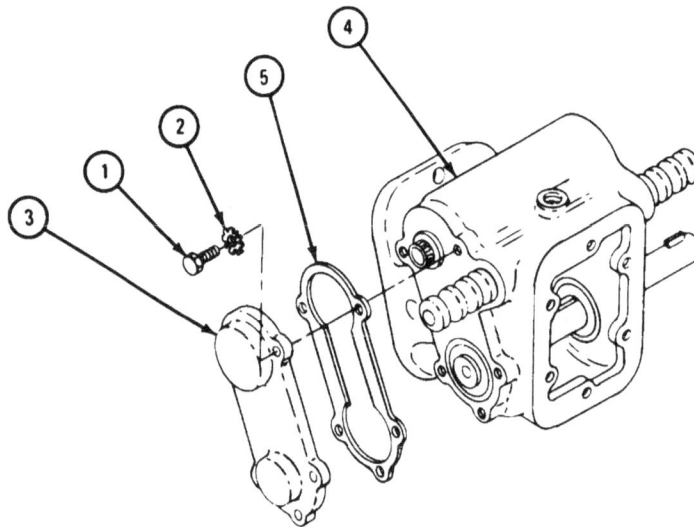

GO TO FRAME 9

TA 087933

FRAME 9

1. Take out key (1).

2. Take out four capscrews (2) and four starwashers (3).

NOTE

If front bearing cap (4) is stuck to power takeoff housing case (5),
tap sides of bearing cap with a soft faced hammer.

3. Take off f rent bearing cap (4) and gasket (6). Throw away gasket.

GO TO FRAME 10

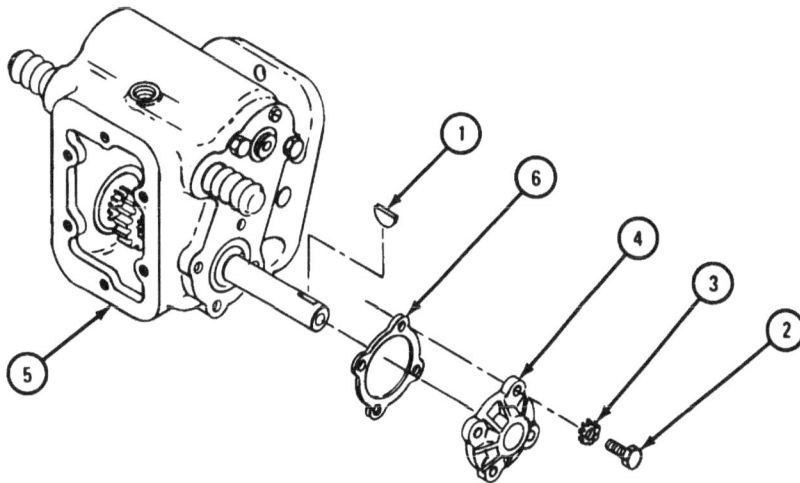

TA 087938

FRAME 10

1. Put front bearing cap (1) on top of vise as shown.

2. Using hammer and brass drift as shown, tap out and throw away seal (2).

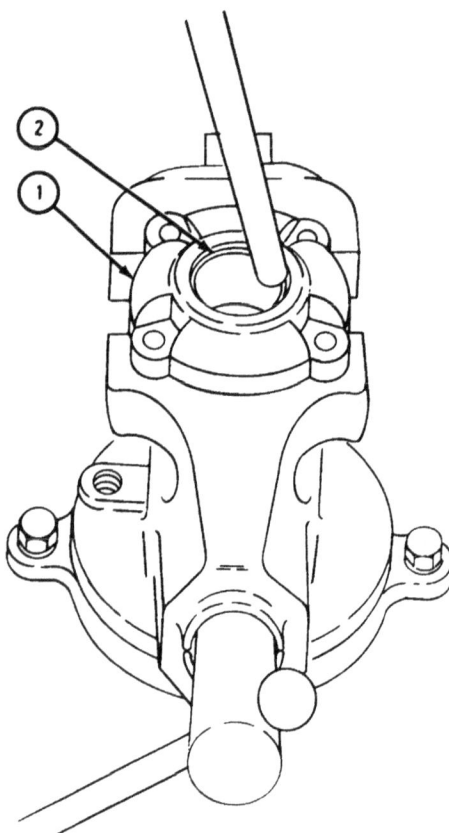

GO TO FRAME 11

TA 087939

FRAME 11

1. Tap output shaft (1) and spacer (2) outward as shown. Spacer will fall off output shaft.

GO TO FRAME 12

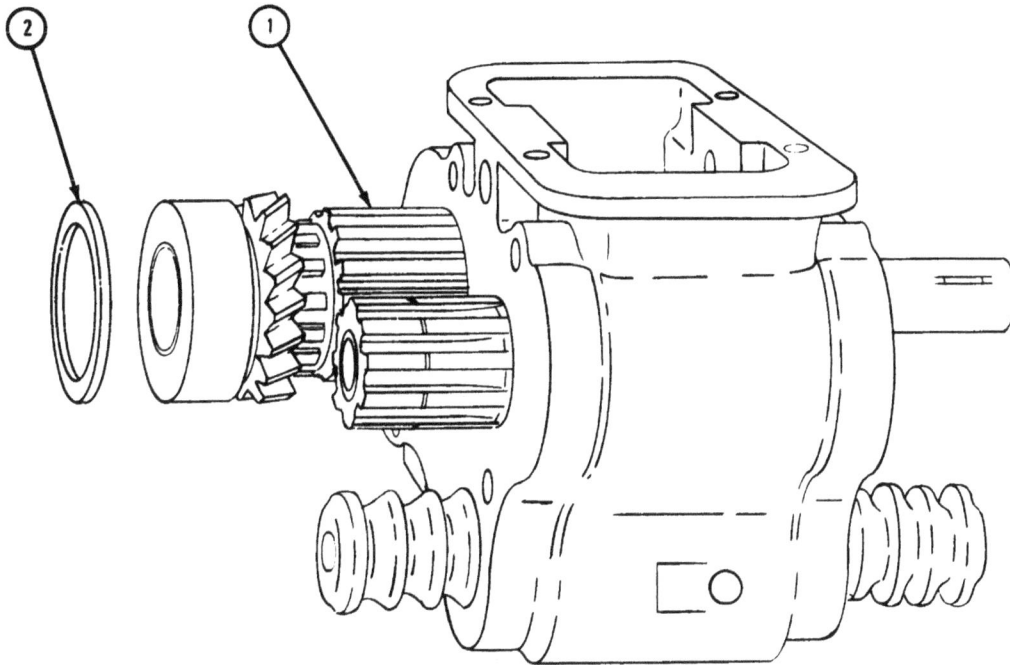

TA 087940

FRAME 12

1. Set up puller as shown and pull off rear bearing (1).
GO TO FRAME 13

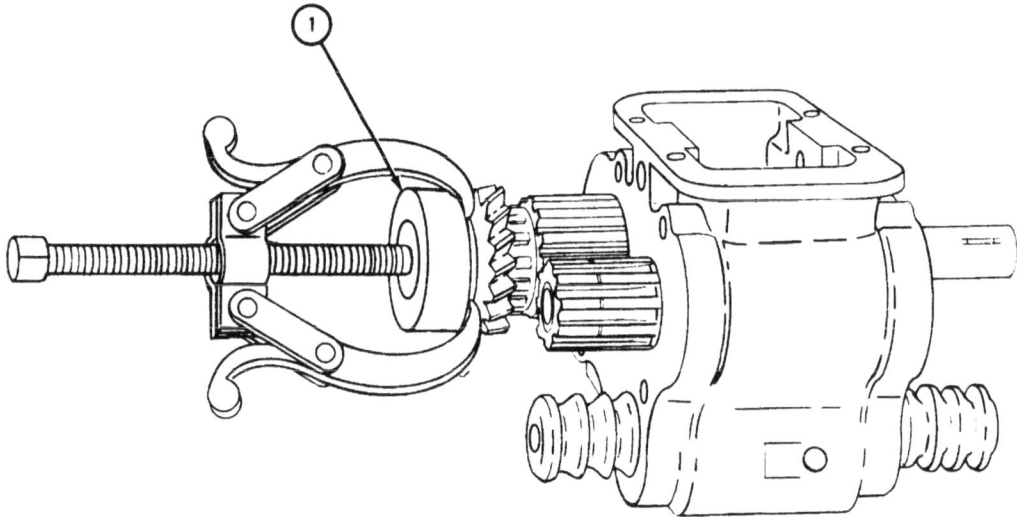

TA 087941

FRAME 13

1. Push output shaft (1) back into place.
2. Take off snapring (2) and spacer (3), and take out helical gear (4).

GO TO FRAME 14

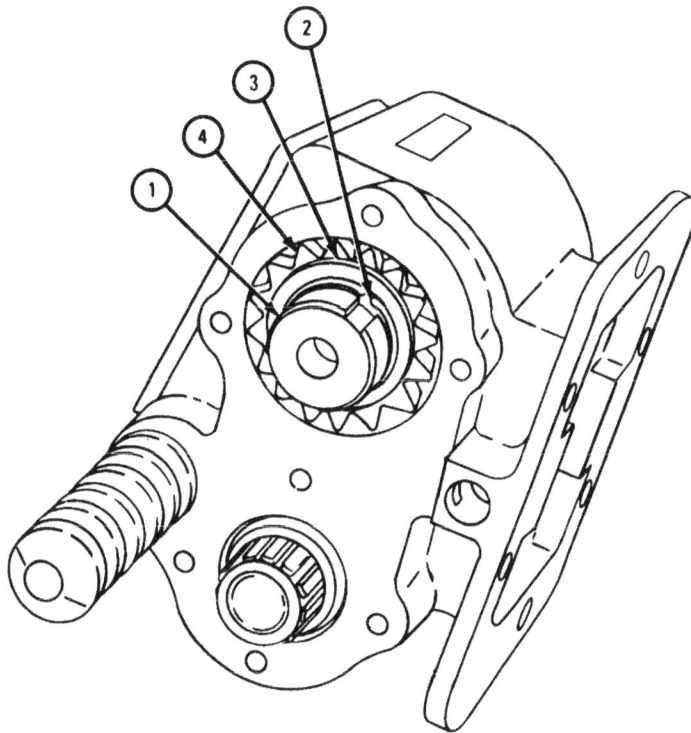

TA 087942

FRAME 14

1. Using hammer and brass drift, tap out output shaft (1) with front bearing (2).
2. Take spur gear (3) out of power takeoff housing (4).

GO TO FRAME 15

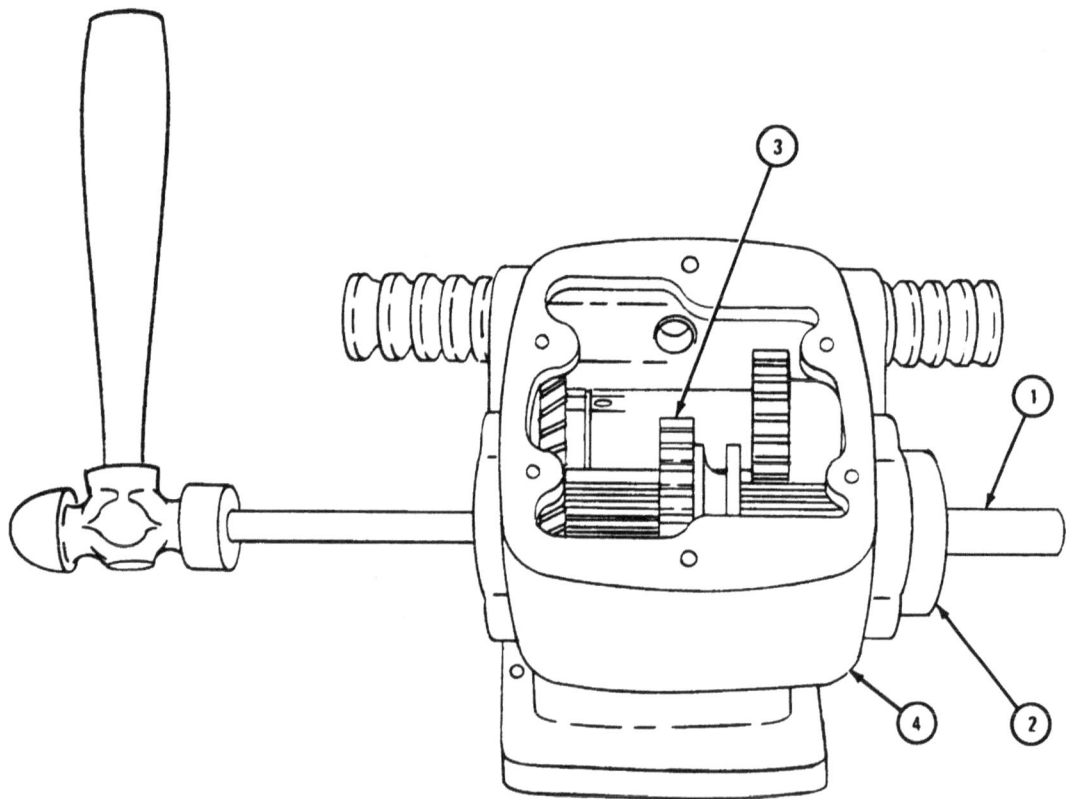

TA 087943

FRAME 15

1. Set up puller as shown, and pull off front bearing (1).

GO TO FRAME 16

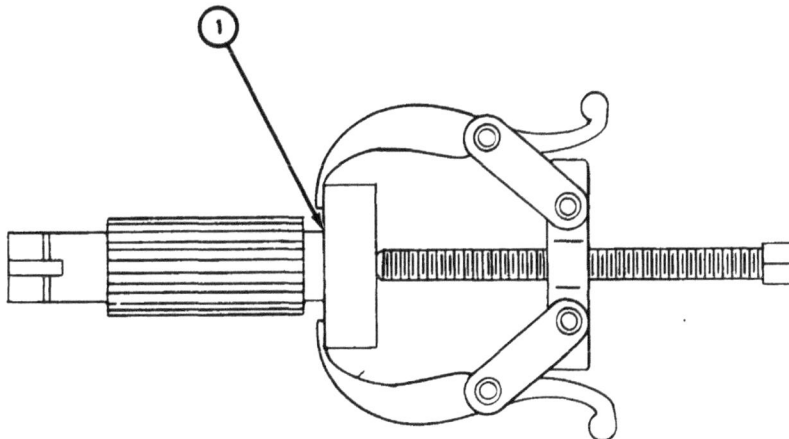

TA 087944

FRAME 16

1. Take out cotter pin (1) and straight headed pin (2).

2. Tap out reverse gear shaft (3), front needle roller bearing (4) and end cap. (5). Take off needle roller bearing.

3. Take reverse gear (6) out of power takeoff housing (7).

4. Take out and throw away two thrust washers (8).

5. Take out woodruff key (9).

GO TO FRAME 17

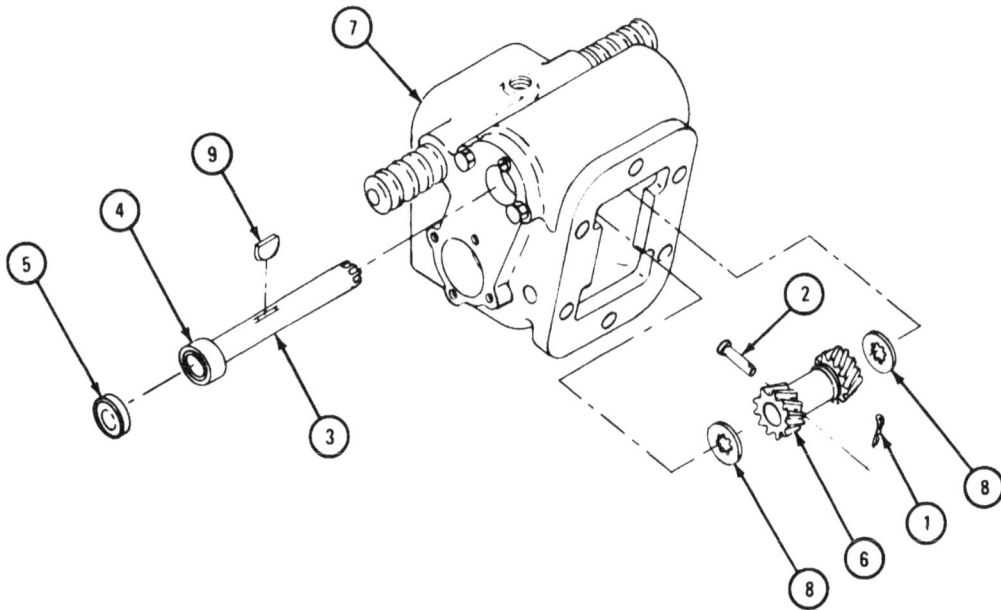

TA 087945

FRAME 17

1. Using hammer and brass drift, tap out bearing sleeve (1).
2. Take out two capscrews (2).
3. Take out plug (3).

GO TO FRAME 18

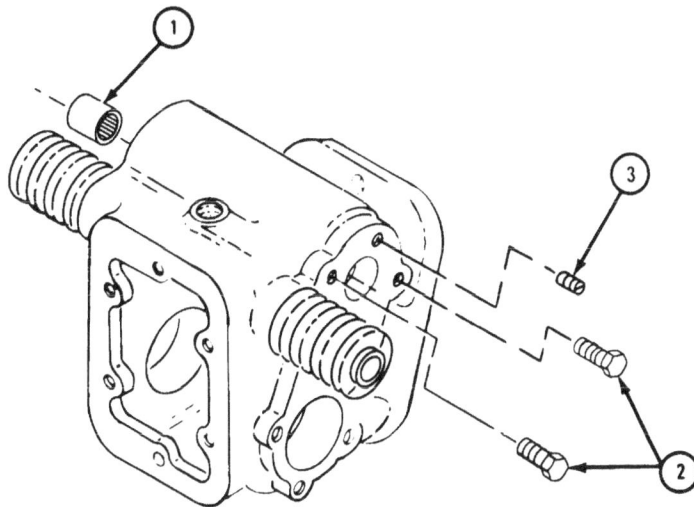

TA 087946

FRAME 18

1. Working from inside of power takeoff housing (1) as shown, tap out and throw away two boots with seals (2).

END OF TASK

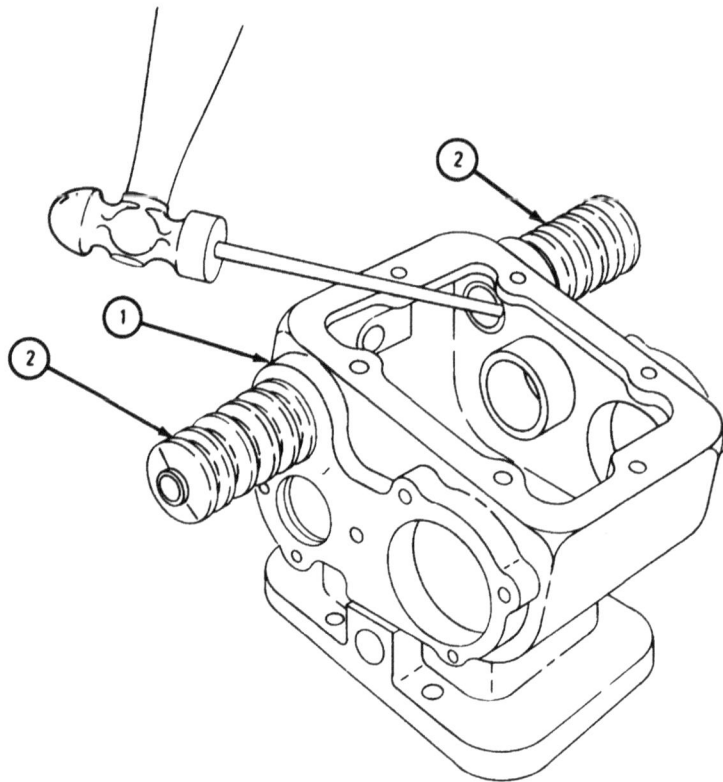

TA 087947

2-26.   CLEANING .   This paragraph gives general instructions for cleaning the transmission power takeoff parts.

   a. Clean all bearing cones and cups. Refer to inspection, care, and maintenance of antifriction bearings, TM 9-214.

WARNING

Dry cleaning solvent is flammable. D   o not use near an open flame.   Keep a fire extinguisher nearby when solvent is used. Use only in well-ventilated places.   Failure to do this may result in injury to personnel and damage to equipment.

Do not use more than 30 psi of air pressure for drying parts.   Eye shields must be worn when using compressed air.   Eye injury can occur if eye shields are not used.

CAUTION

When scraping gasket material from surface of parts, be careful not to scratch or gouge the metal surface.

   b.   Clean all other parts with solvent.   Scrape all gasket material from surface of parts.   Rinse parts in clean solvent and dry with compressed air.

2-27.   GENERAL INSPECTION.   This paragraph gives instructions to check for damage on the transmission power takeoff housings, gear shafts, and gears.

CAUTION

It is easy to damage the equipment if you don't know what you're doing. Do not try to do this task unless you are experienced at it, or you have an experienced person with you.

NOTE

Small chips, burrs or scratches on gears and gear shafts can be repaired. Cracks in housing castings that do not go into screw holes or openings can be repaired. If parts are damaged in any other way, throw away parts and get new ones.

FRAME 1

1. Check all bearings (1 through 6). Refer to TM 9-214 for inspection, care and maintenance of antifriction bearings.

2. Check that housings (7 and 8) do not have any broken bolts or stripped threads. If housings are damaged, mark them for repair.

3. Check that housings (7 and 8) have no cracks, warped gasket surfaces or small holes.

GO TO FRAME 2

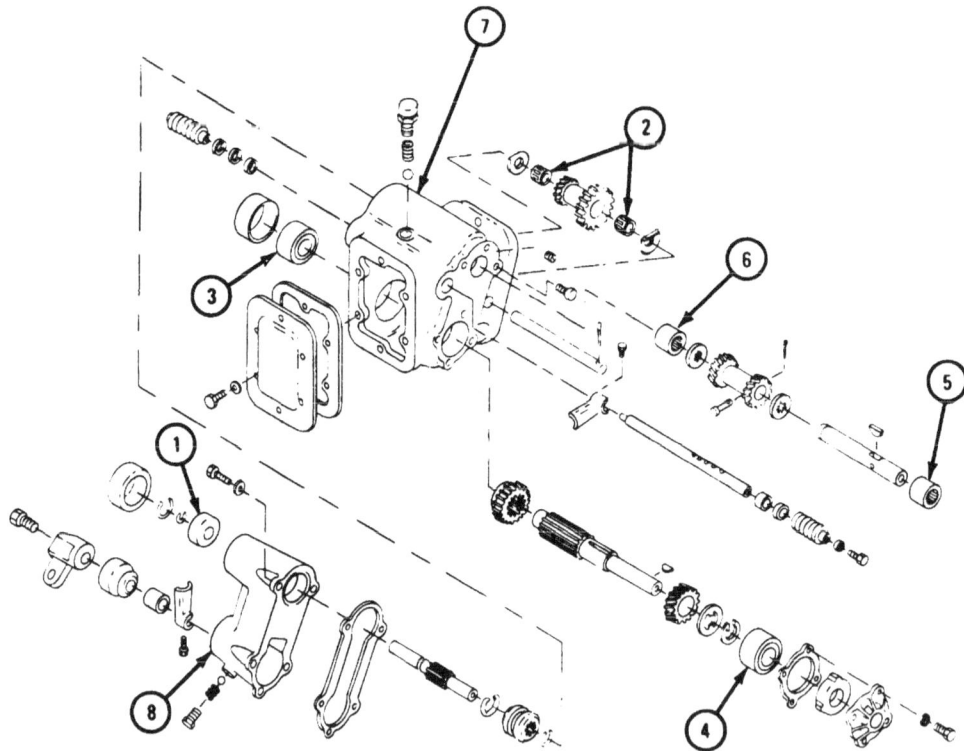

NOTE
CHECK ONLY THOSE PARTS WHICH ARE CALLED OUT IN
THIS FRAME. PARTS WITHOUT CALLOUTS ARE SHOWN
ONLY FOR REFERENCE PURPOSES OR ARE CHECKED IN
ANOTHER FRAME.

TA 087948

FRAME 2

1. Check that shafts (1 through 6) are not scored, pitted or worn.

2. Check that shafts (1, 4, and 5) do not have twisted splines.

3. Check that gears (7, 8, 9, and 10) do not have chipped, cracked or broken teeth. Check that bores of gears are not pitted or scored.

GO TO FRAME 3

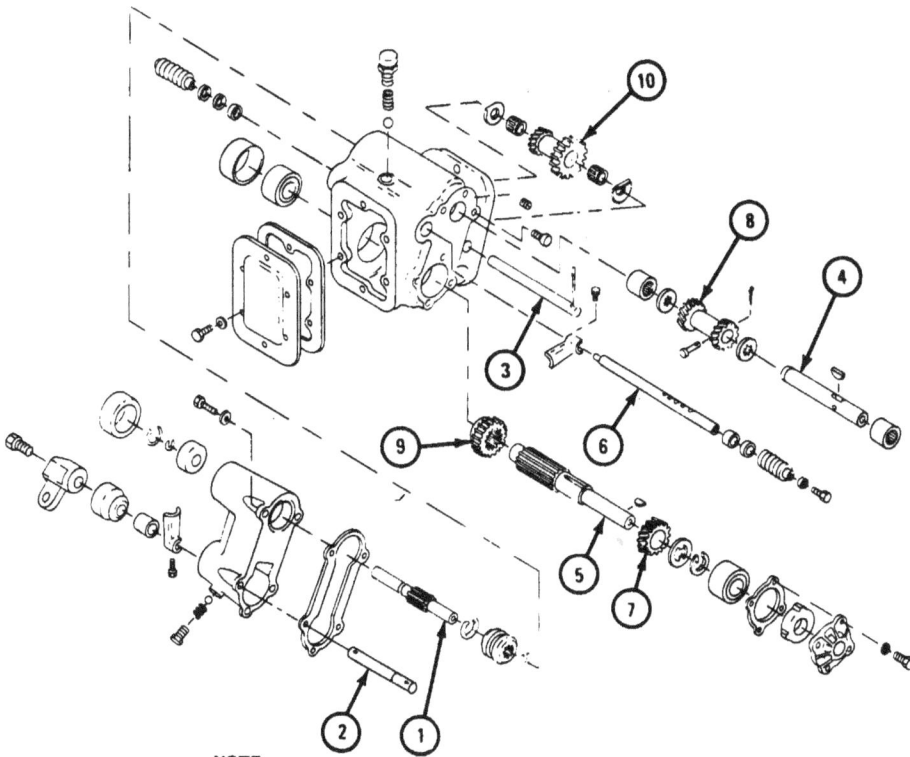

NOTE

CHECK ONLY THOSE PARTS WHICH ARE CALLED OUT IN THIS FRAME. PARTS WITHOUT CALLOUTS ARE SHOWN ONLY FOR REFERENCE PURPOSES OR ARE CHECKED IN ANOTHER FRAME.

TA 087949

FRAME 3

1. Check that ball (1) has no flat spots.
2. Check that spring (2) is not weak or broken.
3. Check that shifter fork (3) is not cracked or bent.

END OF TASK

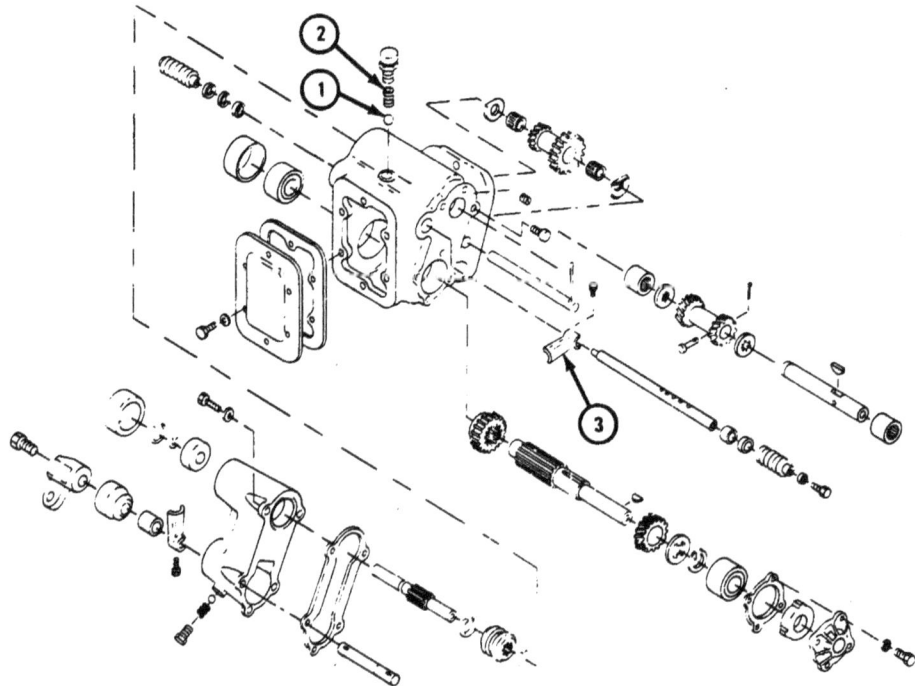

NOTE
CHECK ONLY THOSE PARTS WHICH ARE CALLED OUT IN
THIS FRAME. PARTS WITHOUT CALLOUTS ARE SHOWN
ONLY FOR REFERENCE PURPOSES OR ARE CHECKED IN
ANOTHER FRAME.

TA 087950

2-28. WEAR LIMIT INSPECTION. The following paragraphs give instructions for checking the minimum and maximum wear limits for each subassembly to which a part or parts may be worn before a new part is needed.

    a.   Reverse Gear Shaft Assembly.

FRAME 1

NOTE

Readings must be within limits given in table 2-58. If readings are not within given limits, throw away part and get a new one.

1.   Measure inside diameter of two bearings (1).

2.   Measure outside diameter of reverse gear shaft (2).

3.   Measure thickness of two new thrust washers (3).

4.   Measure inside diameter of reverse gear (4).

GO TO FRAME 2

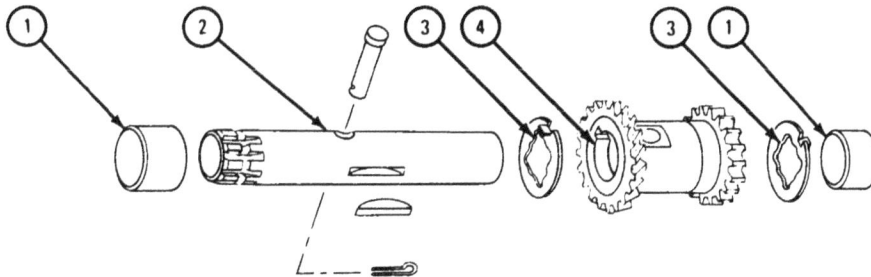

NOTE: CHECK ONLY THOSE PARTS WHICH ARE CALLED OUT IN THIS FRAME. PARTS WITHOUT CALLOUTS ARE SHOWN ONLY FOR REFERENCE PURPOSES OR ARE CHECKED IN ANOTHER FRAME.

TA 087951

Table 2-58. Reverse Gear Shaft Assembly Wear Limits

Index Number	Item/Point of Measurement	Size and Fit of New Parts (inches)	Wear Limit (inches)
1	Bearing inside diameter	1.2510 to 1.2520	0.001
2	Shaft outside diameter	1.2495 to 1.2505	0.003
3	Thrust washer thickness	0.061 to 0.063	None
4	Gear inside diameter	1.252 to 1.253	None

FRAME 2

NOTE

Readings must be within limits given in table 2-59.
The letter L indicates a loose fit.   If readings are
not within given limits,   throw away part and get
a new one.

1.   Measure fit of two bearings (1) on reverse gear shaft (2).

2.   Measure fit of reverse gear (3) on reverse gear shaft (2).

END OF TASK

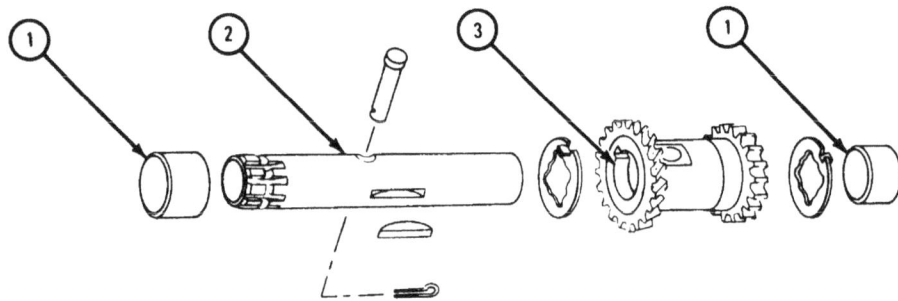

NOTE
CHECK ONLY THOSE PARTS WHICH ARE CALLED OUT IN
THIS FRAME. PARTS WITHOUT CALLOUTS ARE SHOWN
ONLY FOR REFERENCE PURPOSES OR ARE CHECKED IN
ANOTHER FRAME.

TA 087952

Table 2-59. Reverse Gear Shaft Assembly Fits and Tolerances

Index Number	Item/Point of Measurement	Size and Fit of New Parts (inches)	Wear Limit (inches)
1 and 2	Fit of bearings on shaft	0.0005L to 0.0025L	0.004
2 and 3	Fit of reverse gear on shaft	0.0015L to 0.0035L	0.0055L

b. Input Gear Shaft Assembly.

FRAME 1

NOTE

Readings must be within limits given in table 2-60. If readings are not within given limits, throw away part and get a new one.

1. Measure input gear shaft outside diameter (1).

2. Measure thickness of two new thrust washers (2).

3. Measure inside diameter (3) and outside diameter (4) of two roller bearing assemblies.

4. Measure inside diameter of input gear (5).

5. Measure input gearshaft outer end diameter (6).

GO TO FRAME 2

TA 087953

Table 2-60. Input Gear Shaft Assembly Wear Limits

Index Number	Item/Point of Measurement	Size and Fit of New Parts (inches)	Wear Limit (inches)
1	Gear shaft outside diameter	0.7495 to 0.7500	0.003
2	Thrust washer thickness	0.061 to 0.063	None
3	Bearing assembly inside diameter	0.7500	None
4	Bearing assembly outside diameter	1.250	None
5	Input gear inside diameter	1.250 to 1.251	0.007
6	Shaft outer end diameter	0.7510 to 0.7515	None

FRAME 2

NOTE

Readings must be within limits given in table 2-61.
The letter L indicates a loose fit.   If readings are
not within given limits,   throw away part and get
a new one.

1.   Measure fit of two roller bearing assemblies (1) on gear shaft (2).

2.   Measure fit of two roller bearing assemblies (1) in input gear (3).

END OF TASK

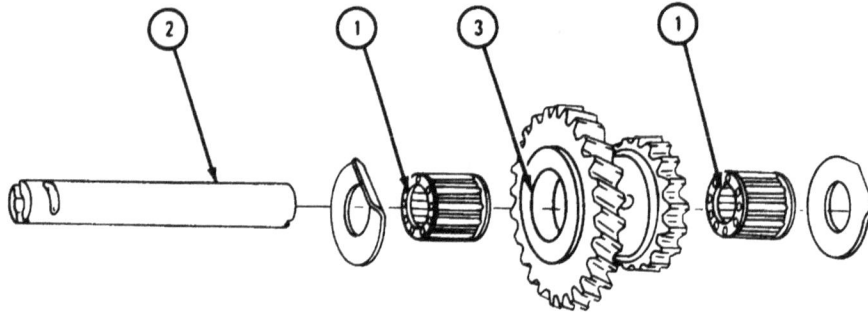

NOTE: CHECK ONLY THOSE PARTS WHICH
ARE CALLED OUT IN THIS FRAME.
PARTS WITHOUT CALLOUTS ARE
SHOWN ONLY FOR REFERENCE
PURPOSES OR ARE CHECKED IN
ANOTHER FRAME.

TA 087954

Table 2-61.   Input Gear Shaft Assembly Fits and Tolerances

Index Number	Item/Point of Measurement	Size and Fit of New Parts (inches)	Wear Limit (inches)
1 and 2	Fit of bearing assemblies on gearshift	0.0000L to 0.0005L	0.004L
1 and 3	Fit of bearing assemblies in input gear	0.000L to 0.00IL	0.004L

c.   Output Shaft Assembly.

FRAME 1

NOTE

Readings must be within limits given in table 2-62.
If readings are not within given limits, throw away
part and get a new one.

1.   Measure two bearing inside diameters (1) and outside diameters (2).

2.   Measure inside diameter of helical gear (3).

3.   Measure output shaft outside diameter (4).

4.   Measure output shaft outside diameter (5).

5.   Measure output shaft outside diameter (6).

GO TO FRAME 2

NOTE: CHECK ONLY THOSE PARTS WHICH
ARE CALLED OUT IN THIS FRAME.
PARTS WITHOUT CALLOUTS ARE
SHOWN ONLY FOR REFERENCE
PURPOSES OR ARE CHECKED IN
ANOTHER FRAME.

Table 2-62.   Output Shaft Assembly Wear Limits

Index Number	Item/Point of Measurement	Size and Fit of New Parts (inches)	Wear Limit (inches)
1	Bearing inside diameter	1.3780	None
2	Bearing outside diameter	2.8346	None
3	Helical gear inside diameter	1.4060 to 1.4065	None
4	Shaft outside diameter	1.3785 to 1.3795	None
5	Shaft outside diameter	1.3985 to 1.4000	0.003
6	Shaft outside diameter	1.3785 to 1.3795	None

TM 9-2520-246-34-1

FRAME 2

NOTE

Readings must be within limits given in table 2-63. The letter L indicates a loose fit and the letter T indicates a tight fit. If readings are not within given limits, throw away part and get a new one.

1. Measure fit of two bearings (1) on output shaft (2).

2. Measure fit of helical gear (3) on output shaft (4).

3. Measure fit of spur gear (5) on output shaft splines (6).

4. Measure fit of helical gear splines (7) in splines of spur gear (5).

GO TO FRAME 3

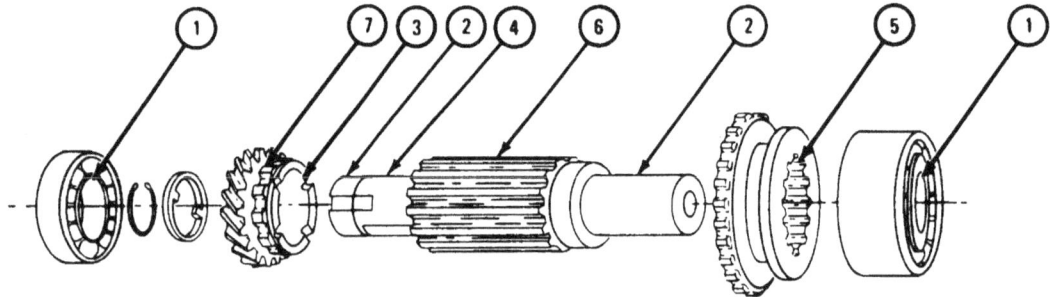

NOTE
CHECK ONLY THOSE PARTS WHICH ARE CALLED OUT IN THIS FRAME. PARTS WITHOUT CALLOUTS ARE SHOWN ONLY FOR REFERENCE PURPOSES OR ARE CHECKED IN ANOTHER FRAME.

TA 087956

Table 2-63. Output Shaft Assembly Fits and Tolerances

Index Number	Item/Point of Measurement	Size and Fit of New Parts (inches)	Wear Limit (inches)
1 and 2	Fit of bearings on shaft	0.0005 to 0.0015T	None
3 and 4	Fit of helical gear on shaft	0.006L to 0.008L	None
5 and 6	Fit of spur gear on shaft splines	0.004L to 0.007L	0.0011
5 and 7	Fit of helical gear splines in spur gear splines	0.004L to 0.007L	None

FRAME 3

NOTE
Readings must be within limits given in table 2-64.
The letter L indicates a loose   fit. If readings are
not within given limits, throw   away part and get
a new one.

1.   Measure diameter of bearing bore (1).

2.   Measure diameter of bearing bore (2).

3.   Measure fit of bearing (3) in bearing bore (1).

4.   Measure fit of bearing (4) in bearing bore (2).

END OF TASK

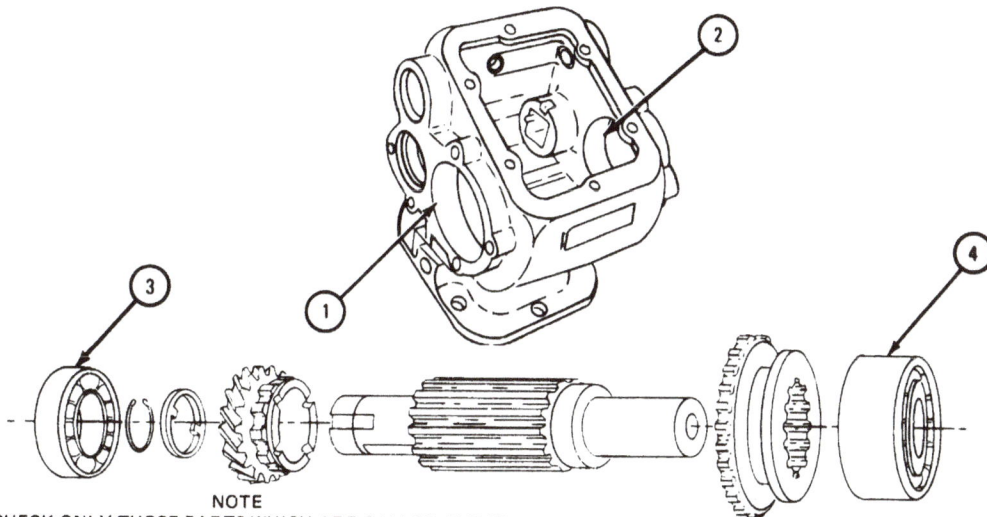

NOTE
CHECK ONLY THOSE PARTS WHICH ARE CALLED OUT IN
THIS FRAME. PARTS WITHOUT CALLOUTS ARE SHOWN
ONLY FOR REFERENCE PURPOSES OR ARE CHECKED IN
ANOTHER FRAME.

TA 087957

Table 2-64. Output Shaft Bearings Fits and Tolerances

Index Number	Item/Point of Measurement	Size and Fit of New Parts (inches)	Wear Limit (inches)
1	Bearing bore diameter	2.8346 to 2.8356	0.001
2	Bearing bore diameter	2.8346 to 2.8356	0.0008
3 and 1	Fit of bearing in bore	0.0000L to 0.0010L	None
4 and 2	Fit of bearing in bore	0.0000L to 0.0010L	0.0018L

d.   Shouldered Shaft Assembly.

FRAME 1

NOTE

Readings must be within limits given in table 2-65. The letter L indicates a loose fit and the letter T indicates a tight fit.   If readings are not within given limits, throw away part and get a new one.

1.   Measure shouldered shaft diameter (1).

2.   Measure shouldered shaft diameter (2).

3.   Measure inside diameter of bearing (3).

4.   Measure fit of bearing (3) on shouldered shaft diameter (2).

END OF TASK

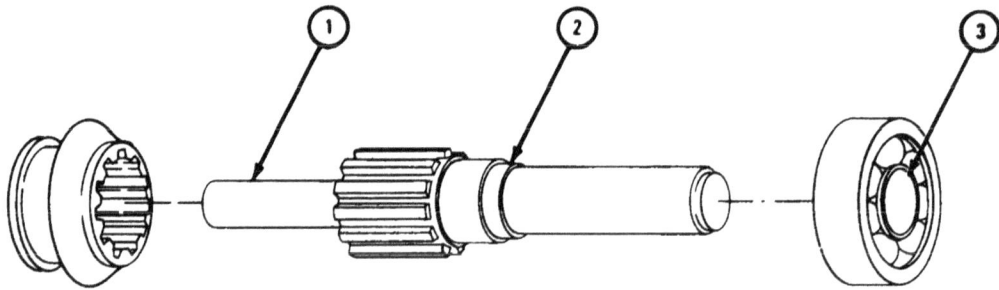

NOTE: CHECK ONLY THOSE PARTS WHICH
ARE CALLED OUT IN THIS FRAME.
PARTS WITHOUT CALLOUTS ARE
SHOWN ONLY FOR REFERENCE
PURPOSES OR ARE CHECKED IN
ANOTHER FRAME.

TA 087958

Table 2-65.   Shouldered Shaft Assembly Wear Limits

Index Number	Item/Point of Measurement	Size and Fit of New Parts (inches)	Wear Limit (inches)
1	Shaft outside of diameter	0.6230 to 0.6235	None
2	Shaft outside diameter	0.9843 to 0.9847	None
3	Bearing inside diameter	0.9840	None
3 and 2	Fit of bearing on shaft	0.003T to 0.007T	None

2-29. REPAIR.   This paragraph gives instructions for repairs that can be done on the transmission power takeoff.

    a.   Smooth out any chips, scratches or burrs on gear shafts and gears with a honing stone.

    b.   Weld cracks and small holes in housing castings. Refer to TM 9-237.

    c.   Drill out any bolts broken off in tapped holes.

    d.   Drill out threaded holes that are stripped or out-of-round to the next larger size and retap them.   When putting transmission power takeoff together, use a bolt the size of the new tapped hole.

2-30. ASSEMBLY.   This paragraph gives instructions for assembling the transmission power takeoff.

NOTE

Keep all parts clean and protected from dust and dirt.   Coat all bearings with multipurpose lubricant during assembly.

Coat all oil seals, gears and shafts with engine lubricating oil during assembly.

Coat shafts and bores of gears with white lead pigment during assembly.

Use new seals, thrust washers, and gaskets during assembly.

FRAME 1

1. Tap two seals (1) in place.

2. Using hammer with wood block, tap two boots with retaining cups (2) into place.

GO TO FRAME 2

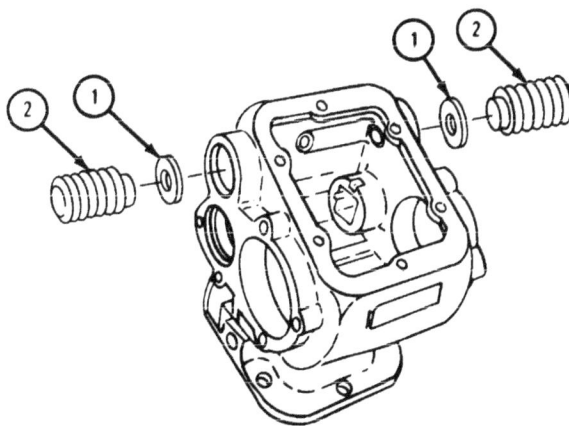

TA 087959

FRAME 2

1.   Using hammer and wood block, tap in bearing sleeve (1).

2.   Put in two capscrews (2).

3.   Put in plug (3).

GO TO FRAME 3

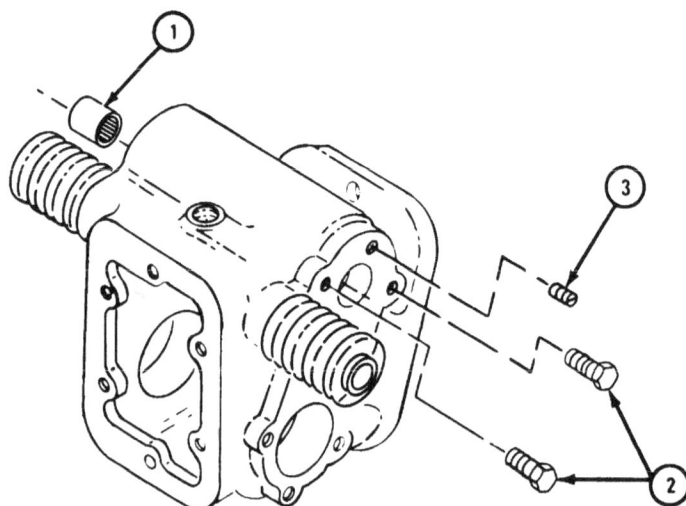

TA 087946

FRAME 3

1. Put in woodruff key (1).

2. Put two thrust washers (2) into place in power takeoff housing (3).

3. Put in and hold reverse gear (4) in place as shown.

4. Aline key (1) with keyway (5) in reverse gear (4) and corner of thrust washer (2) as shown.

5. Tap in reverse gear shaft (6).

GO TO FRAME 4

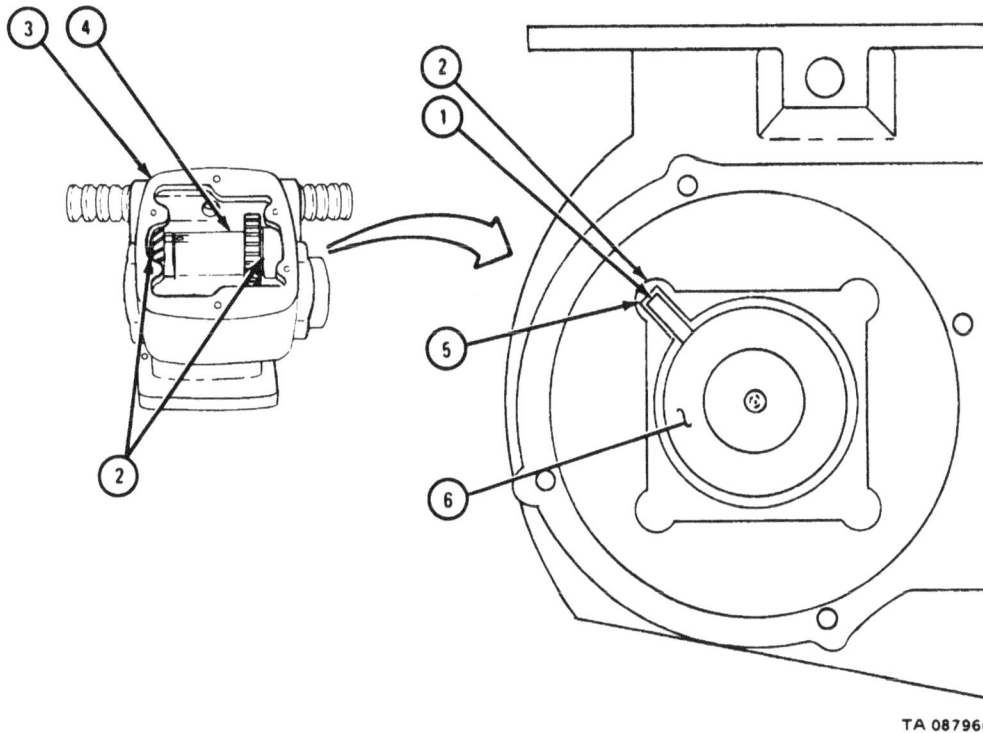

TA 087960

FRAME 4

1. Put rear needle roller bearing (1) on reverse gear shaft (2).
2. Tap in needle roller bearing (1).
3. Put in and hold straight headed pin (3).
4. Turn reverse gear shaft (2) 180° and put in cotter pin (4).
5. Tap end cover (5) into place.
GO TO FRAME 5

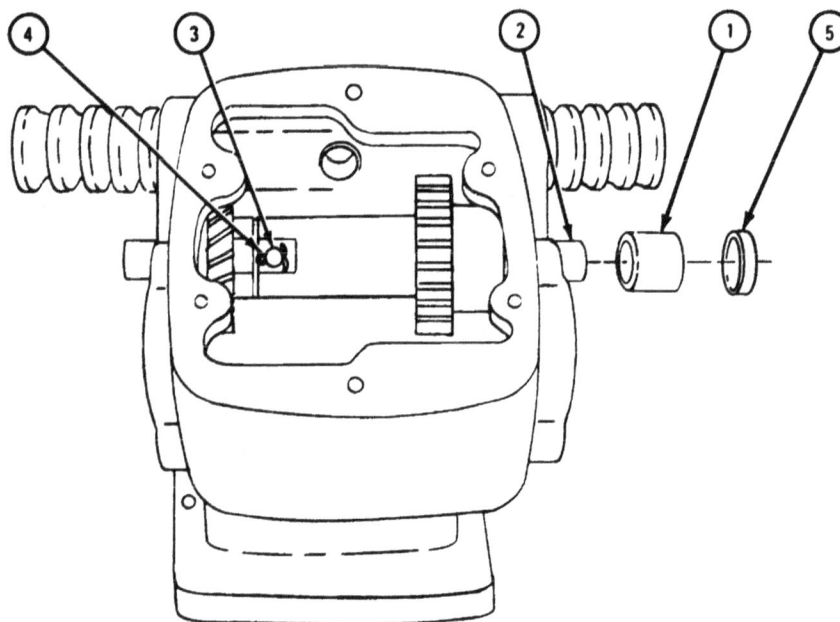

TA 087961

FRAME 5

1. Put in and hold spur gear (1) in place.
2. Tap in output shaft (2).
3. Tap bearing (3) onto output shaft (2) and into housing (4).
GO TO FRAME 6

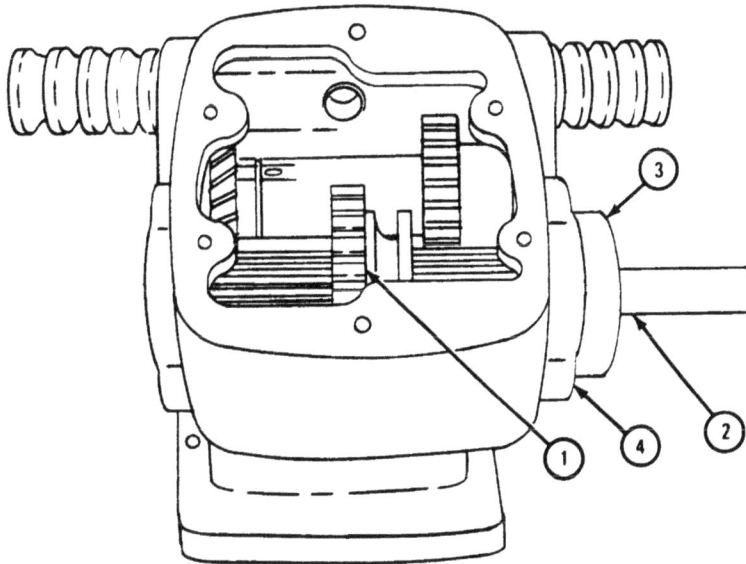

TA 087962

FRAME 6

1. Put on helical gear (1), spacer (2), and snapring (3).
GO TO FRAME 7

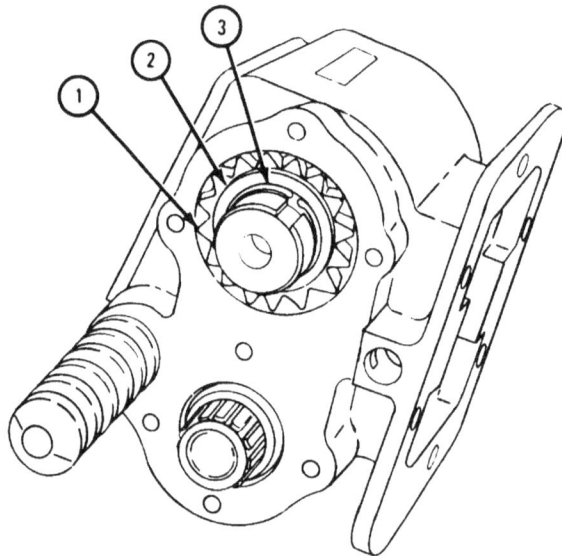

TA 087963

FRAME 7

1. Using hammer and brass drift, tap in bearing (1).

GO TO FRAME 8

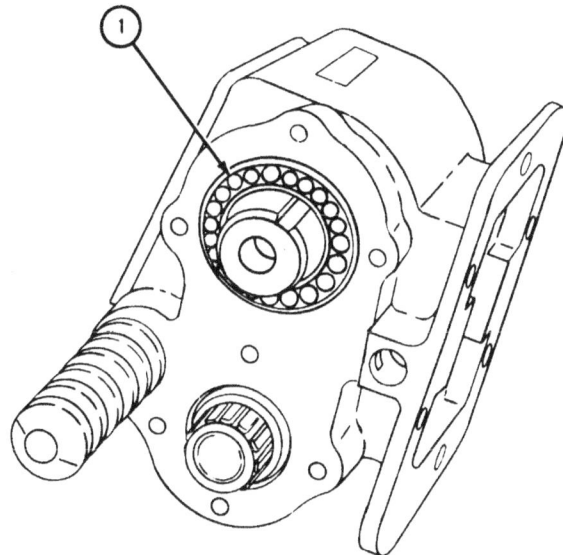

TA 087964

FRAME 8

1. Using hammer and brass drift, tap seal (1) into front bearing cap (2).
GO TO FRAME 9

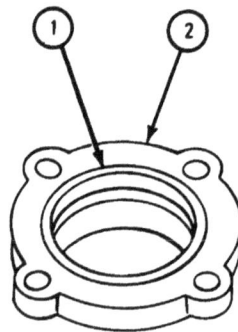

TA 087965

FRAME 9

1. Put on gasket (1) and front bearing cap (2).

2. Put in four capscrews (3) and four starwashers (4).

3. Tap in key (5).

IF WORKING ON DOUBLE OUTPUT POWER TAKEOFF (MODEL WND-7-28), GO TO
FRAME 10.
IF WORKING ON SINGLE OUTPUT POWER TAKEOFF (MODEL WN-7-28), GO TO
FRAME 15

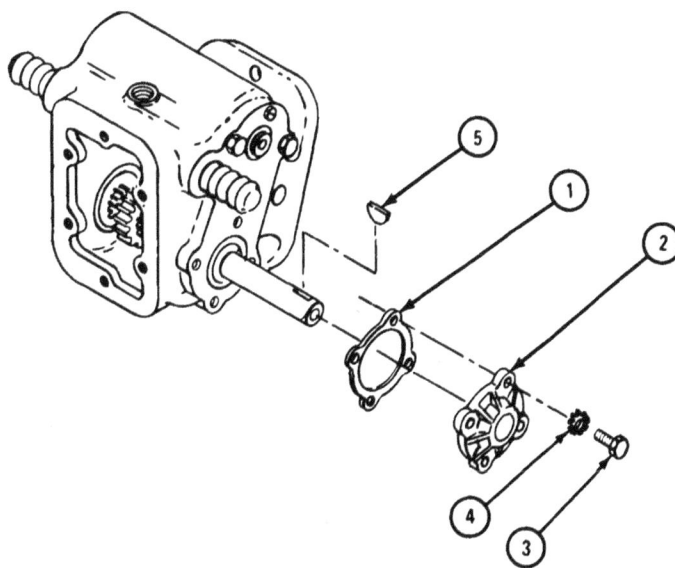

TA 087966

FRAME 10

1. Press bearing (1) on output shaft (2). Put on snapring (3).

2. Put snapring (4) into accessory drive housing (5).

3. Put clutch sliding sleeve (6) into accessory drive housing (5).

4. Put output shaft (2) in through front of accessory drive housing (5) and in clutch sliding sleeve (6).

5. Put in snapring (7).

GO TO FRAME 11

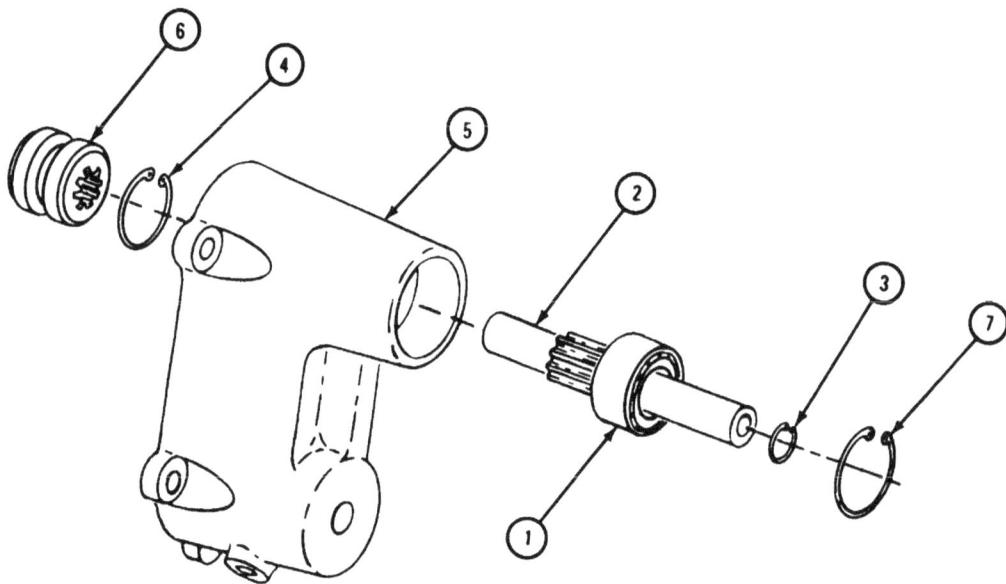

TA087967

FRAME 11

1.   Put in seal (1).
2.   Put in key (2).
GO TO FRAME 12

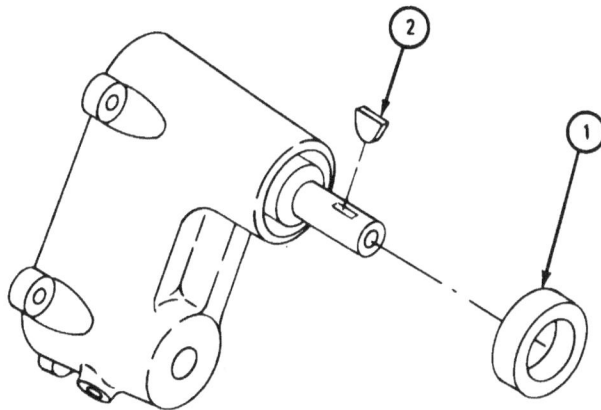

TA 087968

FRAME 12

1. Put shifter fork (1) in accessory drive housing (2). Aline shifter fork with grooves in clutch sliding sleeve (3) and hold it in place.

2. Slide shifter shaft (4) through shifter fork (1).

3. Aline setscrew hole in shifter fork (1) with setscrew hole in shifter shaft (4) and put in setscrew (5).

4. Put safety wire (6) through hole in capscrew (5) and through hole in shifter fork (1).

GO TO FRAME 13

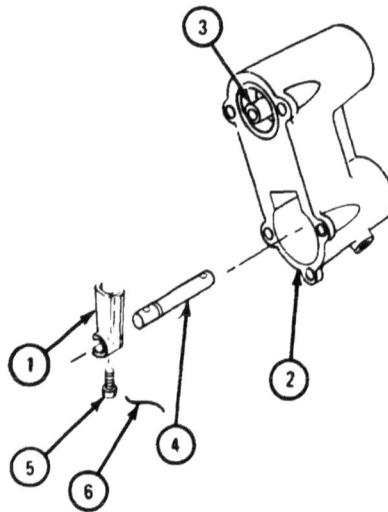

TA 087969

FRAME 13

1. Tap in seal (1) and boot (2).
2. Put in ball (3), spring (4), and plug (5).
3. Put on connector (6) and capscrew (7).

GO TO FRAME 14

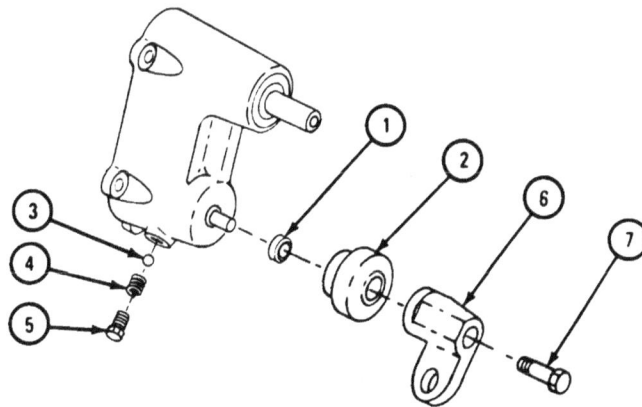

TA 087970

FRAME 14

1. Put on spacer (1),   Put gasket (2) and power accessory drive housing (3) on power takeoff housing (4) and aline screw holes.

2. Put in five capscrews (5) and five starwashers (6).

GO TO FRAME 15

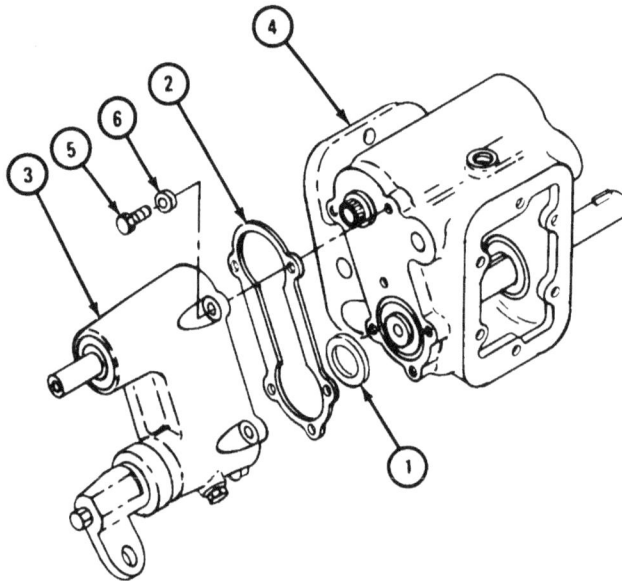

TA 087971

FRAME 15

1. Put gasket (1) and rear bearing cap (2) on power takeoff housing (3).
2. Put in five capscrews (4) and five starwashers (5).
GO TO FRAME 16

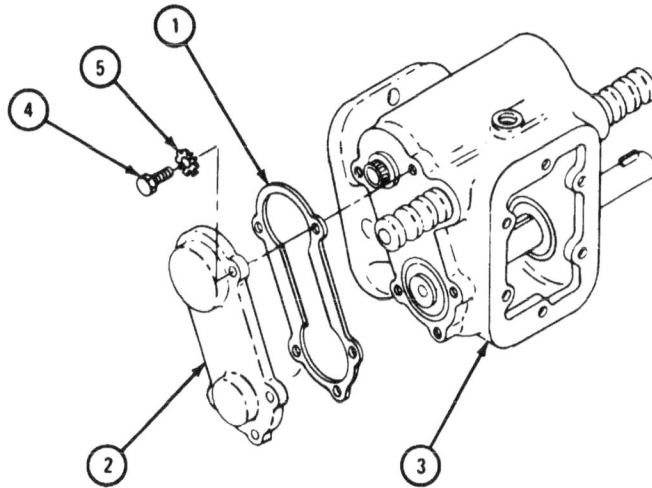

TA 087972

FRAME 16

1. Put grease on the back of two thrust washers (1). Put two thrust washers in place as shown.

2. Put two sets of roller bearing assemblies (2) in input gear (3) as shown.

GO TO FRAME 17

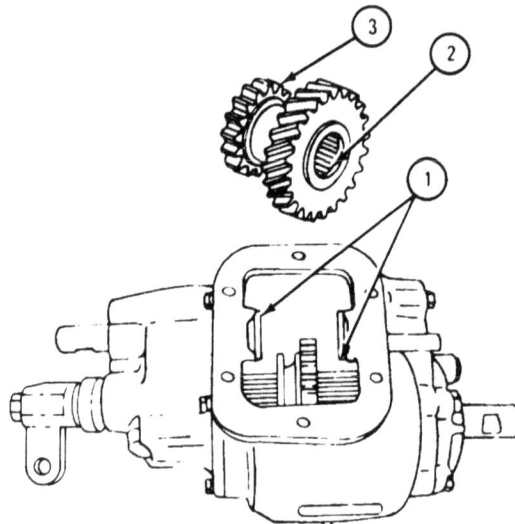

TA 087973

FRAME 17

1. Put in and hold input gear (1) in place.
2. Tap in input gear shaft (2).

GO TO FRAME 18

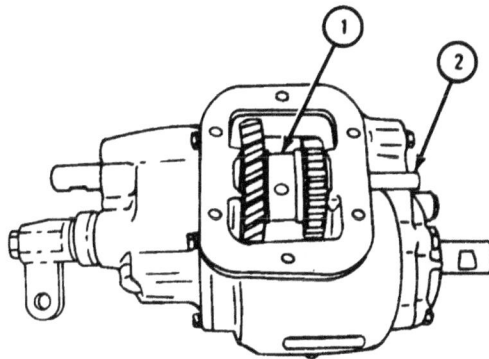

TA 087974

FRAME 18

1. Using ratchet and screwdriver bit, turn input gear shaft (1) and aline cotter pin holes.
2. Put in cotter pin (2).

GO TO FRAME 19

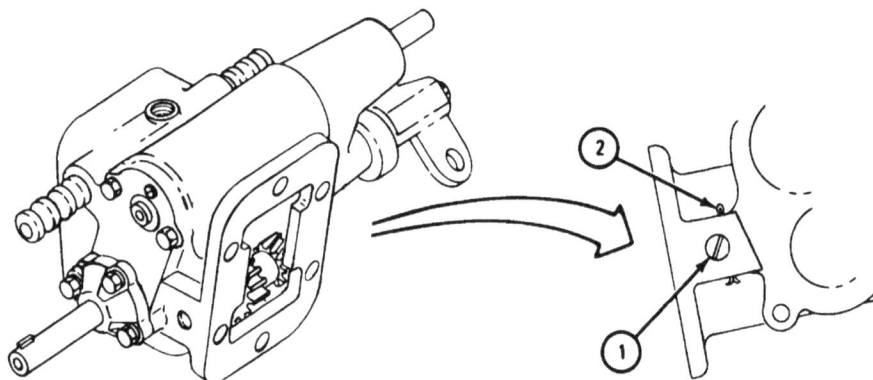

TA087975

FRAME 19

1. Working from inside of power takeoff assembly (1), put in and hold shifter fork (2) in place.

2. Slide in shifter shaft (3) and aline setscrew hole in shifter shaft with setscrew hole in shifter fork (2).

3. Put in setscrew (4).

4. Put safety wire (5) through hole in setscrew (4) and shifter fork (2).

5. Put in eyebolt (6) and locknut (7).

6. Put in capscrew (8) and washer (9).

GO TO FRAME 20

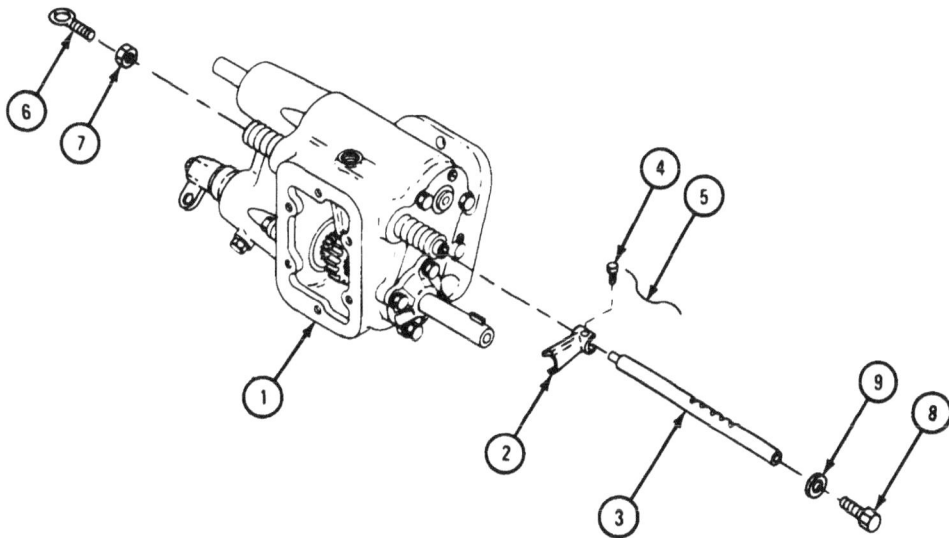

TA 087976

FRAME 20

1. Put in ball (1), poppet spring (2), and poppet retainer (3).
2. Put on gasket (4) and power takeoff cover (5).
3. Put in six capscrews (6) and six starwashers (7).

END OF TASK

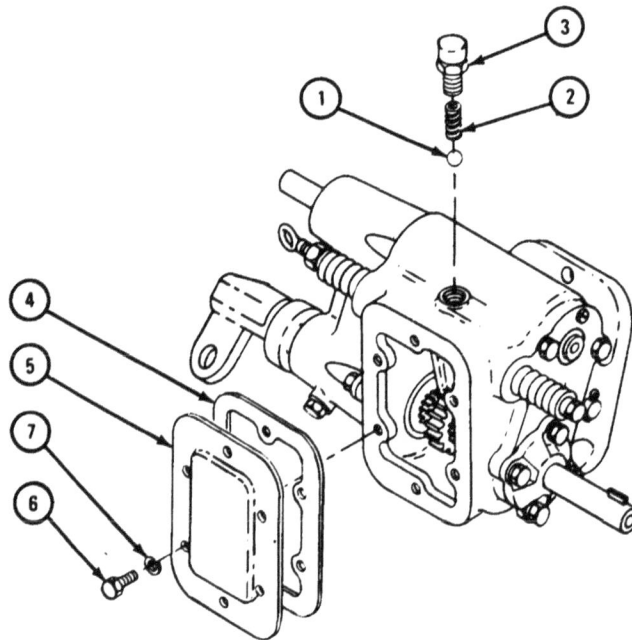

TA 087977

2-31.   SHIFT TEST.   The following paragraphs give instructions for testing the transmission power takeoff for smooth and positive shifting after assembly.

    a.   Neutral Position.

NOTE

There are two neutral positions on the trans-
mission power takeoff, one between the high
speed and low speed positions and one be–
tween the low speed and reverse positions.

FRAME 1

1.   Pull shifter shaft (1) all the way out.

2.   Push shifter shaft (1) in until you feel it go into the first detent.

3.   Hold input gear (2) and turn output shaft (3).   Output shaft should turn freely.

4.   If output shaft (3) does not turn freely, do the following:

    a.   Disassemble transmission power takeoff. Refer to para 2-25, frames 1 and 2.

    b.   Assemble transmission power takeoff.   Refer to para 2-30, frames 19 and 20.

GO TO FRAME 2

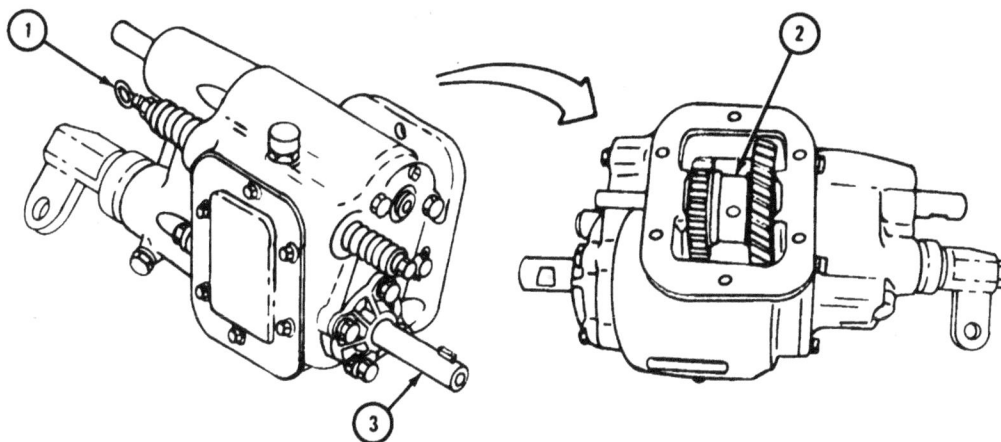

TA 102060

FRAME 2

1. Push shifter shaft (1) all the way in.

2. Pull shifter shaft (1) out until you feel it go into the first detent.

3. Hold input gear (2) and turn output shaft (3). Output shaft should turn freely.

4. If output shaft (3) does not turn freely, do the following:

   (a) Disassemble transmission power takeoff. Refer to para 2-25, frames 1 and 2.

   (b) Assemble transmission power takeoff. Refer to para 2-30, frames 19 and 20.

END OF TASK

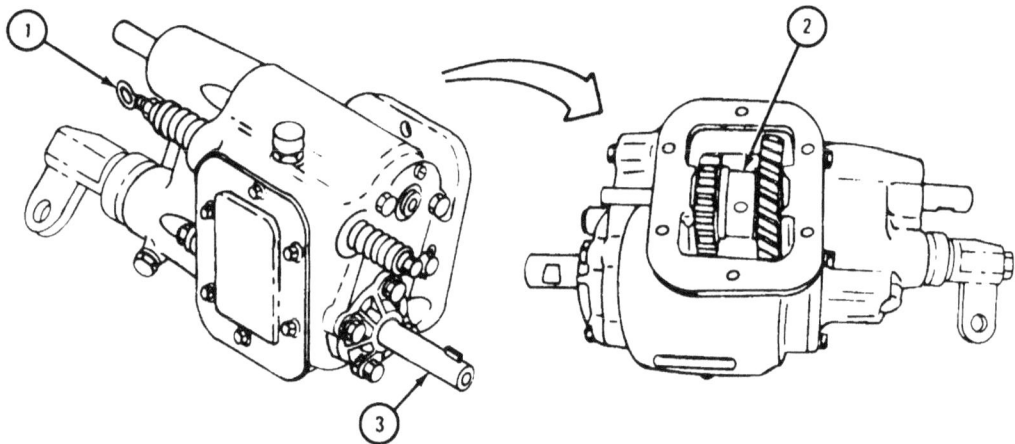

TA 102060

b. High Speed and Reverse Positions.

FRAME 1

1. Pull shifter shaft (1) all the way out.

2. Turn input gear (2). Output shaft (3) should turn. Note direction output shaft turns.

3. Push shifter shaft (1) all the way in.

4. Turn input gear (2). Output shaft (3) should turn in opposite direction to direction noted in step 2.

5. If output shaft (3) did not turn in steps 2 and 4, do the following:

    (a) Disassemble transmission power takeoff. Refer to para 2-25, frames 1 and 2.

    (b) Assemble transmission power takeoff. Refer to para 2-30, frames 19 and 20.

END OF TASK

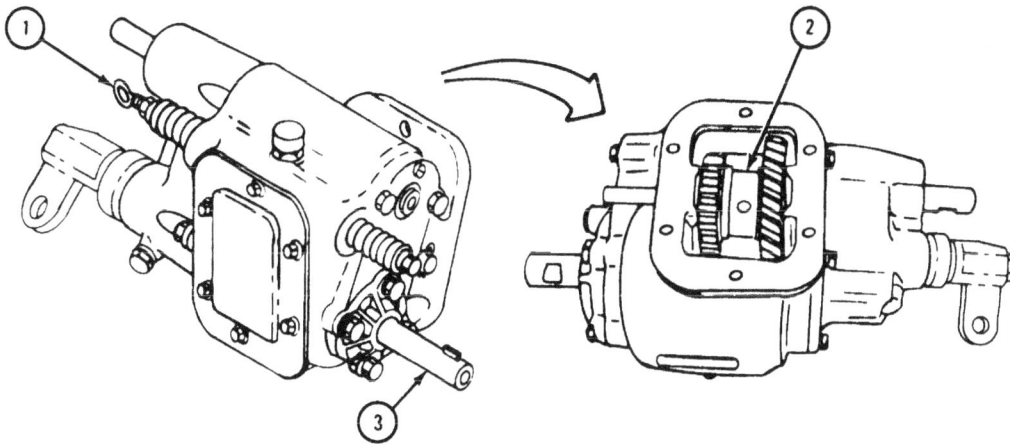

TA 102060

c. Low Speed Position.

FRAME 1

1. Pull shifter shaft (1) all the way out.
2. Push shifter shaft (1) in until you feel detent two times.
3. Turn input gear (2).   Output shaft (3) should turn.
4. If output shaft (3) does not turn, do the following:
   (a) Disassemble transmission power takeoff.   Refer to para 2-25, frames 1 and 2.
   (b) Assemble transmission power takeoff.   Refer to para 2-30, frames 19 and 20.

END OF TASK

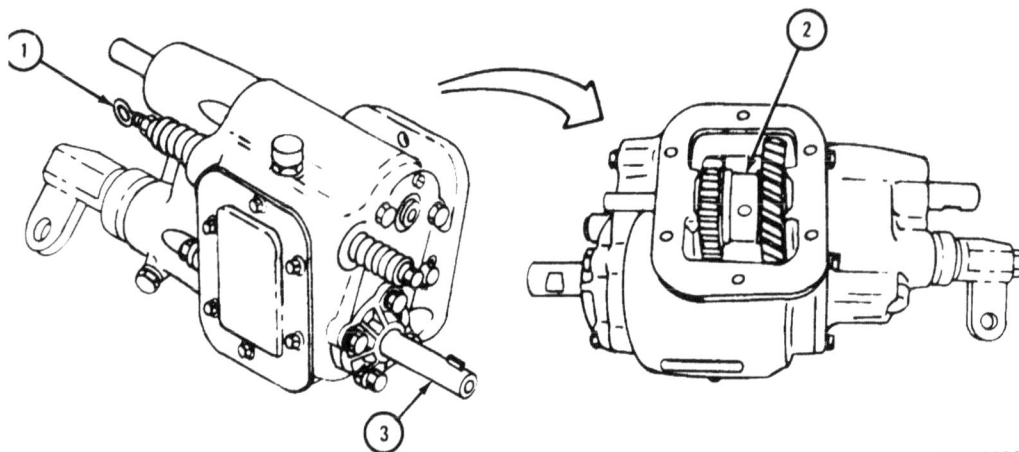

TA 102060

d. Accessory Drive (Model WND-7-28).

FRAME 1

1. Pull shifter shaft (1) all the way out.

2. Push shifter shaft (2) all the way in.

3. Turn input gear (3). Accessory drive   output shaft (4) should turn.

4. If accessory drive output shaft (4)   does not turn, do the following:

   a. Remove and disassemble accessory drive. Refer to para 2-25, frames 4, 5, and 6.

   b. Assemble and replace accessory drive. Refer to para 2-30, frames 12, 13, and 14.

END OF TASK

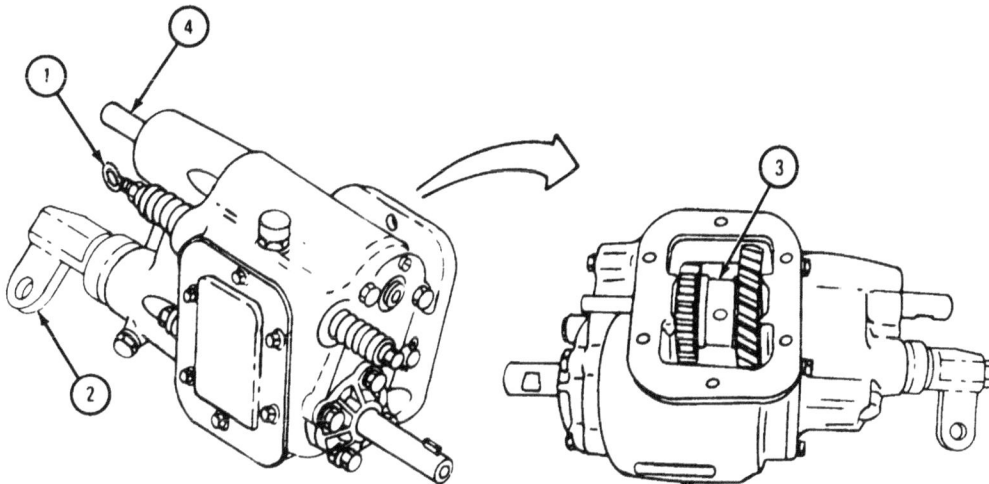

TA 102061

Section V.   MAINTENANCE OF TRANSMISSION TRANSFER POWER TAKEOFF

TOOLS :   Power train rebuild tool kit, pn 7950356

SUPPLIES :   Solvent, dry cleaning, type II (SD-2), Fed. Spec P-D-680
White lead pigment,   NSN 8010-00-290-6643
Universal gear lubricant, GO 80/90, MIL-L-2105
Lubricating oil, ICE,   OE/HDO 10, MIL-L-2104
Safety wire, MS20995F47
Oil pump mounting gasket
Rear bearing retainer gasket
End play shim set
Rear bearing retainer oil seal
Shifter shaft preformed packing
Compressed air source, 30 psi max

PERSONNEL: Tw o

EQUIPMENT CONDITION :   Transmission transfer power takeoff on workbench.

2-32. CLEANING BEFORE DISASSEMBLY.   This paragraph gives instructions for cleaning the transmission transfer power takeoff before disassembly. Note and scribe transmission transfer power takeoff case in places that have oil soaked road mud. It is not necessary to mark oil spots around gaskets or seals since new gaskets and seals will be put in.   Scrape, brush, and steam clean all dirt and road mud from the transmission transfer power takeoff assembly.

2-33. DISASSEMBLY. This paragraph gives instructions to completely disassemble the transmission transfer power takeoff assembly.

FRAME 1

1. Take out drain plug (1).
2. Take out cotter pin (2). Take off washer (3).
3. Take out cotter pin (4).   Take off nut (5) and washer (6).
4. Take off shift lever (7).

GO TO FRAME 2

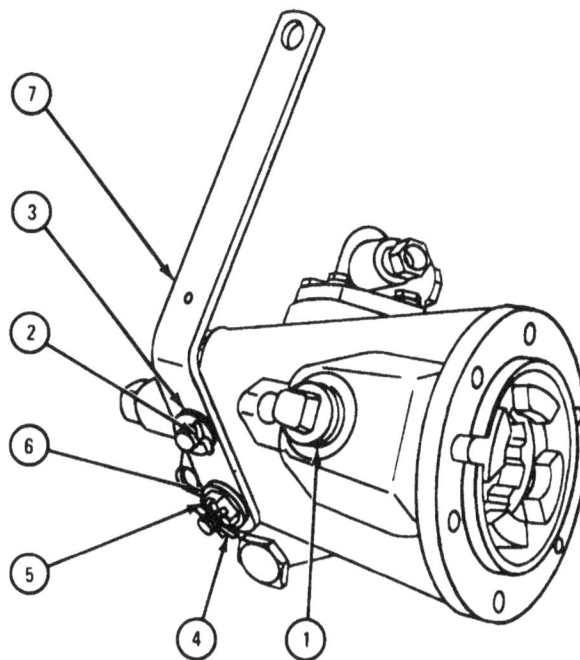

TA 087902

FRAME 2

1. Take out four capscrews (1) and four lockwashers (2).
2. Take off oil pump valve assembly (3).   Take out spring (4) and plunger (5).
3. Take off and throw away gasket (6).

NOTE

Some models do not have fitting (7).

4. Take off fitting (8) and fitting (7).

GO TO FRAME 3

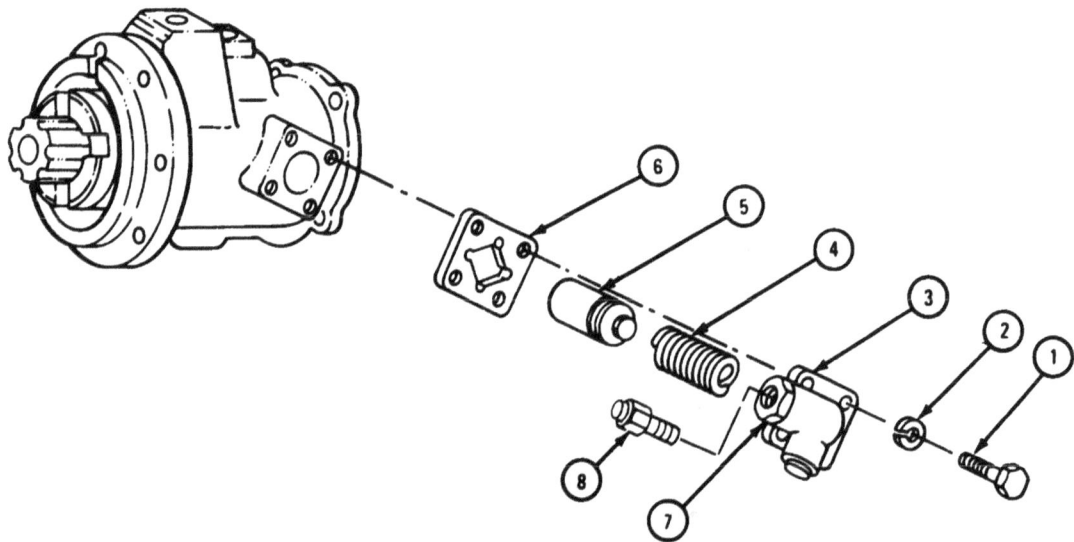

TA 087903

FRAME 3

1. Take out screw (1) and washer (2).
2. Take out spring (3) and ball (4).

GO TO FRAME 4

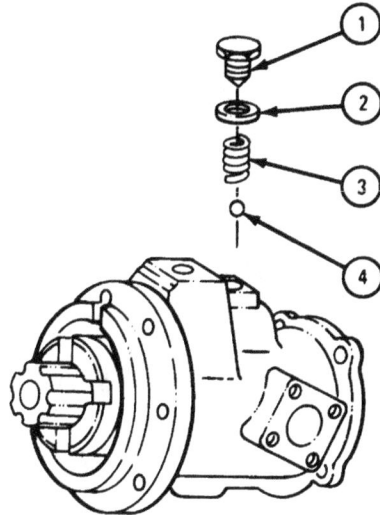

TA 087904

FRAME 4

1. Working through drain plug hole, take out setscrew (1).

2. Pull out shaft (2) and take off and throw away preformed packing (3).

3. Take out sleeve (4). Take out fork (5).

GO TO FRAME 5

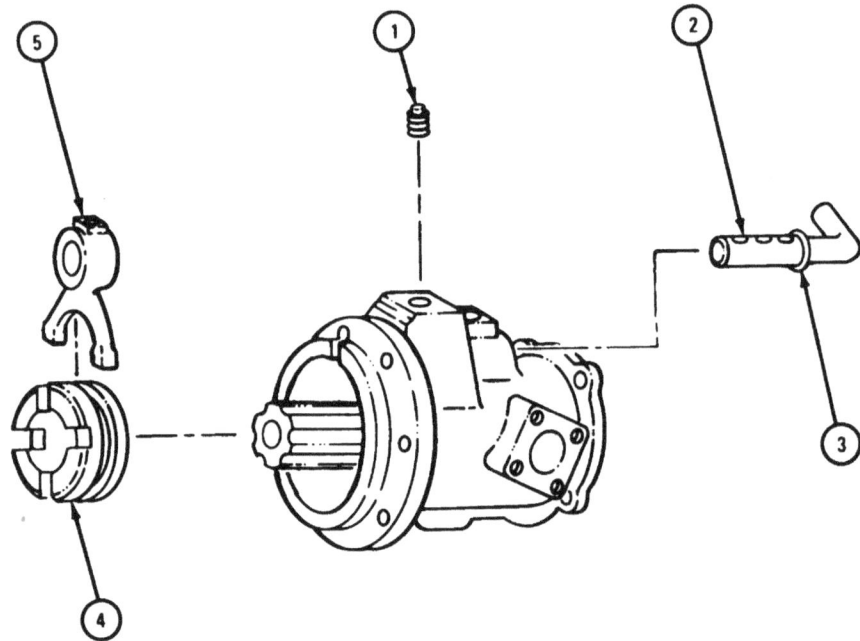

TA 087905

FRAME 5

1. Take out key (1).

2. Take out four capscrews (2) and four lockwashers (3).

NOTE

Some power takeoffs may not have shims (4).

3. Take off retainer (5), gasket (6), and shims (4). Throw away shims and gasket.

GO TO FRAME 6

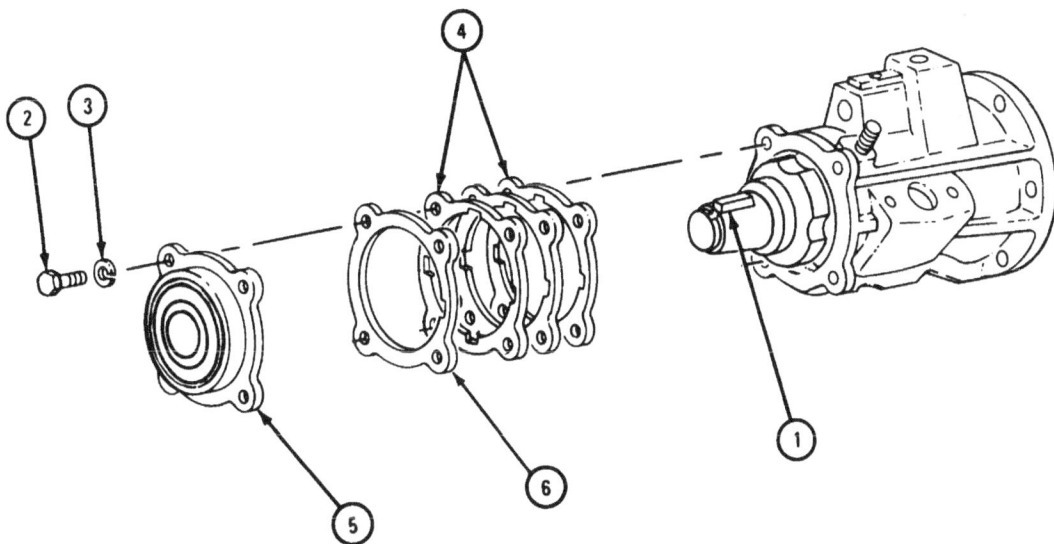

TA 087906

FRAME 6

1. Put retainer (1) on vise as shown.
2. Using hammer and brass drift, tap out and throw away oil seal (2).

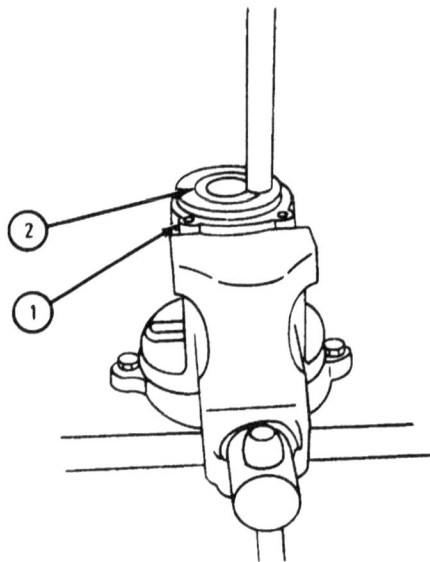

GO TO FRAME 7

TA 087907

FRAME 7

1. Tap out output shaft (1) as shown.

NOTE

Tapered roller bearing cup will come out with output shaft.

GO TO FRAME 8

TA 087908

FRAME 8

NOTE

There are two bearing bores inside carrier (1). Bearing
cup (2) must be driven out of first bearing bore, and then
driven out of second bearing bore.

1. Working through front of carrier (1) and using hammer and brass drift, tap out
   bearing cup (2).

GO TO FRAME 9

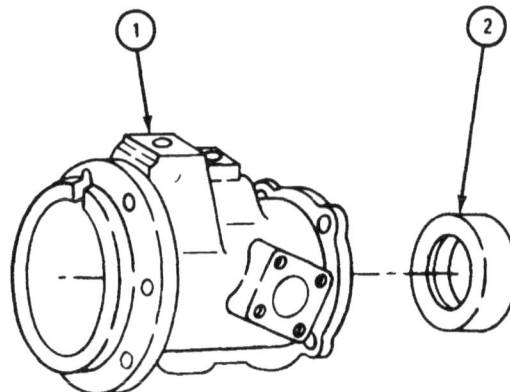

TA 087909

FRAME 9

Soldier A 1.   Put output shaft (1) in hydraulic press as shown.

Soldier B 2.   Working from under press, hold bottom of output shaft (1) to keep it from falling when front bearing (2) is pressed off. Tell soldier A when ready.

Soldier A 3.   Using hydraulic press, press output shaft (1) out of front bearing (2).

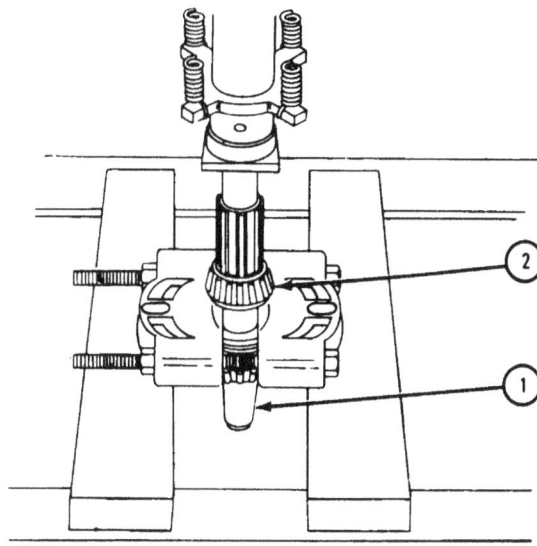

GO TO FRAME 10

TA 087910

FRAME 10

Soldier A 1.   Put output shaft (1) in hydraulic press as shown.

Soldier B 2.   Working from under press, hold bottom of output shaft (1) to keep it from falling when drive gear (2) and rear bearing (3) are pressed off. Tell soldier A when ready.

Soldier A 3.   Using hydraulic press, press output shaft (1) out of drive gear (2) and rear bearing (3).

GO TO FRAME 11

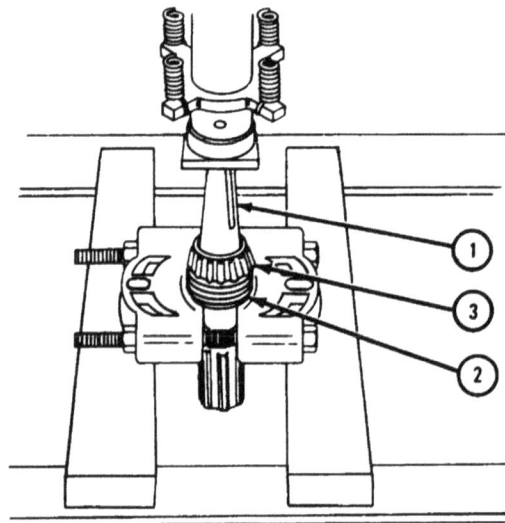

TA 087911

FRAME 11

1.   Take out key (1).
END OF TASK

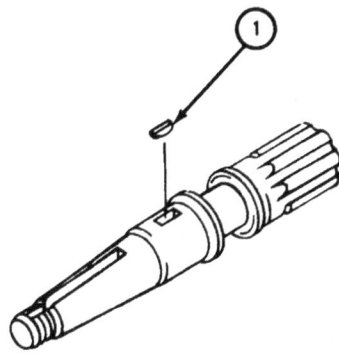

TA 087912

2-34. CLEANING. This paragraph gives general instructions for cleaning the transmission transfer power takeoff parts.

    a.   Clean all bearing cones and cups.   Refer to inspection, care and maintenance of antifriction bearings, TM 9-214.

WARNING

Dry cleaning solvent is flammable. Do not use near an open flame. Keep a fire extinguisher nearby when solvent is used. Use only in well-ventilated places. Failure to do this may result in injury to personnel and damage to equipment.

Do not use more than 30 psi of air pressure for drying parts. Eye shields must be worn when using compressed air. Eye injury can occur if eye shields are not used.

CAUTION

When scraping gasket material from surface of parts, be careful not to scratch or gouge the metal surface.

    b.   Clean all other parts with solvent. Scrape all gasket material from surface of parts. Rinse parts in clean solvent and dry with compressed air.

2-35. GENERAL INSPECTION.   This paragraph gives instructions to check for damage on the transmission transfer power takeoff carrier, gear shafts, and gears.

CAUTION

It is easy to damage the equipment if you don't know what you are doing. Do not try to do this task unless you are experienced at it, or you have an experienced person with you.

NOTE

Small chips, burrs or scratches on gears and gear shafts can be repaired. Cracks in carrier castings that do not go into screw holes or openings can be repaired. If parts are damaged in any other way, throw away parts and get new ones.

FRAME 1

1. Check that all bearing cones (1) and bearing races (2) are not damaged. Refer to inspection, care and maintenance of antifriction bearings, TM 9-214.

2. Check that carrier (3) does not have any broken bolts or stripped threads. Mark them for repair.

3. Check that carrier (3) and pump assembly (4) are not cracked, chipped, warped or have small holes.

4. Check that gear (5) is not cracked or chipped and that it does not have damaged teeth.

GO TO FRAME 2

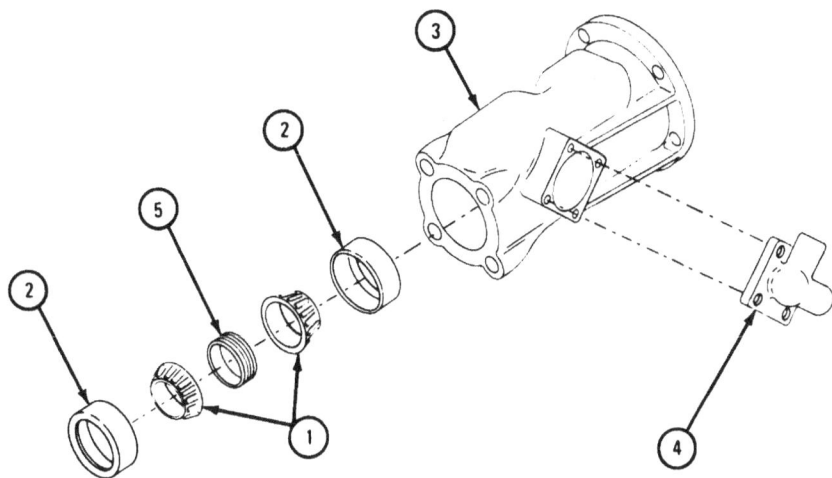

TA 087913

FRAME 2

NOTE

Small chips, burrs or scratches on shafts, gears, sliding
clutch, and shifter fork can be repaired. If parts are
damaged in any other way, throw parts away and get new
ones.

1. Check that shaft (1) is not chipped or cracked.

2. Check that shaft splines (2) are not chipped, cracked, twisted or burred.

3. Check that sliding clutch (3) is not chipped or cracked and that internal splines are
not twisted or burred.

4. Check that shifter fork (4) and shifter shaft (5) are not cracked or bent.

5. Check that all threaded parts are not stripped or crossthreaded.

6. Check that shifting lever lock ball (6) has no flat spots.

7. Check that shifting lever lock spring (7) is not damaged.

GO TO FRAME 3

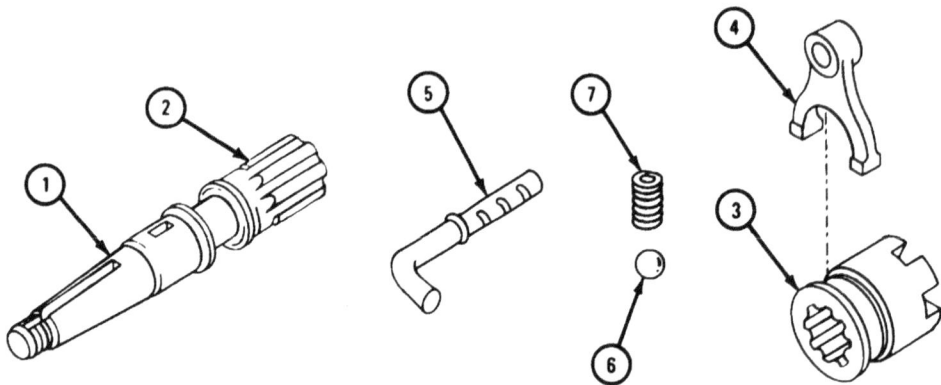

TA 087914

FRAME 3

NOTE

Valve assembly (1) cannot be repaired. If valve assembly
does not pass the test given in the following steps, throw
valve assembly away and get a new one. If a new valve
assembly is used, do this frame again on new part.

1. Make sure that valve assembly (1) is clean.

2. Hold finger over port (2) and blow into port (3). If air goes   through, valve is
    damaged.

3. Hold finger over port (2) and blow into port (4). If air does   not go through, valve
    is damaged.

4. Check that spring (5) is not damaged.

5. Check that plunger (6) has no nicks, burrs or scratches. Small burrs, nicks or
    scratches can be repaired.

END OF TASK

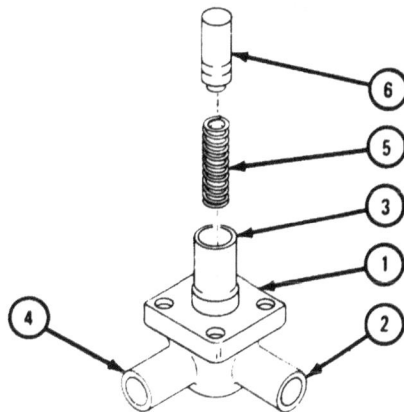

TA 087915

2-36. WEAR LIMIT INSPECTION. The following paragraph gives instructions for checking the minimum and maximum wear limits to which a part or parts may be worn before a new part is needed.

FRAME 1	

NOTE

Readings must be within limits given in table 2-66. The letter L indicates a loose fit and the letter T indicates a tight fit. If readings are not within given limits, throw away part and get a new one.

1. Measure inside diameter of two bearings (1).

2. Measure shaft outside diameter (2).

3. Measure fit of bearings (1) on shaft diameter (2).

4. Measure inside diameter of gear (3).

5. Measure fit of gear (3) on shaft diameter (2).

6. Measure output shaft cam section (4).

END OF TASK

TA 087916

Table 2-66. Output Shaft Wear Limits

Index Number	Item/Point of Measurement	Size and Fit of New Parts (inches)	Wear Limit (inches)
1	Bearing inside diameters	1.688	None
2	Shaft outside diameters	1.6885 to 1.6895	None
1 and 2	Fit of bearings on shaft	0.005 to 0.0015T	None
3	Gear inside diameter	1.6900 to 1.6915	None
3 and 2	Fit of gear on shaft	0.0005L to 0.003L	None
4	Cam section of shaft	1.685 to 1.690	None

2-37. REPAIR. This paragraph gives instructions to repair the transmission transfer power takeoff parts.

    a.   Smooth out any chips, scratches, or burrs on shafts and gear with a honing stone.

    b.   Weld cracks and small holes in carrier casting. Refer to TM 9-237.

    c.   Drill out any bolts that are broken off in tapped holes.

    d.   Drill out threaded holes that are stripped or out-of-round to the next larger size and retap them.   When putting parts together, use a bolt the size of the new tapped hole.

2-38. ASSEMBLY. This paragraph gives instructions for putting the transmission transfer power takeoff together and also has end play adjustment needed.

NOTE

Keep all parts clean and protected from dust and dirt.
Coat all bearings with multipurpose lubricant during
assembly. Coat oil seal, gear and shafts with engine
lubricating oil during assembly. Coat shafts and
bore of gear with white lead pigment during assembly.
Use new seal and gasket during assembly.

FRAME 1

1.   Put in key (1).

2.   Put bearing cone (2) and shaft (3) in hydraulic press as shown.

NOTE

Make sure that tapered end of bearing cone (2) is facing splined
end of shaft (3).

3.   Press shaft (3) into bearing cone (2).

GO TO FRAME 2

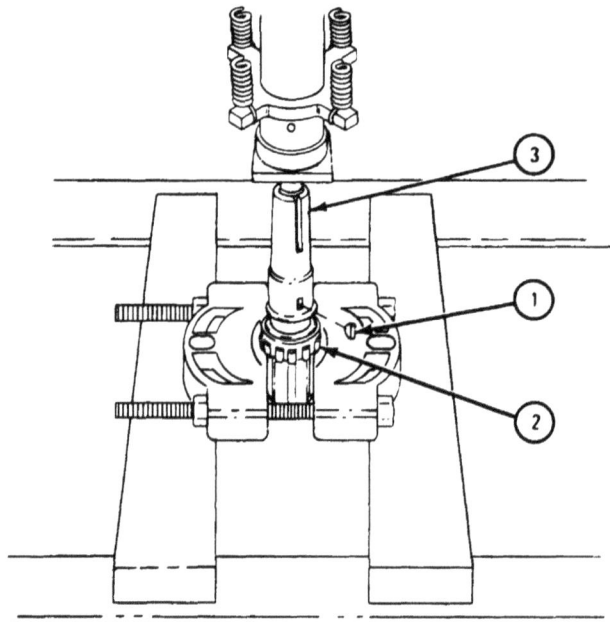

TA 087917

FRAME 2

1. Aline keyway in gear (1) with key in shaft (2) and set up in hydraulic press as shown.

2. Press gear (1) onto shaft (2).

GO TO FRAME 3

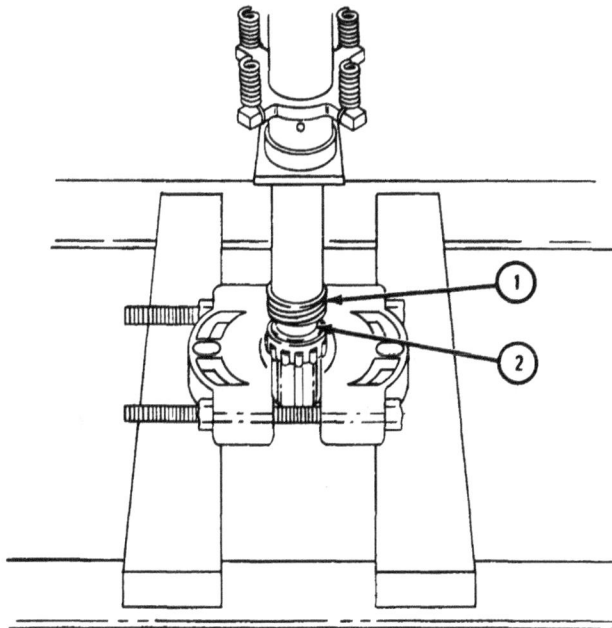

TA 087918

FRAME 3

1. Put bearing cone (1) and shaft (2) in hydraulic press as shown.

NOTE

Make sure that tapered end of bearing cone (1) faces threaded end of shaft (2).

2. Press bearing cone (1) on shaft (2).

GO TO FRAME 4

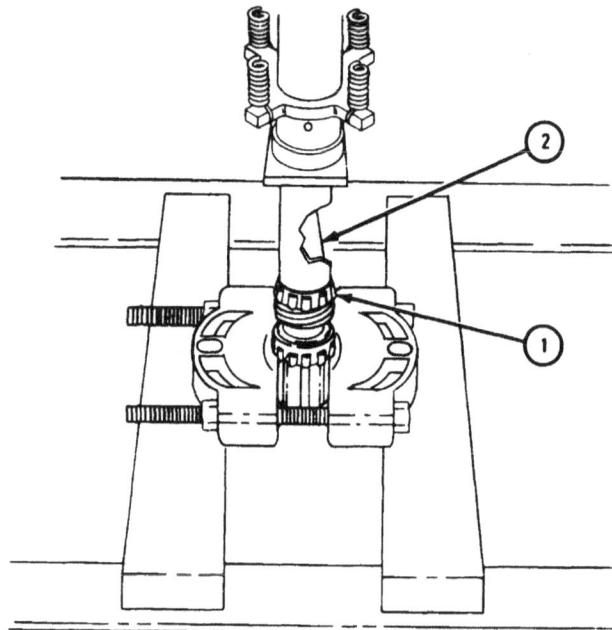

TA 087919

FRAME 4

NOTE

There are two bearing bores inside carrier (1). Bearing cup (2) must be pressed past the first bore and into the second bore.

1. Set up hydraulic press with tapered end of bearing cup (2) facing arbor.

2. Press bearing cup (2) through first bearing bore in carrier (1).

3. Aline bearing cup (2) with second bearing bore and press in bearing cup until it is seated in carrier (1).

GO TO FRAME 5

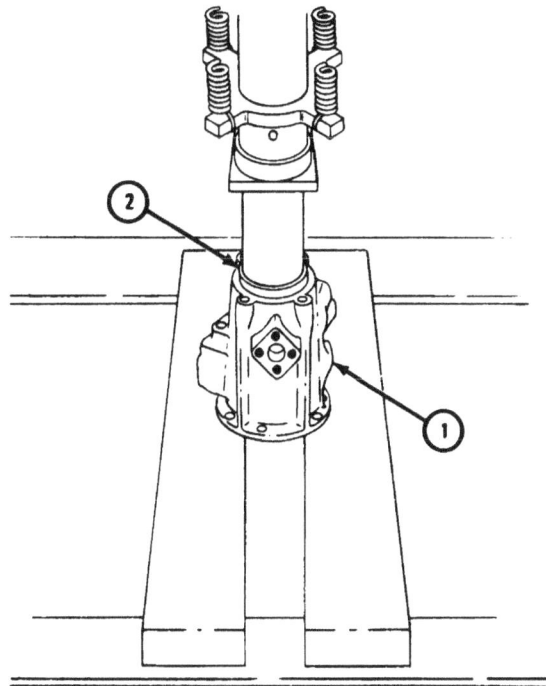

TA 087920

FRAME 5

1. Put output shaft assembly (1) through rear of carrier (2) as shown.

2. Using hammer and brass drift and working through rear of carrier (2), tap in bearing cup (3).

GO TO FRAME 6

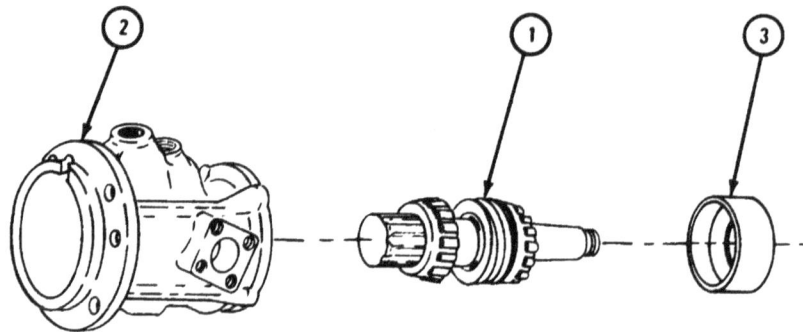

TA 087921

FRAME 6

1. Using hammer and brass drift, tap oil seal (1) into back of retainer (2) as shown. Make sure that oil seal is seated straight in retainer.

GO TO FRAME 7

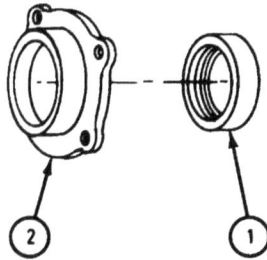

TA 087922

FRAME 7

1. Put on retainer (1) and gasket (2).

2. Put in four capscrews (3) and four lockwashers (4). Evenly tighten four capscrews to 20 pound-feet.

3. Put dial indicator on front output shaft (5) as shown.

GO TO FRAME 8

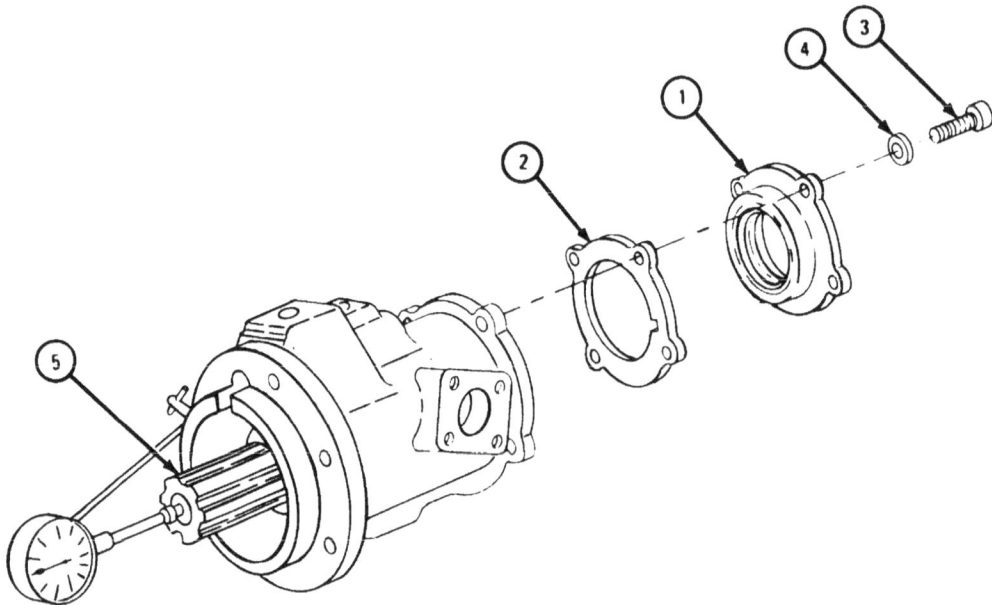

TA 087923

FRAME 8

1. Push output shaft (1) back away from dial indicator and set dial to 0.

2. Pull output shaft (1) forward against dial indicator and note dial reading.

3. If reading in step 2 is 0.002 to 0.006-inch, take off dial indicator and go to frame 10.

4. If reading in step 2 is more than 0.006-inch, go to frame 9.

5. If reading in step 2 is less than 0.002-inch, do the following:

    a. Disassemble transmission transfer power takeoff. Refer to para 2-33, frame 5.

    b. Seat bearing cup and assemble transmission transfer power takeoff. Refer to para 2-38, frame 5.

GO TO FRAME 9

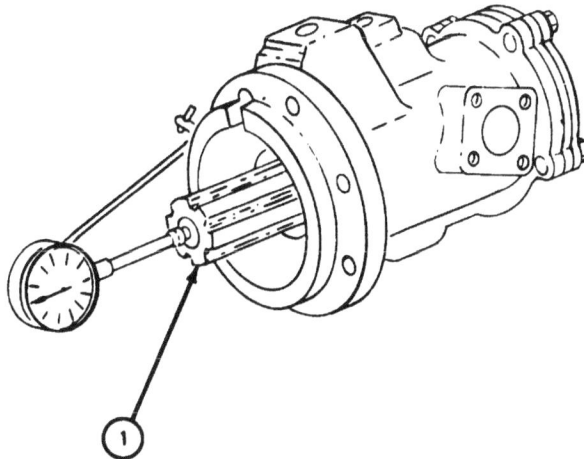

TA 087924

FRAME 9

1. Take out four capscrews (1) and four lockwashers (2).

2. Take off retainer (3) and gasket (4).

3. Subtract 0.006-inch from dial indicator reading in frame 8. Difference is thickness of shims (5) needed.

4. Put on needed thickness of shims (5), gasket (4), and retainer (3). Aline all screw holes and put in four capscrews (1) and four lockwashers (2).

5. Evenly tighten capscrews (1) to 20 pound-feet.

6. Do frame 8 again.

GO TO FRAME 10

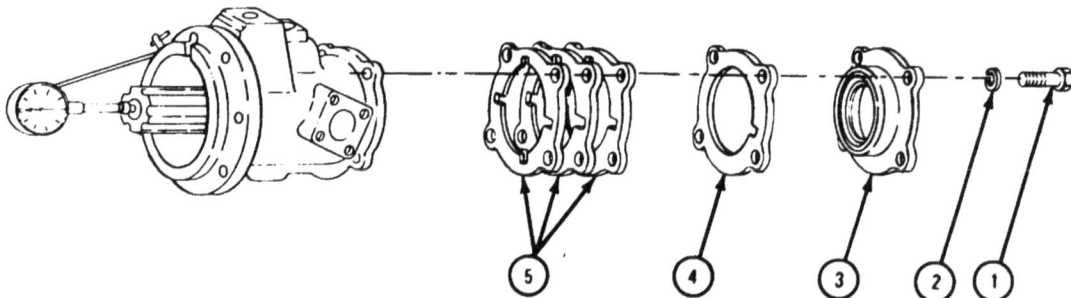

TA 087925

FRAME 10

1. Put fork (1) inside carrier (2).

2. Aline splines in sliding clutch (3) with splines on output shaft (4) and put sliding clutch on output shaft and inside carrier (2). Slide fork (1) on sliding clutch.

3. Put preformed packing (5) on shifter shaft (6). Put shifter shaft in carrier (2) and through hole in fork (1).

4. Working through drain plug hole (7), aline setscrew hole in fork (1) with setscrew hole in shifter shaft (6). Put in setscrew (8).

GO TO FRAME 11

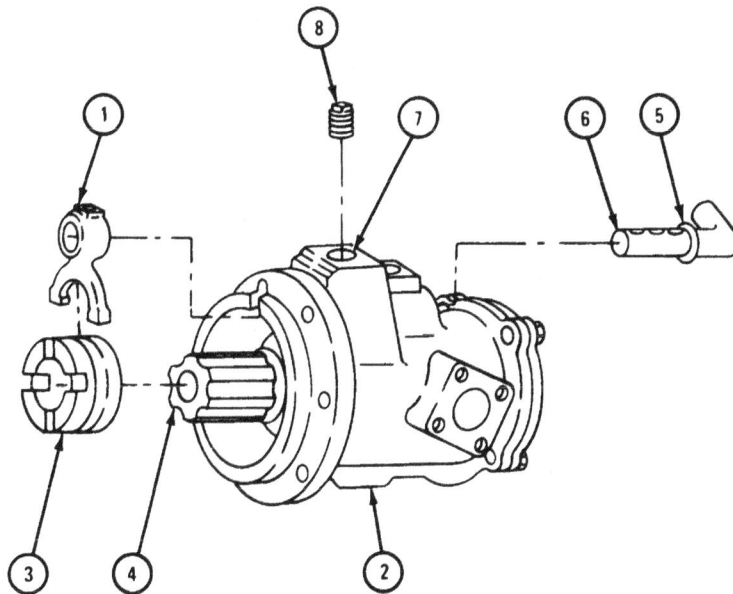

TA 087926

2-373

FRAME 11

1. Put in ball (1) and spring (2).
2. Put screw (3) and washer (4) into housing (5).

GO TO FRAME 12

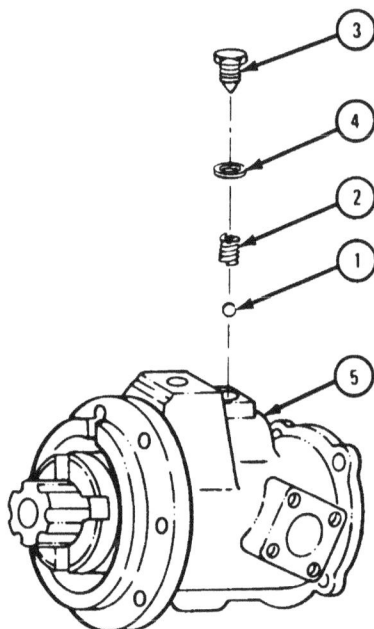

TA 087927

FRAME 12

1.   Put on gasket (1).

2.   Put spring (2) and plunger (3)   in valve assembly   (4).

3.   Put on valve assembly (4).

4.   Put in four capscrews (5) and four washers (6).

NOTE

Some models may not have fitting (7).

5.   Put in fitting (7) and fitting (8).

GO TO FRAME 13

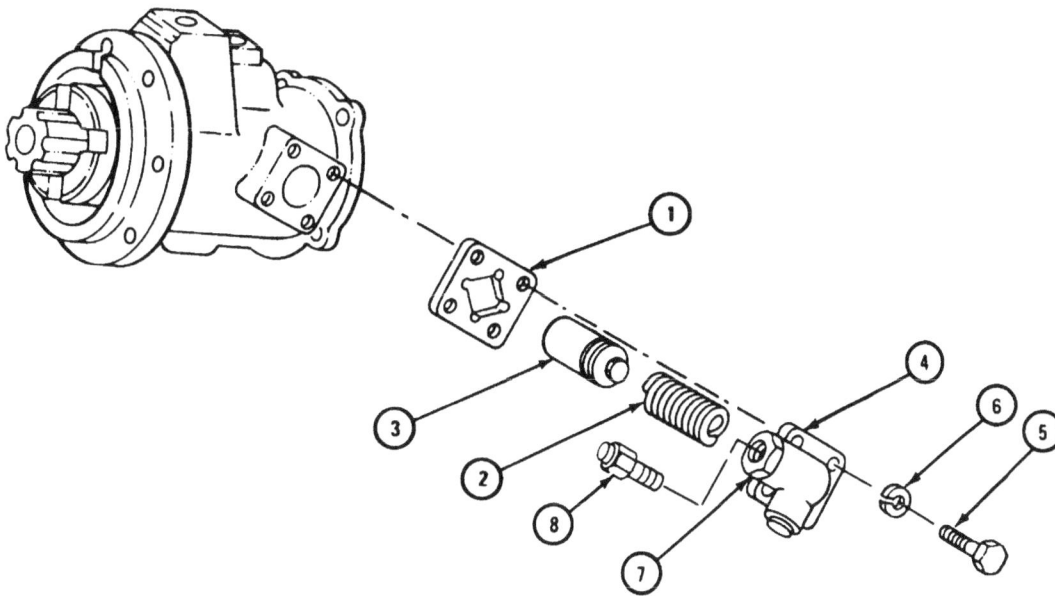

TA 087928

FRAME 13

1. Put in key (1).

2. Put shift lever (2) in place.

3. Put on washer (3).   Put in cotter pin (4).

4. Put on washer (5) and nut (6). Put in cotter pin (7).

5. Put in drain plug (8).

END OF TASK

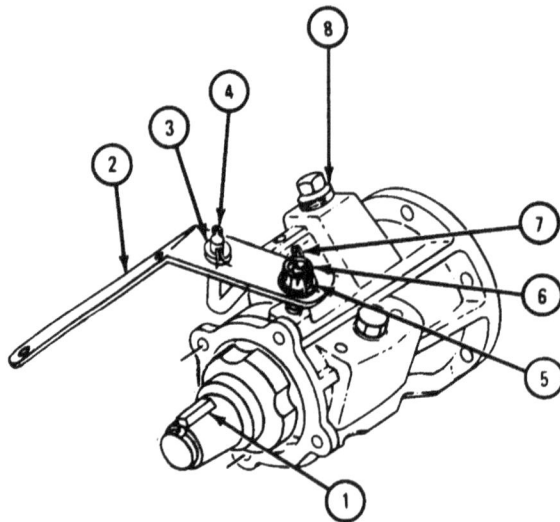

TA 087929

2-39. SHIFT TEST. This transfer power takeoff for paragraph gives instructions for testing the transmission positive shifting.

FRAME 1

NOTE

Turn output shaft (1) after shifting in each position to make sure it turns freely.

1. Push shift lever (2) forward.   Clutch sleeve (3) should move forward.

2. Pull shift lever (2) back until you feel it go into the first detent. Clutch sleeve (3) should move back.

3. Pull shift lever (2) all the way back.   Clutch sleeve (3) should move all the way back.

4. If transmission transfer assembly (4) does not shift into all three positions or does not turn freely, do the following:

   (a) Disassemble transmission transfer power takeoff. Refer to para 2-33, frames 1 through 5.

   (b) Check that all parts move freely and that there are no damaged parts. Refer to para 2-35, frame 2.

   (c) Assemble transmission transfer power takeoff. Refer to para 2-38, frames 10 through 13.

END OF TASK

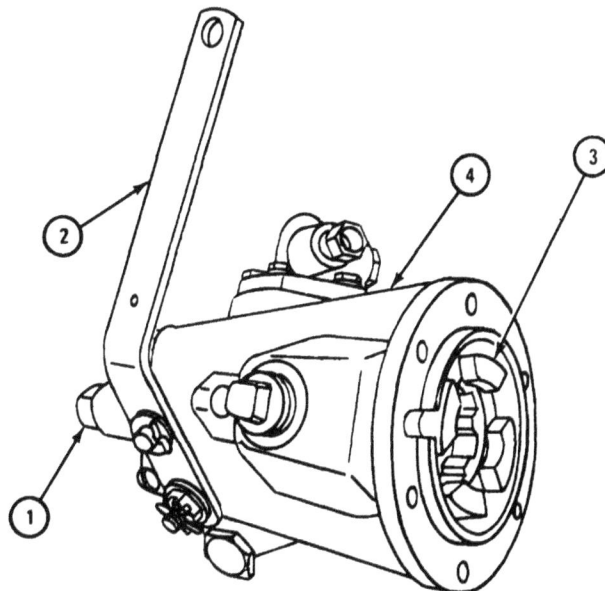

TA 102062

APPENDIX A

REFERENCES

A-1. PUBLICATION INDEXES AND GENERAL REFERENCE.

Indexes should be checked often for the latest changes or revisions of references given in this appendix and for new publications on materiel covered in this technical manual.

a. Military Publications Indexes.

Index of Army Motion Pictures and Related Audio-Visual Aids	DA Pam 108-1
Index of Administrative Publications	DA Pam 310-1
Index of Blank Forms	DA Pam 310-2
Index of Doctrinal Training and Organizational Publications	DA Pam 310-3

Military Publications:

Index of Technical Manuals, Technical Bulletins, Supply Bulletins, and Lubrications Orders . . . , ...	DA Pam 310-4
Index of Supply Catalogs and Supply Manuals (excluding types 7, 8 and 9)	DA Pam 310-6
Index of Modification Work Orders	DA Pam 310-7

Common Tools and Equipment
Supply Manuals . . . . , . DA Supply Manuals
SC 4910-95-CL-A01, SC 4910-95-CL-A02, SC 4910-95-CL-A31, SC 4910-95-CL-A33, SC 4910-95-CL-A50, SC 4910-95-CL-A63, SC 4910-95-CL-A67, SC 4910-95-CL-A68, SC 4910-95-CL-A72, SC 4910-95-CL-A73, and SC 4910-95-CL-A74.

b. General References.

Dictionary of United States Army Terms	AR 310-25
Authorized Abbreviations and Brevity Codes	AR 310-50

A-2. FORMS.

The following forms are for this materiel.   (Refer to DA pamphlet 310-2 for index of blank forms and to TM 38-750 for explanation of their use.)

Recommended Changes to Equipment
  Publications . . . .   DA Form 2028

Maintenance Request . . . .   DA Form 2407

Equipment Log Assembly (Records) . . . .   DA Form 2408

A-3. OTHER PUBLICATIONS.

a. Vehicle.

Lubrication Order . . . .   LO 9-2320-209-12/1

Operator's Manual . . . .   TM 9-2320-209-10

Organizational Maintenance Manual
  (Multifuel Engine) . . . .   TM 9-2320-209-20

Direct Support and General Support
  Maintenance Manual
  (Multifuel Engine) . . . .   TM 9-2320-209-34

Organizational Maintenance Repair Parts
  and Special Tool List . . . .   TM 9-2320-209-20P

Direct Support and General Support
  Maintenance Repair Parts and
  Special Tool List . . . .   TM 9-2320-209-34P

b. General.

Safety Inspection and Testing of Lifting
  Devices . . . .   TB 43-0142

Inspection, Care, and Maintenance of
  Antifriction Bearings . . . .   TM 9-214

Welding Theory and Application . . . .   TM 9-237

Materials Used for Cleaning, Preserving,
  Abrading, and Cementing Ordnance
  Material and Related Materials Including
  Chemicals . . . .   TM 9-247

Army Maintenance Management System . . . . .   TM 38-750

Painting Instructions for Field Use . . . .   TM 43-0139

Equipment Improvement Report and Maintenance
  Summary for TARCOM Equipment; Tank-
  Automotive, Commercial, Construction and
  Material Handling Equipment (MHE) . . . . .   TM 43-0143

A-3.   OTHER PUBLICATIONS. (Cont)

Equipment Improvement Report and Maintenance
  Digest (EIR MD) and Equipment Improvement
  Report and Maintenance Summary (EIR MS) . . . . . .   TB 43-0001-39 Series

INDEX

INDEX-CONT

INDEX-CONT

INDEX-CONT

INDEX-CONT

INDEX-CONT

By Order of the Secretaries of the Army and the Air Force:

E. C. MEYER
General, United States Army
Chief of Staff

Official:

J.C. PENNINGTON
Major General, United States Army
The Adjutant General

LEW ALLEN, JR., *General, USAF*
Chief of Staff

Official:

VAN L. CRAWFORD, JR., *Colonel, USAF*
Director of Administration

Distribution:

To be distributed in accordance with DA Form 12-38, Direct and General Support Maintenance requirements for 2½-Ton Truck Cargo, etc.

⚫ **U.S. GOVERNMENT PRINTING OFFICE:** 1994-300-421/82321

RECOMMENDED CHANGES TO EQUIPMENT TECHNICAL PUBLICATIONS

SOMETHING WRONG WITH THIS PUBLICATION?

THEN... JOT DOWN THE DOPE ABOUT IT ON THIS FORM, CAREFULLY TEAR IT OUT, FOLD IT AND DROP IT IN THE MAIL!

FROM (PRINT YOUR UNIT'S COMPLETE ADDRESS)
CDR, 1st Bn, 65th ADA
Attn: SP4 Jane Idone
Key West, FL 33040

DATE SENT
27 July 1980

PUBLICATION NUMBER	PUBLICATION DATE	PUBLICATION TITLE
TM 9-2520-246-34-1	15 June 1980	Transm., Transm. Transfers, & PTO's, Dir & Gen Support Maintenance

BE EXACT PIN-POINT WHERE IT IS				IN THIS SPACE TELL WHAT IS WRONG AND WHAT SHOULD BE DONE ABOUT IT:
PAGE NO	PARA GRAPH	FIGURE NO	TABLE NO	
2-14	2-4 9			FRAME 1, step 1 reads "Take off five capscrews (1) with lockwashers (2)." Should read "Take off six capscrews (1) with six lockwashers (2)."
2-51			2-5	Item 5, Third gem bore - Wear Limit reads "0.009." Should read "0.004."
2-189	2-18 9			FRAME 2 - Change illustration callouts. Reason: callouts for countershaft front bearing outer race (5) and front output shaft front bearing outer race (7) are reversed.

SAMPLE

PRINTED NAME GRADE OR TITLE AND TELEPHONE NUMBER	SIGN HERE
SP4 Jane Idone Autovon 222-2224	Jane Idone

DA FORM 2028-2
1 JUL 79

PREVIOUS EDITIONS ARE OBSOLETE

P.S. IF YOUR OUTFIT WANTS TO KNOW ABOUT YOUR RECOMMENDATION MAKE A CARBON COPY OF THIS AND GIVE IT TO YOUR HEADQUARTERS

TEAR ALONG PERFORATED LINE

REVERSE OF DA FORM 2028-2

TEAR ALONG PERFORATED LINE

FILL IN YOUR UNIT'S ADDRESS

FOLD BACK

DEPARTMENT OF THE ARMY

OFFICIAL BUSINESS

COMMANDER
US ARMY TANK-AUTOMOTIVE
MATERIEL READINESS COMMAND
ATTN: DRSTA-MB
WARREN, MI 48090

RECOMMENDED CHANGES TO EQUIPMENT TECHNICAL PUBLICATIONS

SOMETHING WRONG WITH THIS PUBLICATION?

THEN. . JOT DOWN THE DOPE ABOUT IT ON THIS FORM, CAREFULLY TEAR IT OUT, FOLD IT AND DROP IT IN THE MAIL!

FROM: (PRINT YOUR UNIT'S COMPLETE ADDRESS)

DATE SENT

PUBLICATION NUMBER	PUBLICATION DATE	PUBLICATION TITLE Transm., Transm., Transfers, PTO's, Dir. & Gen. Support Maint.
TM 9-2320-246-34-1	30 JANUARY 1981	

BE EXACT. . .PIN-POINT WHERE IT IS

PAGE NO.	PARA-GRAPH	FIGURE NO.	TABLE NO.	IN THIS SPACE TELL WHAT IS WRONG AND WHAT SHOULD BE DONE ABOUT IT:

PRINTED NAME, GRADE OR TITLE, AND TELEPHONE NUMBER

SIGN HERE:

DA FORM 1 JUL 79 **2028-2**

PREVIOUS EDITIONS ARE OBSOLETE.

P.S.—IF YOUR OUTFIT WANTS TO KNOW ABOUT YOUR RECOMMENDATION MAKE A CARBON COPY OF THIS AND GIVE IT TO YOUR HEADQUARTERS.

TEAR ALONG PERFORATED LINE

REVERSE OF DA FORM 2028-2

FILL IN YOUR UNIT'S ADDRESS

FOLD BACK

DEPARTMENT OF THE ARMY

OFFICIAL BUSINESS

TEAR ALONG PERFORATED LINE

COMMANDER
U.S. ARMY TANK – AUTOMOTIVE
MATERIEL READINESS COMMAND
ATTN: DRSTA-MB
WARREN, MI 48090

SOMETHING WRONG WITH THIS PUBLICATION?

THEN. . . JOT DOWN THE DOPE ABOUT IT ON THIS FORM, CAREFULLY TEAR IT OUT, FOLD IT AND DROP IT IN THE MAIL!

FROM (PRINT YOUR UNIT'S COMPLETE ADDRESS)

DATE SENT

PUBLICATION NUMBER	PUBLICATION DATE	PUBLICATION TITLE
TM 9-2520-246-34-1	30 JANUARY 1981	Transm., Transm. Transfers, & PTO's, Dir. & Gen. Support Maint.

BE EXACT   PIN-POINT WHERE IT IS

PAGE NO	PARA- GRAPH	FIGURE NO	TABLE NO

IN THIS SPACE TELL WHAT IS WRONG AND WHAT SHOULD BE DONE ABOUT IT:

PRINTED NAME GRADE OR TITLE AND TELEPHONE NUMBER

SIGN HERE

DA FORM 1 JUL 79 **2028-2**

PREVIOUS EDITIONS ARE OBSOLETE.

P.S.--IF YOUR OUTFIT WANTS TO KNOW ABOUT YOUR RECOMMENDATION MAKE A CARBON COPY OF THIS AND GIVE IT TO YOUR HEADQUARTERS

TEAR ALONG PERFORATED LINE

REVERSE OF DA FORM 2028-2

FILL IN YOUR UNIT'S ADDRESS

FOLD BACK

TEAR ALONG PERFORATED LINE

DEPARTMENT OF THE ARMY

OFFICIAL BUSINESS

COMMANDER
U.S. ARMY TANK-AUTOMOTIVE
MATERIEL READINESS COMMAND
ATTN: DRSTA-MB
WARREN, MI 48090

SOMETHING WRONG WITH THIS PUBLICATION?

THEN. . JOT DOWN THE DOPE ABOUT IT ON THIS FORM, CAREFULLY TEAR IT OUT, FOLD IT AND DROP IT IN THE MAIL!

FROM: (PRINT YOUR UNIT'S COMPLETE ADDRESS)

DATE SENT

PUBLICATION NUMBER	PUBLICATION DATE	PUBLICATION TITLE
TM 9-2320-246-34-1	30 JANUARY 1981	Transm., Transm. Transfers, & PTO's, Dir. & Gen. Support Maint.

BE EXACT. . .PIN-POINT WHERE IT IS

PAGE NO.	PARA-GRAPH	FIGURE NO.	TABLE NO	IN THIS SPACE TELL WHAT IS WRONG AND WHAT SHOULD BE DONE ABOUT IT:

PRINTED NAME, GRADE OR TITLE. AND TELEPHONE NUMBER

SIGN HERE:

DA FORM 1 JUL 79 **2028-2**

PREVIOUS EDITIONS ARE OBSOLETE.

P.S.–IF YOUR OUTFIT WANTS TO KNOW ABOUT YOUR RECOMMENDATION MAKE A CARBON COPY OF THIS AND GIVE IT TO YOUR HEADQUARTERS.

TEAR ALONG PERFORATED LINE

REVERSE OF DA FORM 2028-2

FILL IN YOUR UNIT'S ADDRESS

FOLD BACK

DEPARTMENT OF THE ARMY

OFFICIAL BUSINESS

TEAR ALONG PERFORATED LINE

COMMANDER
U.S. ARMY TANK-AUTOMOTIVE
MATERIEL READINESS COMMAND
ATTN: DRSTA-MB
WARREN, MI 48090

048083

TECHNICAL MANUAL

DIRECT SUPPORT AND GENERAL SUPPOR T

REPAIR PARTS AND SPECIAL TOOLS LIS T

F O R

T R A N S M I S S I O N S

(2520-00-347-4520) (2520-00-884-4833)

T R A N S F E R S

(2520-00-089-8287) (2520-00-001-7855)

P O W E R   T A K E - O F   F

(2520-00-229-5673) (2520-00-706-1136)

(2520-00-706-1137)

HEADQUARTERS, DEPARTMENT OF THE ARMY

MAY 1976

Technical Manual

TM 9-2520-246-34P

HEADQUARTERS
DEPARTMENT OF THE ARMY
Washington, DC, 28 May 1976

DIRECT SUPPORT AND GENERAL SUPPORT

REPAIR PARTS AND SPECIAL TOOLS LIST

FOR

TRANSMISSIONS

(2520-00-347-4520) (2520-00-884-4833)

TRANSFERS

(2520-00-089-8287) (2520-00-001-7855)

POWER TAKE-OFF.

(2520-00-229-5673)(2520-00-706-1136)

(2520-00-706-1137)

Current as of 30 October 1975

This manual supersedes TM9-2520-246-35P, 11 July 1962, including all changes.

SECTION I

INTRODUCTION

1. Scope

This manual lists repair parts with special tools authorized for direct support and general support of: Transmission Assemblies, 2520-00-347-4520 and 2520-00-884-4833; Transfer Assemblies, 2520-00-001-7855, 2520-00-089-8287 and Power Take-Off Assemblies, 2520-00-229-5673, 2520-00-706-1136, 2520-00-706-1137.

2. General

This Repair Parts and Special Tools List is divided into the following sections:

a. *Section II- Repair Parts List.* A list of repair parts authorized for use in the performance of maintenance. The list also includes parts which must be removed for replacement of the authorized parts. Parts lists are composed for functional groups in ascending numerical sequence, with the parts in each group listed in figure and item number sequence. Bulk materials are listed in NSN sequence.

b. *Section III–Special Tools List.* A list of special tools, TMDE, end support equipment authorized for the performance of maintenance at the organizational level.

c. *Section IV—National Stock Number and Part Number Index.* A list, in ascending numerical sequence, of all National stock numbers appearing in the listings, followed by a list, in alphanumeric sequence, of all part numbers appearing in the listings. Federal stock number and part numbers are cross-referenced to each illustration figure and item number appearance. This index is followed by a cross-reference list of reference designations to figure and item numbers when applicable.

3. Explanation of Columns

The following provides an explanation of columns found in the tabular listings:

a. *Illustration.* This column is divided as follows:

(1) *Figure number.* Indicates the figure number of the illustration in which the item is shown.

(2) *Item number.* The number used to identify each item called out in the illustration.

b. *Source, Maintenance, and Recoverability Codes (SMR).*

(1) *Source code. Source codes are* assigned to support items to indicate the manner of acquiring support items for maintenance, repair, or overhaul of end items. Source codes are entered in the first and second positions of the Uniform SMR Code format as follows:

Code	Definition
PA	Item procured and stocked for anticipated or known usage.
PB	Item procured and stocked for insurance purpose because essentiality dictates that minimum quantity be available in the supply systems.
PC	Item procured and stocked and which otherwise would be coded PA except that it is deteriorative in nature.
PD	Support item, excluding support equipment, procured for initial issue or outfitting and stocked only for subsequent or additional initial issues or outfittings. Not subject to automatic replenishment.
PE	Support equipment procured and stocked for initial issue or outfitting to specified maintenance repair activities.
PF	Support equipment which will not be stocked but which will be centrally procured on demand.
PC	Item procured and stocked to provide for sustained support for the life of the equipment. It is applied to an item peculiar to the equipment which, because of probable discontinuance or shutdown of production facilities, would prove uneconomical to reproduce at a later time.
KD	An item of a depot overhaul/repair kit and not purchased separately. Depot kit defined as a kit that provides items required at the time of overhaul or repair.
KF	An item of a maintenance kit and not purchased separately. Maintenance kit defined as a kit that provides an item that can be replaced at organizational or intermediate levels of maintenance.
KB	Item included in both a depot overhaul/repair kit and a maintenance kit.
MO	Item to be manufactured or fabricated at organizational level.
MF	Item to be manufactured or fabricated at the direct support maintenance level.
MH	Item to be manufactured or fabricated at the general support maintenance level.
MD	Item to be manufactured or fabricated at the depot maintenance level.
AO	Item to be assembled at organizational level.
AF	Item to be assembled at direct support maintenance level.
AH	Item to be assembled at general support maintenance level.
AD	Item to be assembled at depot maintenance level.
XA	Item is not procured or stocked because the requirements for the item will result in the replacement of the next higher assembly.
XB	Item is not procured or stocked. If not available through salvage, requisition.
XD	A support item that is not stocked. When required, item will be procured through normal supply channels.

Cannibalization or salvage may be used as a source of supply for any items source coded above excpet those coded XA, XD, and aircraft support items as restricted by AR 700-42.

(2) *Maintenance code.* Maintenance codes are gned to indicate the levels of maintenance **horized** to USE and REPAIR support items. The ntenance codes are entered in the third and th positions of the Uniform SMR Code format as ows:

(a) The maintenance code entered in the d position will indicate the lowest maintenance d authorized to remove, replace, and use the port item. The maintenance code entered in the d position will indicate one of the following levels maintenance:

Application/Explanation

Crew or operator maintenance performed within organizational maintenance.

Support item is removed, replaced, used at the organizational level.

Support item is removed, replaced, used by the direct support element of integrated direct support maintenance.

Support item is removed, replaced used at the direct support level.

Support item is removed, replaced, used at the general support level.

Support items that are removed, replaced, used at depot, mobile depot, specialized repair activity only.

NOTE
Codes "I" and "F" will be considered the same by direct support units.

(b) The maintenance code entered in the th position indicates whether the item is to be iired and identifies the lowest maintenance level 1 the capability to perform complete repair (i. e., mthorized maintenance functions). This position contain one of the following maintenance codes:

Application/Explanation

The lowest maintenance level capable of complete repair of the support item is the organizational level.

The lowest maintenance level capable of complete repair of the support item is the direct support level.

The lowest maintenance level capable of complete repair of the support item is the general support level.

The lowest maintenance level capable of complete repair of the support item is the depot level, performed by depot, mobile depot or specialized repair activity.

Repair restricted to designated specialized repair activity.

Nonreparable. No repair is authorized.

No repair is authorized. The item may be reconditioned by adjusting, lubricating, etc., at the user level. No parts or special tools are procured for the maintenance of this item.

(3) *Recoverability code.* Recoverability codes are assigned to support items to indicate the disposition action on unserviceable items. The recoverability code is entered in the fifth position of the Uniform SFR Code format as follows:

Recoverability
code   Definition

z   Nonreparable item. When unserviceable, condemn and dispose at the level indicated in position 3.

o   Reparable item. When uneconomically reparable, condemn and dispose at organizational level.

F   Reparable item. When uneconomically reparable, condemn and dispose at the direct support level.

H   Reparable item. When uneconomically reparable, condemn and dispose at the general support level.

D   Reparable item. When beyond lower level repair capability, return to depot. Condemnation and disposal not authorized below depot level.

L   Reparable item. Repair, condemnation, and disposal not authorized below depot/specialized repair activity level.

A   Item requires special handling or condemnation procedures because of specific reasons (i.e., precious metal content, high dollar value, critical material or hazardous material). Refer to appropriate manuals/directives for specific instructions.

c. *National Stock Number.* Indicates the National stock number assigned to the item and will be used for requisitioning purposes.

d. *Part Number.* Indicates the primary number used by the manufacturer (individual, company, firm, corporation, or Government activity), which controls the design and characteristics of the item by means of its engineering drawings, specifications standards, and inspection requirements, to identify an item or range of items.

NOTE
When a stock numbered item i s requisitioned, the repair part received may have a different part number than the part being replaced.

e. *Federal Supply Code for Manufacturer (FSCM).* The FSCM is a 5-digit numeric code listed in SB708-42 which is used to identify the manufacturer, distributor, or Government agency, etc.

f. *Unit of Measure (U/M).* Indicates the standard of the basic quantity of the listed item as used in performing the actual maintenance function. This measure is expressed by a two-character alphabetical abbreviation (e.g., ea, in, pr, etc). When the unit of measure differs from the unit of issue, the lowest unit of issue that will satisfy the required units of measure will be requisitioned.

g. *Quantity Furnished with Equipment.* (Basic Issue Items Only). Indicates the quantity of the basic issue item furnished with the equipment.

h. *Quantity Authorized.* (Items Troop Installed or Authorized Only). Indicates the quantity of the item authorized to be used with the equipment.

i. *Quantity Incorporated in Unit.* Indicates the

quantity of the item used in the breakout shown on the illustration figure, which is prepared for a functional group, subfunctional group, or an assembly. A "V" appearing in this column in lieu of a quantity indicates that no specific quantity is applicable, (e.g., shims, spacers, etc.).

4. Special Information

a Components contained in this manual are applicable to the Truck, Cargo: 2 1/2 Ton, 6x6, M44 Series (TM9-2320-209).

Component	End item		
Transmission (2520-00-347-4520)	M44 Series (gasoline)		
Transmission (2520-00-884-4833)	M44 Series (multifuel)		
Transfer (2520-00-089-8287)	M352A	M45A2G	M29A2
	M35A2C	M46A2	M342A2
	M36A2	M46A2C	M621
	M36A2C	M109A3	M622
	M44A2	M185A3	M623
	M45A2	M275A2	M624
Transfer (2520-00-001-7855)	M44		
Power Take-Off (2520-00-229-5673)	M621	M623	
	M622	M624	
Power Take-Off (2520-00-706-1136)	M47	M58	M59
	M57	M58A1	M352
Power Take-Off (2520-00-706-1137)	M34	M46	M108
	M35	M46A1	M109
	M35A1	M46C	M109A1
	M35A2	M48	M109A2
	M36A2	M49	M109A3
	M36C	M49A1C	M109C
	M44	M49A2C	M109D
	M44A1	M49C	M275
	M44A2	M50	M275A1
	M45	M50A1	M275A2
	M45A1	M50A2	VT17-A/MTQ
	M45A2	M60	VT18-A/MTQ

b. Usable on codes are shown in the description column. Uncoded items are applicable to all models. Identification of the usable on codes used in this publication are:

Code	Used on
A	Transmission (2520-00-347-4520) — Gasoline Engine
B	Transmission (2520-00-884-4833) — Multifuel Engine
C	Power Take-off (2520-00-706-1136) — Front and Rear Output Shafts
D	Power Take-off (2520-00-706-1137) — Front Output Shaft only
E	Transfer (2520-00-001-7855) — Spring type (w/improved oiling)
F	Transfer (2520-00-089-8287) — Air lock-up
H	Power Take-off (2520-00-229-5673) —

c. Detailed manufacturing instructions for items source coded to be manufactured or fabricated are found in TM9-2815-246-35.

d. Detailed assembly instructions for items source coded to be assembled are found in TM 9-2815-246-35.

e. Repair parts kits and gasket set appear as the last entries in the repair parts listing for the figure in which its parts are listed as repair parts. Kits or Sets whose separate components appear as repair parts in more than one figure will be found immediately following the repair parts listing for the last figure containing parts of the Kit.

f. The following publication pertains to transmission, transfers and power take-offs installed on M44 series vehicles.

Publication No.	Title
TM 9-2520-246-35	Direct Support, General Support and Depot Maintenance for Transmissions, Transfers and Power Take-offs.

5. How to Locate Repair Parts

a. When National Stock Number or Part Number is Unknown:

(1) **First.** Using the table of contents, determine the functional subgroup within which the repair part belongs. This is necessary since illustrations are prepared for subgroups, and listings are divided into the same group.

(2) **Second.** Find the illustration covering the functional subgroup to which the repair part belongs.

(3) **Third.** Identify the repair part on the illustration and note the illustration figure and item number of the repair part.

(4) Fourth. Using the Repair Parts Listing, the figure and item number noted on the tration.

When National Stock Number or Part Number nown:

(1) **First. Using** the Index of National Stock ıbers and Part Numbers, find the pertinent lonal Stock Number or part number. This index ascending NSN sequence followed by a list of numbers in ascending alphanumeric sequence, s-referenced to the illustration figure number item number.

(2) **Second.** After finding the figure and item number, locate the figure and item number in the repair parts list.

6. Reporting of Errors

You can improve this manual by recommending improvements. Mail your comments direct to Commander, US Army Tank-Automotive Command, ATTN: DRSTA-MTP, Warren, MI 48090, using DA Form 2028, (Recommended Changes to Publications and Blank Forms). A reply will be sent direct to you.

REPAIR PARTS LIST

(1) ILLUSTRATION		(2)	(3)	(4)	(5)	(6)	(7)	(8) QTY
(a) FIG. NO.	(b) ITEM NO.	SMR CODE	NATIONAL STOCK NUMBER	PART NUMBER	FSCM	DESCRIPTION USABLE ON CODE	U/M	INC IN UNIT
						GROUP 07-TRANSMISSION		
						0700-TRANSMISSION ASSEMBLY		
1	1	PAFZZ	5305-00-317-3102	594119	21450	SCREW,ASSEMBLED LOCKWASHER:TRANSMISSION ASSEMBLY	EA	12
1	2	PAOZZ	4820-00-726-4719	5196397	19207	VALVE,AIR PRESSURE,ASSEMBLY:TRAMSMISSION ASSEMBLY	EA	1
1	3	PAFHH	2520-00-347-4520	7520995	19207	TRANSMISSION,MECHANICAL:ASSEMBLY (GAS) A	EA	1
1	3	PAFHH	2520-00-884-4833	10872096	19207	TRANSMISSION, MECHANICAL:ASSEMBLY (SPICER MODEL 3052 (MULTIFUEL) B	EA	1
2	1	PAFZZ	3110-00-198-1502	2135-1	08162	BEARING,BALL,THRUST:RELEASE,TRANSMISSION CLUTCH HOUSING	EA	1
2	2	PAFZZ	2520-00-752-0966	7520966	19207	CARRIER:RELEASE BEARING,TRANSMISSION CLUTCH HOUSING	EA	1
2	3	PAFZZ	5306-00-174-4202	5214807	19207	BOLT,MACHINE:RELEASE FORK,TRANSMISSION CLUTCH HOUSING	EA	2
2	4	PAFZZ	5310-00-209-0965	MS35338-47	96906	WASHER,LOCK:RELEASE FORK TRANSMISSION CLUTCH HOUSING	EA	2
2	5	PAFZZ	5310-00-661-9568	5226033	19207	WASHER,THRUST:KEY RELEASE FORK,TRANSMISSION CLUTCH HOUSING	EA	2
2	6	PAFZZ	2520-00-752-0967	7520967	19207	FORK,RELEASE:TRANSMISSION CLUTCH HOUSING	EA	1
2	7	PAFZZ	5330-00-752-0959	7520959	19207	GASKET:CLUTCH HOUSING MOUNTING	EA	1
2	8	PAFZZ	5340-00-752-0976	7520976	19207	PLUG,EXPANSION:SHAFT HOLE,THROW-OUT LEVER	EA	1
2	9	PAFZZ	2520-00-752-0952	7520952	19207	HOUSING,CLUTCH	EA	1
2	10	PAOZZ	4730-00-172-0034	MS15003-7	19207	FITTING,LUBRICATION:CLUTCH HOUSING	EA	1
2	11	PAFZZ	5330-00-286-6815	500021	19207	SEAL,PLAIN,ENCASED:CLUTCH HOUSING	EA	1
2	12	PAOZZ	5306-00-044-5319	8712073	19207	BOLT,SPECIAL:TRANSMISSION CASE INPUT SHAFT BEARING COVER(4)INSPECTION HOLE COVER(6)	EA	10
2	13	PAHZZ	2520-00-508-4664	7520989	19207	COVER,TRANSFER INPUT SHAFT BEARING FRONT	EA	1
2	14	PAHZZ	5330-00-594-8953	7520992	19207	GASKET:INPUT SHAFT BEARING(PART OF SET 2520-00-752-0987)	EA	1
2	15	PAFZZ	5330-00-752-0958	7520958	19207	GASKET:CLUTCH HOUSING(PART OF SET 2520-00-752-0987)	EA	1
2	16	XAHZZ		7520988	19207	CASE:TRANSMISSION HOUSING	EA	1
2	17	PAFZZ	5330-00-752-1061	7521061		GASKET:TRANSMISSION (PART OF SET 2520-00-752-0987)		
2	18	KFHZZ		7521004	19207	GASKET:BEARING RETAINER(PART OF SET 2520-00-752-0987)		
2	19	PAHZZ	2520-00-692-6065	7521007	19207	RETAINER WITH OIL SEAL:REAR BEARING	EA	1
2	20	XAHZZ		7748571	19207	CAP:RETAINER TO TRANSMISSION CASE,MAIN SHAFT REAP BEARING	EA	1
2	21	PAHZZ	5330-00-350-9958	583509	21450	SEAL,PLAIN,ENCASED:REAR BERAING MAIN SHAFT	EA	1
2	22	PAHZZ	5306-00-021-4077	8743049	19207	SCREW,ASSEMBLED W/LOCKWASHER:BEARING COVER,REAR MAINSHAFT (6), COUNTERSHAFT (4), POWER TALE-OFF ACCESS COVER (6)	EA	16

(1) ILLUSTRATION		(2)	(3)	(4)	(5)	(6)	(7)	(8) QTY.
(a) FIG. NO.	(b) ITEM NO.	SMR CODE	NATIONAL STOCK NUMBER	PART NUMBER	FSCM	DESCRIPTION USABLE ON CODE	U/M	INC. IN UNIT
						0700-TRANSMISSION ASSEMBLY-CONTINUED		
2	23	PAHZZ	2520-00-692-6070	7521038	19207	COVER:TRANSFER,COUNTERSHAFT REAR BEARING	EA	1
2	24	PAOZZ	4730-00-278-3380	MS49005-10	96906	PLUG,PIPE:OIL FILLER AND DRAIN	EA	2
2	25	PAHZZ	2520-00-752-1039	7521039	19207	GASKET:COUNTERSHAFT REAR BEARING COVER(PART OF SET 2520-00-752-0987)		
2	26	PAHZZ	5330-00-234-3317	8327322	19207	GASKET:ACCESS HOLE,POWER TAKE-OFF (PART OF SET 5330-00-752-0987)	EA	1
2	27	PAHZZ	2520-00-354-0771	5323473	19207	COVER:ACCESS,TRANSMISSION CASE,POWER TAKE-OFF	EA	1
2	28	PAHZZ	2520-00-752-1032	7521032	19207	RETAINER:COUNTERSHAFT FRONT BEARING	EA	1
2	29	PAFZZ	2520-00-752-0979	7520979	19207	SHAFT:CLUTCH THROW-OUT LEVER	EA	1
2	30	PAOZZ	5330-00-752-0957	7520957	19207	GASKET:INSPECTION HOLE COVER(PART OF SET 5330-00-752-0987)	EA	1
2	31	PAOZZ	2520-00-752-0956	7520956	19207	COVER PLATE:INSPECTION HOLE ACCESS	EA	1
2	32	PAFZZ	5310-00-584-5272	MS35338-48	96906	WASHER,LOCK:CLUTCH HOUSING	EA	5
2	33	PAFZZ	5305-00-716-8184	MS90726-112	96906	SCREW,CAP,HEXAGON HEAD:CLUTCH HOUSING	EA	5
2	34	PAFZZ	5340-00-392-4017	7520980	19207	SPRING:CLUTCH RELEASE BEARING CARRIER	EA	2
2	35	PAFZZ	2520-00-752-0965	7520965	19207	BUTTON:CLUTCH RELEASE BEARING CARRIER	EA	2
2	36	PAOZZ	4730-00-050-4208	MS15003-1	96906	FITTING:LUBRICATION:CLUTCH RELEASE BEARING CARRIER	EA	1
		PAHZZ	2520-00-752-0987	7520987	19207	GASKET SET:TRANSMISSION ASSEMBLY	EA	1
2	14					GASKET:INPUT SHAFT BEARING COVER	EA	1
2	15					GASKET:CLUTCH HOUSING	EA	1
2	17					GASKET:TRANSMISSION CASE	EA	1
2	18					GASKET:BEARING RETAINER	EA	1
2	25					GASKET:COUNTERSHAFT REAR BEARING COVER	EA	1
2	26					GASKET:ACCESS HOLE,POWER TAKE-OFF	EA	1
2	30					GASKET:INSPECTION HOLE COVER	EA	1
						0701-TRANSMISSION INPUT AND OUTPUT SHAFTS, COUNTERSHAFTS AND IDLER SHAFTS WITH GEARS		
3	1	PAHZZ	5365-00-508-4675	7520993	19207	RING,RETAINING:SNAP,INPUT SHAFT BEARING	EA	1
3	2	PAHZZ	3110-00-554-5960	700955	21450	BEARING,BALL,ANNULAR,INPUT SHAFT,FRONT	EA	1
3	3	PAHZZ	2520-00-752-0994	7520994	19207	SHAFT:INPUT WITH INTEGRAL GEAR(GAS) A	EA	1
3	3	PAHZZ	2520-00-885-3423	10914635	19207	SHAFT:INPUT,WITH INTEGRAL GEAR(MULTIFUEL) B	EA	1
3	4	PAHZZ	3110-00-227-4123	7521010	19207	ROLLER BEARING:INPUT SHAFT	EA	14
3	5	PAHZZ	2520-00-752-1022	7521022	19207	SYNCHRONIZER,TRANSMISSION:FOURTH AND FIFTH SPEEDS	EA	1
3	6	PAHZZ	5365-00-699-8456	7521008	19207	RING,RETAINING:SNAP FOURTH SPEED GEAR	EA	1
3	7	PAHZZ	5310-00-285-2169	7521024	19207	WASHER,KEYWAY:INPUT SHAFT	EA	1
3	8	PAHZZ	3020-00-752-1015	7521015	19207	GEAR,HELICAL:FOURTH SPEED (GAS) A	EA	1
3	8	PAHZZ	4730-00-884-4838	10914637	19207	GEAR,HELICAL:MAINSHAFT,FOURTH SPEED (MULTIFUEL) B	EA	1
3	9	PAHZZ	5315-00-699-8458	7521018	19207	PIN:STRAIGHT HEADED,FOURTH SPEED GEAR(SLEEVE)	EA	1
3	10	PAHZZ	3120-00-752-1020	7521020	19207	BEARING,SLEEVE:FOURTH SPEED GEAR	EA	1
3	11	PAHZZ	3020-00-752-1016	7521016	19207	GEAR,HELICAL:THIRD SPEED	EA	1
3	12	PAHZZ	2520-00-752-1023	7521023	19207	SYCHRONIZER,TRANSMISSION:SECOND AND THIRD SPEED	EA	1
3	13	PAHZZ	5365-00-699-8457	7521009	19207	RING:RETAINING:CLUTCH GEAR	EA	1
3	14	PAHZZ	3020-00-752-1005	7521005	19207	GEAR:CLUTCH,SECOND AND THIRD SPEED	EA	1

(1) ILLUSTRATION		(2)	(3)	(4)	(5)	(6)		(7)	(8) QTY.
(a) FIG. NO.	(b) ITEM NO.	SMR CODE	NATIONAL STOCK NUMBER	PART NUMBER	FSCM	DESCRIPTION USABLE ON CODE		U/M	INC. IN UNIT
						0701-TRANSMISSION INPUT AND OUTPUT SHAFTS, COUNTERSHAFTS AND IDLER SHAFTS WITH GEARS-CONTINUED			
3	15	PAHZZ	3020-00-752-1017	7521017	19207	GEAR,HELICAL:SECOND SPEED		EA	1
3	16	PAHZZ	5315-00-012-4553	MS35756-17	96906	KEY,WOODRUFF:SYCHRONIZER SLEEVE		EA	2
3	17	PAHZZ	2520-00-752-1019	7521019	19207	SHAFT:MAIN		EA	1
3	18	PAHZZ	3020-00-752-1006	7521006	19207	GEAR,SPUR:FIRST SPEED AND REVERSE		EA	1
3	19	PAHZZ	3110-00-155-6706	700772	00000	BEARING,BALL,ANNULAR:REAR,MAIN SHAFT		EA	1
3	20	PAHZZ	5365-00-752-1021	7521021	19207	SPACER,SLEEVE:REAR BEARING,MAIN SHAFT		EA	1
3	21	PAHZZ	5315-00-298-1481	MS24665-357	96906	PIN,COTTER:SLOTTED NUT		EA	2
3	22	PAHZZ	5310-00-529-4103	5294103	19207	NUT,PLAIN,SLOTTED:MAIN SHAFT(1),COUNTER(1)		EA	2
3	23	PAHZZ	2520-00-752-1003	7521003	19207	FLANGE,COMPANION,UNION:MAIN SHAFT		EA	1
3	24	PAHZZ	3020-00-752-1031	7521031	19207	GEAR,HELICAL:COUNTERSHAFT DRIVE (GAS) A		EA	1
3	24	PAHZZ	2520-00-884-4835	10914636		GEAR,HELICAL:COUNTERSHAFT DRIVE(MULTIFUEL) B		EA	1
3	25	PAHZZ	5365-00-752-1046	7521046	19207	RING:RETAINING:DRIVE GEAR(1)FOURTH SPEED GEAR(1)		EA	2
3	26	PAHZZ	3020-00-752-1029	7521029	19207	GEAR,HELICAL:COUNTERSHAFT FOURTH SPEED (GAS) A		EA	1
3	26	PAHZZ	2520-00-884-4832	10914638	19207	GEAR,HELICAL:COUNTERSHAFT FOURTH SPEED (MULTIFUEL) B		EA	1
3	27	PAHZZ	5315-00-058-8583	MS35756-110	96906	KEY,WOODRUFF:GEAR RETAING,HELICAL DRIVE AND FOURTH SPEED COUNTERSHAFT		EA	2
3	28	PAHZZ	2520-00-752-1037	7521037	19207	COUNTERSHAFT:WITH INTEGRAL GEARS		EA	1
3	29	PAHZZ	3110-00-155-6686	700771	21450	BEARING,BALL,ANNULAR:COUNTERSHAFT REAR		EA	1
3	30	PAHZZ	5310-00-752-1036	7521036	19207	WASHER,FLAT:COUNTERSHAFT NUT		EA	1
3	31	PAHZZ	2520-00-752-1047	7521047	19207	SHAFT:REVERSE IDLER GEAR		EA	1
3	32	PAHZZ	3120-00-752-1049	7521049	19207	WASHER,THRUST:REVERSE IDLER GEAR		EA	1
3	33	PAHZZ	3110-00-120-4276	708231	00000	ROLLER ASSEMBLY:BEARING,IDLER GEAR		EA	2
3	34	PAHZZ	3020-00-752-1030	7521030	19207	GEAR CLUSTER:HELICAL:REVERSE IDLER		EA	1
3	35	PAHZZ	5365-00-752-1045	7521048	19207	RING,RETAING:DRIVE GEAR,COUNTERSHAFT		EA	1
3	36	PAHZZ	5310-00-752-1048	7521048	19207	WASHER,FLAT:FRONT BEARING,COUNTERSHAFT		EA	1
3	37	PAHZZ	3110-00-117-0693	707617	00000	BEARING,ROLLER CYLINDRICAL:COUNTERSHAFT,FRONT		EA	1
						0704-TRANSMISSION TOP COVER ASSEMBLY			
4	1	PAFZZ	2520-00-347-4591	7521056	19207	COVER ASSEMBLY:GEAR SHIFT:TRAMSMISSION		EA	1
4	2	PAOZZ	5355-00-962-3018	6184279	19207	KNOB;GEAR SHIFT HAND LEVER		EA	1
4	3	PAOZZ	2520-00-692-6072	7521062	19207	BOOT,DUST AND MOISTURE:GEAR SHIFT HAND LEVER		EA	1
4	4	PAFZZ	2520-00-930-3138	10937790	19207	LEVER:HAND GEAR SHIFT		EA	1
4	5	PAFZZ	5310-00-209-0965	MS35338-47	96906	WASHER,LOCK:HAND LEVER CLAMP		EA	1
4	6	PAFZZ	5310-00-880-7745	MS51968-11	96906	NUT,PLAIN HEXAGON,HEAND LEVER CLAMP		EA	1
4	7	PAFZZ	5310-00-692-6101	7539146	19207	PIN,STRAIGHT HEADED,PIVOT,GEAR SHAFT		EA	2
4	8	PAFZZ	2520-00-627-5978	7529023	19207	COVER:TRANSMISSION		EA	1
4	9	PAFZZ	5330-00-752-7750	MS35769-4	96906	GASKET:SHIFTER SHAFT,INTERLOCK ATTCAHMENT		EA	4
4	10	PAFZZ	5310-00-984-3807	MS51922-13	96906	NUT:SHIFTER SHAFT ATTACHING		EA	2
4	11	PAFZZ	5306-00-021-4077	8743049	19207	SCREW,ASSEMBLED LOCKWASHER:TRANSMISSION COVER		EA	8
4	12	PAFZZ	5330-00-454-0338	MS51915-21-1	96906	SEAL,PLAIN ENCASED:SHIFTER SHAFT		EA	1

0704-TRANSMISSION TOP COVER ASSEMBLY-CONTINUED

'ZZ	2520-00-752-1094	7521094	19207	BRACKET,SHIFTER SHAFT:REVERSE GEAR	EA	1
'ZZ	5305-00-752-1058	7521058	19207	SCREW:LOCK,SHIFTER FORK	EA	4
'ZZ	2520-00-752-1060	7521060	19207	FORK:SHIFTER,REVERSE GEAR	EA	1
'ZZ	2520-00-752-1097	7521097	19207	SHAFT,STRAIGHT:REVERSE GEAR SHIFTER	EA	1
'ZZ	9605-00-248-9850	MS20995F47	96906	LOCKWIRE:SETSCREW,SHIFTER FORK	EA	V
'ZZ	5360-00-347-4563	7521100	19207	SPRING,HELICAL,COMPRESSION:SHIFTER SHAFT DETENT BALL	EA	3
'ZZ	3110-00-142-6040	MS19061-11	96906	BALL,BEARING:SHIFTER SHAFT DETENTS	EA	3
'ZZ	2520-00-752-1093	7521093	19207	LEVER:HAND GEAR SHIFT EXTENSION	EA	1
'ZZ	2520-00-752-1096	7521096	19207	SEAT,HELICAL,COMPRESSION;GEAR SHIFT LEVER SUPPORT SPRING	EA	1
'ZZ	5360-00-692-6075	7521102	19207	SPRING,HELICAL,COMPRESSION:SUPPORT SHIFTER LEVER	EA	1
'ZZ	5365-00-699-8459	7521095	19207	RING,RETAINING;SHIFT LEVER SUPPORT SPRING	EA	1
'ZZ	2520-00-933-3112	10937974	19207	PLATE,INTERLOCK;SHIFTER SHAFT	EA	1
'ZZ	5360-00-692-6074	7521101	19207	SPRING,HELICAL,COMPRESSION:SHIFTER SHAFT INTERLOCK	EA	2
'ZZ	5310-00-167-0820	48488	62983	WASHER,FLAT:SHIFTER SHAFT INTERLOCK	EA	2
'ZZ	5340-00-050-1593	MS35648-5	96906	PLUG,EXPANSION:SHIFTER SHAFT	EA	3
'ZZ	2520-00-752-1099	7521099	19207	SHAFT,SHIFTER:SECOND,THIRD,FOURTH AND FIFTH SPEED GEARS	EA	2
'ZZ	2520-00-752-1057	7521057	19207	FORK,SHIFTER:FOURTH AND FIFTH GEARS	EA	1
'ZZ	2520-00-752-1059	7521059	19207	FORK,SHIFTER:SECOND AND THIRD GEARS	EA	1
'ZZ	5306-00-752-1055	7521055	19207	BOLT,SHOULDER:TOP COVER ASSEMBLY	EA	2
'ZZ	5305-00-710-4195	MS90726-90	96906	SCREW:SHIFT LEVER CLAMP	EA	1

GROUP 0800-TRANSFER
0801-POWER TRANSFER

'ZZ	5310-00-241-6664	MS51943-44	96906	NUT,SELF-LOCKING,HEXAGON;SUPPORT STUD, TRANSFER ASSEMBLY	EA	10
'ZZ	5307-00-695-7213	8331791	19207	STUD,PLAIN:SUPPORT,TRANSFER ASSEMBLY	EA	10
'HH	2520-00-001-7855	11609226	19207	TRANSFER,TRANSMISSION;ASSEMBLY WITH FLANGE E (ROCKWELL-STANDARD MODEL T-136-21)	EA	1
'HH	2520-00-089-8287	11609224	19207	TRANSFER,TRANSMISSION;ASSEMBLY WITH FLANGE F (W-AIR SHIFT) (ROCKWELL STANDARD MODEL,T-136-27)		
IZZ		7521363-1	19207	CASE WITH COVER ASSEMBLY:TRANSFER TRANSMISSION	EA	1
'ZZ	5305-00-071-1788	MS90728-87	96906	SCREW,CAP,HEXAGON HEAD:REAR BEARING RETAINER, OUTPUT SHAFT	EA	6
IZZ	5310-00-209-0965	MS35338-47	96906	WASHER,LOCK:TRANSFER CASE FRONT COVER(39), TOP COVER(4),BEARING RETAINER CAP(6), BEARING COVER(6)	EA	55
IZZ	5305-00-071-2056	MS90728-90	96906	SCREW,CAP,HEXAGON HEAD:BEARING COVER E COUNTERSHAFT	EA	4

10

| (1) ILLUSTRATION | | (2) | (3) | (4) | (5) | (6) | (7) | (8) QTY. |
(a) FIG. NO.	(b) ITEM NO.	SMR CODE	NATIONAL STOCK NUMBER	PART NUMBER	FSCM	DESCRIPTION / USABLE ON CODE	U/M	INC. IN UNIT
						0801-POWER TRANSFER-CONTINUED		
6	5	PAHZZ	5305-00-165-8250	11609328-1	19207	SCREW,CAP,HEXAGON HEAD:BEARING COVER, COUNTERSHAFT	EA	1
6	6	PAHZZ	5305-00-071-2055	MS90728-89	96906	SCREW,CAP,HEXAGON HEAD;BEARING COVER, COUNTERSHAFT	EA	2
6	7	PAHZZ	2520-00-752-1439	7521439	19207	COVER,ACCESS:REAR BEARING,COUNTERSHAFT	EA	1
6	8	PAHZZ	2520-00-752-1441	7521441	19207	SHIM:REAR BEARING COVER (PART OF SET 5330-00-752-1437)	EA	1
6	9	PAHZZ	2520-00-752-1446	7521446	19207	SHIM:REAR BEARING COVER PART OF SET 5330-00-752-1437)	EA	1
6	10	PAHZZ	5305-00-071-1785	MS90728-84	96906	SCREW,CAP,HEXAGON HEAD:COVER BEARING, INPUT SHAFT	EA	6
6	11	PAHZZ	2520-00-752-1585	7521585	19207	COVER:REAR BEARING,INPUT SHAFT	EA	1
6	12	PAHZZ	5330-00-585-7502	8344200-1	19207	GASKET:REAR BEARING COVER,INPUT SHAFT (PART OF SET 5330-00-752-1437)	EA	1
6	13	PAHZZ	5340-00-752-1372	7521372	19207	PLUG EXPANSION:SHIFTER SHAFT	EA	1
6	14	PAHZZ	4820-00-808-7442	11621193	19207	VALVE,DIAPHRAM:HOUSING	EA	1
6	15	XAHZZ		11621147	19207	HOUSING:TRANSFER CASE AND COVER ASSEMBLY	EA	1
6	16	PAFZZ	5330-00-311-7774	7971280	19207	GASKET:TOP ACCESS COVER (PART OF SET 5330-00-752-1437)	EA	1
6	17	PAFZZ	2520-00-692-5753	7735641	19207	COVER:TRANSFER CASE,TOP ACCESS	EA	1
6	18	PAOZZ	4820-00-726-4719	5196397	19207	VALVE,AIR PRESSURE,ASSEMBLY:BREATHER TRANSFER CASE	EA	1
6	19	PAFZZ	5305-00-543-4372	MS90728-58	96906	SCREW,CAP,HEXAGON HEAD:TOP COVER	EA	4
6	20	KFHZZ		7521364	19207	GASKET:TRANSFER CASE HOUSING (PART OF SET 5330-00-752-1437)	EA	1
6	21	PAHZZ	2520-00-781-6264	8675959	19207	COVER,TRANSFER,TRANSMISSION:FRONT	EA	1
6	22	PAHZZ	5305-00-071-1789	MS90728-88	96906	SCREW,CAP,HEXAGON HEAD:TRANSFER COVER,FRONT	EA	22
6	23	PAHZZ		10938287	19207	SEAL,PLAIN,ENCASED:SHIFTER SHAFT	EA	1
6	24	PAHZZ	5305-00-139-4620	11609328-2	19207	SCREW,CAP,HEXAGON HEAD:FRONT COVER	EA	12
6	25	PAOZZ	4730-00-737-5249	7375249	19207	PLUG,PIPE,MAGNETIC:HOUSING	EA	1
6	26	PAHZZ	5310-00-880-7745	MS51968-11	96906	NUT,PLAIN,HEXAGON:HOUSING FRONT COVER	EA	21
6	27	PAHZZ	5315-00-699-8460	7521237	19207	PIN,STRAIGHT,HEADLESS:BEARING RETAINER, REAR INPUT SHAFT	EA	1
6	28	PAOZZ	4730-00-044-4655	444655	19207	PLUG,PIPE:HOUSING	EA	1
6	29	KFHZZ	5365-00-752-1277	7521277	19207	SHIM:REAR BEARING RETAINING (PART OF SET 5330-00-752-1437)	EA	V
6	30	PAHZZ	5330-00-143-8666	7521241	19207	SEAL,PLAIN,ENCASED,OUTPUT SHAFT	EA	1
6	31	PAHZZ	2520-00-024-6540	7521275	19207	CAP,REAR,BEARING:RETAINER	EA	1
7	1	KFHZZ		7521230	19207	GASKET,COVER:FRONT PUTPUT DRIVE CASE (PART OF SET 5330-00-752-1437)	EA	1
7	2	XAHZZ		10948109	19207	CASE FRONT DRIVE OUTPUT:TRANSFER CASE F	EA	1
7	3	PAOZZ	4730-00-737-5248	8757704	19207	PLUG.PIPE:FILLER,HOUSING,OUTPUT SHAFT	EA	1
7	4	XAHZZ		8757713	19207	CASE FRONT DRIVE,OUTPUT E	EA	1

(1)		(2)	(3)	(4)	(5)	(6)	(7)	(8)
ILLUSTRATION								QTY.
(a)	(b)	SMR	NATIONAL	PART	FSCM	DESCRIPTION	U/M	INC.
FIG.	ITEM	CODE	STOCK	NUMBER				IN
NO.	NO.		NUMBER			USABLE ON CODE		UNIT

						0801-POWER TRANSFER-CONTINUED		
7	5	PAHZZ	5330-00-579-8156	MS28775-212	96906	SEAL:SHIFTER SHAFT E	EA	1
7	6	PAHZZ	5330-00-285-5121	5164515	19207	WASHER,NON-METALLIC:SHIFTER SHAFT E	EA	1
7	7	PAHZZ	5330-00-353-2465	7702649	19207	RETAINER,PACKING:SHIFTER SHAFT E	EA	1
7	8	KFHZZ		7521574	19207	GASKET:FRONT BEARING RETAINER	EA	1
						(PART OF SET 5330-00-752-1437)		
7	9	PAHZZ	2520-00-692-6091	7521573	19207	RETAINER,TRANSFER BEARING ASSEMBLY:FRONT BEARING,	EA	1
						INPUT SHAFT		
7	10	PAHZZ	5330-00-143-8666	7521241	19207	SEAL,OIL,FRONT BEARING,INPUT SHAFT	EA	1
7	11	PAHZZ	2520-00-781-6265	7700135	19207	RETAINER,TRANSFER BEARING:FRONT BEARING,	EA	1
						OUTPUT SHAFT		
7	12	PAHZZ	2520-00-752-1276	7521276	19207	FLANGE,COMPANION W/DEFLECTOR ASSEMBLY:INPUT	EA	2
						SHAFT(1) OUTPUT SHAFT(1)		
7	13	PAHZZ	2520-00-692-6080	7521242	19207	DEFLECTOR:DIRT AND DUST COMPANION FLANGE	EA	1
7	14	PAHZZ	2520-00-134-5124	8757674	19207	FLANGE,DIFFERENTIAL:COMPANION FLANGE	EA	1
7	15	PAHZZ	5315-00-013-7214	MS24665-359	96906	PIN,COTTER:NUT RETAINING,INPUT AND OUTPUT SHAFTS	EA	2
7	16	PAHZZ	5310-00-752-1234	7521234	19207	NUT,SLOTTED HEXAGON:FLANGE INPUT AND OUTPUT SHAFTS	EA	2
7	17	PAHZZ	5310-00-752-1235	7521235	19207	WASHER,FLAT:FLANGE	EA	2
7	18	PAHZZ	5305-00-071-1788	MS90728-87	96906	SCREW,CAP,HEXAGON:RETAINER AND GEAR COVER	EA	15
7	19	PAHZZ	5310-00-209-0965	MS35338-47	96906	WASHER,LOCK:BEARING RETAINER AND GEAR COVER	EA	19
8	1	PAHZZ	5365-00-752-1584	7521584	96906	RING,RETAINING:REAR BERAING,INPUT SHAFT	EA	1
8	2	PAHZZ		713925	21450	BEARING,BALL,ANNULAR:REAR,INPUT SHAFT,TRANSFER	EA	1
						GEARS		
8	3	PAHZZ	3120-00-752-1583	7521583	19207	BEARING WASHER,THRUST:GEAR,HELICAL,INPUT SHAFT	EA	2
8	4	PAHZZ	3020-00-786-0211	7363017	19207	GEAR,HELICAL:TRANSFER INPUT SHAFT	EA	1
8	5	PAHZZ	3110-00-554-3078	700603	21450	BEARING,BALL,ANNULAR:TRANSFER GEARS,INPUT SHAFT	EA	4
8	6	PAHZZ	5365-00-752-1575	7521575	19207	SPACER,SLEEVE:TRANSFER GEAR BEARINGS,INPUT SHAFT	EA	1
8	7	PAHZZ	3040-00-808-7441	11621145	19207	SHAFT,SHOULDERED:INPUT TRANSFER GEARS	EA	1
8	8	PAHZZ	2520-00-752-1385	7521385	19207	SHAFT:SHIFTER,LOW RANGE	EA	1
8	9	PAHZZ	3110-00-100-6156	MS19059-55	96906	BALL,BEARING:TRANSFER GEARS,SHIFTER SHAFT	EA	1
8	10	PAHZZ	2520-00-514-3782	5143782	19207	PLUNGER,SHIFTER SHAFT:TRANSFER GEARS	EA	1
8	11	PAHZZ	5360-00-692-6089	7521387	19207	SPRING,HELICAL COMPRESSION:SHIFTER SHAFT	EA	1
8	12	PAHZZ	5305-00-695-7174	7521390	19207	SETSCREWA:SHIFTER FORK	EA	1
8	13	PAHZZ	2520-00-786-0209	7363018	19207	FORK,SHIFTER SHAFT:TRANSFER GEARS	EA	1
8	14	PAHZZ	2520-00-752-1581	7521581	19207	SYNCHRONIZER ASSEMBLY:GEAR SHIFTING INPUT SHAFT	EA	1
8	15	PAHZZ	3120-00-753-9011	7539011	19207	SPACER,WASHER,THRUST:GEAR,HELICAL	EA	1
8	16	PAHZZ	3030-00-786-0213	7363016	19207	GEAR,HELICAL: INPUT SHAFT, LOW RANGE	EA	1
8	17	PAHZZ	3110-00-293-9305	700795	00000	BEARING,BALL,ANNULAR:FRONT INPUT SHAFT	EA	1
8	18	PAHZZ	3120-00-752-1487	7521487	19207	BEARING,WASHER,THRUST:FRONT,INPUT SHAFT	EA	1
8	19	PAHZZ	5305-00-071-1786	MS90728-85	96906	SCREW,CAP,HEXAGON HEAD:RETAINER DRIVE GEAR,COUN-	EA	4
						TERSHAFT		
8	20	PAHZZ	5310-00-832-7904	8327904	19207	LOCK ASSEMBLY:COUNTERSHAFT BEARING AND DRIVE GEAR	EA	2
						RETAINER		
8	21	PAHZZ	2520-00-752-1373	7521373	19207	PLATE:RETAINER,BEARING AND COUNTERSHAFT	EA	2

(1)		(2)	(3)	(4)	(5)	(6)	(7)	(8)
ILLUSTRATION								QTY.
(a)	(b)	SMR	NATIONAL	PART	FSCM	DESCRIPTION	U/M	INC.
FIG.	ITEM	CODE	STOCK	NUMBER				IN
NO.	NO.		NUMBER			USABLE ON CODE		UNIT

0801-POWER TRANSFER-CONTINUED

8	22	PAHZZ	3020-00-303-5106	7521438	19207	GEAR,HELICAL;DRIVE,FRONT,COUNTERSHAFT E	EA	1
8	23	PAHZZ	5365-00-752-1374	7521374	19207	RING,RETAINING:FRONT BEARING CUP,COUNTERSHAFT E	EA	1
8	24	PAFZZ	3110-00-013-7743	10948082	19207	BEARING,CUP,CONE AND ROLLER,TAPERED:	EA	2
						FRONT AND REAR,COUNTERSHAFT		
8	25	PAFZZ	3020-00-312-8348	7521376	19207	GEAR,HELICAL:LOW RANGE,COUNTERSHAFT	EA	1
8	26	PAFZZ	5315-00-850-7038	MS35756-105	96906	KEY,WOODRUFF:COUNTERSHAFT	EA	2
8	27	PAFZZ	2520-00-752-1340	7521340	19207	COUNTERSHAFT:TRANSFER E	EA	1
8	28	PAFZZ	3040-00-808-7448	11621191	19207	SHAFT SHOULDERED:TRANFER COUNTERSHAFT F	EA	1
8	NI	PAFZZ	3020-00-692-6084	7521375	19207	GEAR,HELICAL:HIGH RANGE COUNTERSHAFT	EA	1
8	NI	PAHZZ	5365-00-808-7423	10948103	19207	SPACER,SLEEVE:COUNTERSHAFT STEEL TUBE F	EA	1
8	NI	PAHZZ		10924755	19207	RETAINER:COUNTERSHAFT F	EA	1
9	1	PAHZZ	3110-00-165-6329	10948080	19207	BEARING ASSEMBLY ROLLER:REAR OUTPUT	EA	1
9	2	PAHZZ	3020-00-312-8350	7521240	19207	GEAR,HELICAL:DRIVEN,OUTPUT SHAFT,REAR	EA	1
9	3	PAHZZ	5315-00-850-7038	MS35756-105	96906	KEY,WOODRUFF:OUTPUT SHAFT,REAR	EA	1
9	4	PAHZZ	2520-00-301-7734	8376462	19207	SHAFT,SHOULDERED,OUTPUT SHAFT,REAR E	EA	1
9	5	PAHZZ	3110-00-101-6480	712663	00000	BEARING,ROLLER:REAR OUTPUT SHAFT FRONT	EA	1
9	6	PAHZZ	5365-00-696-0268	7521434	19207	RING,RETAINING:FRONT BEARING CUP,	EA	1
						REAR OUTPUT SHAFT		
9	7	PAHZZ	2520-00-737-2011	7372011	19207	SHAFT,OUTPUT,FRONT E	EA	1
9	8	PAHZZ	5315-00-281-7939	8329891	19207	KEY,WOODRUFF:OUTPUT SHAFT,FRONT E	EA	1
9	9	PAHZZ	2520-00-737-2012	7372012	19207	GEAR,TRANSMISSION:DRIVEN,OUTPUT SHAFT,FRONT E	EA	1
9	10	PAHZZ	2520-00-737-2010	7372010	19207	COLLAR,TRANSMISSION:CLUTCH,SHIFT E	EA	1
						OUTPUT SHAFT		
9	11	PAHZZ	2520-00-737-1995	7371995	19207	SHIFTER FORK:DE-CLUTCH SHAFT E	EA	1
9	12	PAHZZ	5305-00-752-1391	7521391	19207	SETSCREW:FORK SHIFTER	EA	1
9	13	PAHZZ	2520-00-752-1341	7521341	19207	SHAFT:SHIFTER,REVERSE E	EA	1
9	14	PAHZZ	5360-00-737-2020	7372020	19207	SPRING,RETAINING:OIL SEAL AND BEARING E	EA	1
						FRONT OUTPUT SHAFT		
9	15	PAHZZ	2520-00-513-5811	8329892	19207	CLUTCH ASSEMBLY:SPRAG UNIT E	EA	1
9	16	PAHZZ		8329893	19207	WASHER,FLAT,INNER RACE,SPRAG UNIT E	EA	1
9	17	PAHZZ	5365-00-696-0267	7521433	19207	RING,RETAINING;OIL SEAL AND BEARING	EA	3
						FRONT OUTPUT SHAFT		
9	18	PAHZZ	3110-00-516-5491	ST202	31007	BEARING,BALL,ANNULAR:FRONT OUTPUT SHAFT	EA	1
9	219	PAHZZ	3120-00-752-1487	7521487	19207	BEARING,WASHER,THRUST:FRONT OUTPUT SHAFT	EA	1
10	1	PAHZZ	5305-00-071-2235	MS90725-16	96906	SCREW,CAP,HEXAGON HEAD:AIR CYLINDER COVER F	EA	4
10	2	PAHZZ	5310-00-145-2085	8327009	19207	WASHER,KEY:AIR CYLINDER COVER SCREW F	EA	4
10	3	PAHZZ	2520-00-808-7445	10948042	19207	COVER,AIR CYLINDER:SHIFTER SHAFT F	EA	1
10	4	PAHZZ	5365-00-832-7774	10913209	19207	GASKET:AIR CYLINDER TUBE F	EA	1
10	5	PAHZZ	3120-00-832-7334	8758292-2	19207	BUSHING,SLEEVE:AIR CYLINDER F	EA	1
10	6	PAHZZ	5310-00-225-6993	MS51922-33	96906	NUT,SELF-LOCKING,HEXAGON:PISTON RETAINING, F	EA	1
						AIR CYLINDER		
10	7	PAHZZ	5310-00-177-0973	11609221	19207	WASHER,FLAT;PISTON RETAINING F	EA	1
10	8	PAHZZ	2520-00-832-8236	10948043	19207	PISTON,TRANSFER:AIR CYLINDER F	EA	1

(1)		(2)	(3)	(4)	(5)	(6)	(7)	(8)
ILLUSTRATION								QTY.
(a)	(b)	SMR	NATIONAL	PART	FSCM	DESCRIPTION	U/M	INC.
FIG.	ITEM	CODE	STOCK	NUMBER				IN
NO.	NO.		NUMBER			USABLE ON CODE		UNIT

						0801-POWER TRANSFER-CONTINUED		
10	9	PAHZZ	2520-00-832-8235	8758294	19207	SEAL,TRANSFER:PISTON,AIR CYLINDER F	EA	1
10	10	PAHZZ	5310-00-527-4222	8758299	19207	WASHER,FLAT:SHAFT,AIR CYLINDER F	EA	1
10	11	PAHZZ	2520-00-832-8074	10948099	19207	SHAFT,SHIFTER:CLUTCH,AIR CYLINDER F	EA	1
10	12	PAHZZ	9505-00-248-9850	MS20995-47	96906	WIRE,SAFETY:SHIFTER FORK SETSCREW F	FT	V
10	13	PAHZZ	5305-00-752-1391	7521391	19207	SETSCREW:SHIFTER FORK	EA	1
10	14	PAHZZ	2520-00-832-8071	10948100	19207	SHIFTER FORK:CLUTCH SHAFT,AIR CYLINDER F	EA	1
10	15	PAHZZ	2520-00-808-7428	10948108	19207	SEAT,HELICAL SPRING:SHIFTER SHAFT AIR CYLINDER F	EA	2
10	16	PAHZZ	5360-00-832-8072	10945186	19207	SPRING,HELICAL COMPRESSION:SHIFTER SHAFT, F AIR CYLINDER	EA	1
10	17	PAHZZ	3110-00-165-6329	10948080	19207	BEARING,ROLLER,TAPERED:REAR OUTPUT SHAFT	EA	1
10	18	PAHZZ	5315-00-850-7038	MS35756-105	96906	KEY.WOODRUFF:REAR OUTPUT SHAFT	EA	1
10	19	PAHZZ	2520-00-832-7794	10937638	19207	SHAFT,REAR AXLE:REAR OUTPUT F	EA	1
10	20	PAHZZ	2520-00-832-8075	10948101	19207	SHAFT,AXLE,AUTOMOTIVE:FRONT OUTPUT F	EA	1
10	21	PAHZZ	3110-00-101-6480	712663	00000	BEARING,ROLLER,TAPERED:FRONT OUTPUT SHAFT	EA	1
10	22	PAHZZ	5305-00-165-8248	10937391	19207	SETSCREW:CLUTCH,DRIVING,OUTPUT SHAFT F	EA	1
10	23	PAHZZ	2520-00-832-8237	10948106	19207	CLUTCH,DRIVING,FRONT:OUTPUT SHAFT F	EA	1
10	24	PAHZZ	2520-00-808-7424	10948105	19207	CLUTCH,SLIDING,SLEEVE:OUTPUT SHAFT F	EA	1
10	25	PAHZZ	5365-00-696-0267	7521433	19207	RING,RETAINING:OIL SEAL AND BEARING	EA	3
10	26	PAHZZ	3110-00-516-5491	ST202	31007	BEARING,BALL,ANNULAR:FRONT OUTPUT SHAFT	EA	1
10	27	PAHZZ	3120-00-752-1487	7521487	19207	BEARING,WASHER,THRUST:FRONT OUTPUT SHAFT	EA	1
10	28	PAHZZ	5330-00-143-8666	7521241	19207	SEAL,PLAIN ENCASED:FRONT OUTPUT SHAFT	EA	1
		PAHZZ	5330-00-752-1437	7521437	19207	GASKET AND SHIM SET POWER TRANSFER	EA	1
10	29	PAHZZ	2520-00-394-9718	5704159	19207	KIT,TRANSFER:AIR CYLINDER	EA	1
6	8					SHIM:REAR BEARING COVER	EA	1
6	9					SHIM:REAR BEARING COVER	EA	1
6	12					GASKET:REAR BEARING COVER	EA	1
6	16					GASKET:TOP ACCESS COVER	EA	1
6	20					GASKET:HOUSING	EA	1
6	29					SHIM:REAR BERAING RETAINER	EA	1
7	1					GASKET:COVER,TRANSFER CASE	EA	1
7	8					GASKET:FRONT,BEARING RETAINER	EA	1
						GROUP 12 BRAKES		
						1201-PARKING BRAKE DRUM AND SHOES		
						AND RELATED PARTS		
11	1	PAOZZ	5306-00-752-1014	7521014	19207	BOLT,RIBBED NECK:PARKING BRAKE DRUM	EA	4
11	2	PAOZZ	2530-00-752-1490	7521490	19207	BRAKEDRUM:PARKING BRAKE	EA	1
11	3	XBFZZ	2530-00-622-3949	7397808	19207	BRACKET:BRAKE CABLE	EA	1
11	4	PAFZZ	5310-00-514-6674	MS35335-34	96906	WASHER,LOCK,EXTERNAL TOOTH:SHOE STOP BRACKET	EA	1
11	5	PAFZZ	5306-00-089-3837	8757599	19207	BOLT,MACHINE:BRAKE CABLE BRACKET	EA	1
11	6	PAOZZ	5305-00-071-1788	MS90728-87	96906	SCREW,CAP,HEXHEAD		
11	7	PAOZZ	5310-00-209-0965	MS35338-47	19207	WASHER,LOCK SPLIT:SHOE STOP BRACKET(1)	EA	1
11	8	XBOZZ	2530-00-752-9383	8757711	19207	SHIELD,GREASE:PARKING BRAKE DRUM	EA	1

| (1) ILLUSTRATION | | (2) | (3) | (4) | (5) | (6) | (7) | (8) QTY. |
(a) FIG. NO.	(b) ITEM NO.	SMR CODE	NATIONAL STOCK NUMBER	PART NUMBER	FSCM	DESCRIPTION USABLE ON CODE	U/M	INC. IN UNIT
						1201-PARKING BRAKE DRUM AND SHOES AND RELATED PARTS-CONTINUED		
11	9	PAFZZ	2520-00-752-1276	7521276	19207	FLANGE,COMPANION W/DEFLECTOR ASSEMBLY: TRANSFER REAR SHAFT	EA	1
11	10	XAFZZ		7521242	19207	DEFLECTOR,DIRT AND LIQUID:COMPANION FLANGE	EA	1
11	11	PAOZZ	2520-00-134-5124	8757674	19207	FLANGE,COMPANION:BRAKE DRUM	EA	1
11	12	PAOZZ	5310-00-584-5272	MS35338-10	96906	WASHER,LOCK:PARKING BRAKE DRUM	EA	4
11	13	PAOZZ	5310-00-732-0560	MS51968-14	96906	NUT,HEXAGON:PARKING BRAKE DRUM	EA	4
11	14	PAOZZ	5315-00-013-7214	MS34665-359	96906	PIN,COTTER,SPLIT:BRAKE DRUM FLANGE	EA	1
11	15	PAOZZ	5310-00-752-1234	7521234	19207	NUT,SLOTTED,HEX HEAD:BRAKE DRUM FLANGE	EA	1
11	16	PAOZZ	5310-00-752-1235	7521235	19207	WASHER,FLAT:BRAKE DRUM FLANGE	EA	1
11	17	PAOZZ	5360-00-693-0615	7521284	19207	SPRING,HELICAL,EXTENSION:BRAKE SHOE ADJUSTING ANGLE RETURN	EA	1
11	18	PAFZZ	5310-00-926-5916	MS51968-12	96906	NUT,PLAIN,HEX:BRAKESHOE ADJUSTING SPRING SCREW	EA	1
11	19	XBFZZ	2530-00-752-1488	7521488	19207	BRACKET,ANGLE:BRAKESHOE ADJUSTING	EA	1
11	20	PAFZZ	5305-00-139-4620	11609328-2	19207	SCREW,CAP,HEX HEAD:BRAKESHOE ADJUSTING RETURN SPRING	EA	1
11	21	PAFZZ	5305-00-165-9250	11609328-1	19207	SCREW:BRAKESHOE ADJUSTING ANGLE	EA	1
11	22	PAFZZ	5305-00-071-1789	MS90728-87	96906	SCREW:BRAKESHOE ADJUSTING ANGLE	EA	1
11	23	PAOFF	2530-00-693-0680	8380559	19207	BRAKESHOE: ASSEMBLY:OUTER PARKING,W/LINING	EA	1
11	24	PAFZZ	5320-00-443-5065	10896748	19207	RIVET,TUBULAR:BRAKESHOE LINING	EA	8
11	25	PAFZZ	2530-00-289-7207	7368681	19207	LINING,FRICTION:BRAKESHOE,PARKING,INNER	EA	1
11	26	XAFZZ		8757706	19207	SHOE,BRAKE:OUTER,PARKING	EA	1
11	27	PAOZZ	4730-00-050-4208	MS15003-1	96906	FITTING,LUBRICATION:INNER AND OUTER BRAKESHOE	EA	2
11	28	PAOZZ	5310-00-333-7519	7064468	19207	WASHER,SLOTTED:BRAKE PIN	EA	2
11	29	PAOFF	2530-00-693-0679	8380558	19207	BRAKESHOE ASSEMBLY:INNER PARKING,W/LINING	EA	1
11	30	XAFZZ		8757707	19207	SHOE,BRAKE:INNER,PARKING	EA	1
11	31	PAFZZ	2530-00-736-8682	7368682	19207	LINING,FRICTION:BRAKESHOE,PARKING,INNER	EA	1
11	32	PAFZZ	5320-00-443-5065	10896748	19207	RIVET,TUBULAR:BRAKESHOE LINING	EA	8
11	33	PAOZZ	5315-00-316-0992	7064470	19207	PIN,SHOULDER,HEADLESS:INNER BRAKESHOE	EA	1
11	34	PAOZZ	5360-00-342-7071	7373243	19207	SPRING,HELICAL,EXTENSION:LEVER RETURN	EA	1
11	35	PAOFF	2530-00-736-8683	7368683	19207	LEVER AND PIN ASSEMBLY:PARKING BRAKE ACTUATING	EA	1
11	36	XAFZZ		8757709	19207	LEVER:PARKING BRAKE	EA	1
11	37	PAFZZ	5315-00-312-0831	7521448	19207	PIN,GROOVED,HEADLESS:PARKING BRAKE LEVER	EA	2
11	38	PAOZZ	5365-00-664-2779	MS16633-1098	96906	WASHER,RETAINING LEVER PIN	EA	2
11	39	PAOZZ	5340-00-321-6375	7064466	19207	SPRING:STABILIZER,BRAKESHOE	EA	1
11	40	PAOZZ	5310-00-279-3314	7064467	19207	WASHER,FLAT:BRAKE SPACER TIE RING	EA	4
11	41	PAOZZ	5305-00-206-0932	7064469	19207	SCREW,SHOULDER:OUTER BRAKESHOE STABILIZER	EA	1
11	42	PAOZZ	5306-00-312-0845	7521283	9207	BOLT,MACHINE:ANCHOR,OUTER BRAKESHOE	EA	1
11	43	XBOZZ		8757705	19207	WASHER FLAT:BRAKESHOE ANCHOR BOLT	EA	1
11	44	PAOZZ	5310-00-753-9169	7539169	19207	NUT,PLAIN,HEX:BRAKESHOE ANCHOR BOLT	EA	1
11	45	PAOZZ	5310-00-637-9541	MS122036	96906	WASHER,LOCK,SPLIT:BRAKESHOE STABILIZER SCREW	EA	1
11	46	PAOZZ	5310-00-732-0558	MS51967-8	96906	NUT,PLAIN,HEX:BRAKESHOE STABILIZER SCREW	EA	1
11	47	PAFZZ	5310-00-209-0965	MS35338-47	96906	WASHER,LOCK:SPLIT HELICAL	EA	2

15

(1)		(2)	(3)	(4)	(5)	(6)	(7)	(8)
ILLUSTRATION								QTY.
(a)	(b)	SMR	NATIONAL	PART	FSCM	DESCRIPTION	U/M	INC.
FIG.	ITEM	CODE	STOCK	NUMBER				IN
NO.	NO.		NUMBER			USABLE ON CODE		UNIT

						GROUP 20-POWER TAKE-OFF		
						2004-POWER TAKE-OFF ASSEMBLY		
12	1	PAFHH	2520-00-706-1136	7061136	19207	POWER TAKE-OFF TRANSMISSION:WITH ACCESSORY DRIVE, C	EA	1
						SPICER MODEL WND-7-28 (P2)		
12	2	PAFHH	2520-00-706-1137	7061137	19207	POWER TAKE-OFF TRANSMISSION:SPICER MODEL D	EA	1
						WN-7-28- (P1)		
12	3	PAHZZ	5306-00-404-0075	8344155	19207	BOLT,MACHINE:SHIFTER FORK,REVERSE GEAR SHAFT	EA	1
12	4	PAHZZ	2520-00-706-1273	7061273	19207	SHIFTER FORK:SHAFT,REVERSE GEAR	EA	1
12	5	PAHZZ	3110-00-902-1686	713421	00000	BEARING,ROLLER,NEEDLE:REVERSE GEAR SHAFT,FRONT	EA	1
12	6	PAHZZ	3040-00-706-1159	7061159	19207	SHAFT ASSEMBLY POWER TAKE-OFF:REVERSE GEAR	EA	1
12	7	XAHZZ		8743044	19207	SHAFT:REVERSE GEAR	EA	1
12	8	PAHZZ	5315-00-058-8581	MS35756-109	96906	KEY,WOODRUFF:SHAFT,REVERSE GEAR	EA	1
12	9	PAHZZ	3120-00-737-3115	7373115	19207	BEARING,SLEEVE:REVERSE GEAR SHAFT	EA	1
12	10	PAHZZ	5365-00-706-1276	7061276	19207	RING,RETAINING:DRIVE CLUTCH GEAR	EA	1
12	11	PAHZZ	5306-00-225-8494	MS90725-31	96906	BOLT:SHIFTER SHAFT	EA	1
12	12	PAHZZ	5310-00-080-6004	MS27183-14	96906	WASHER:BOLT,SHIFTER SHAFT	EA	2
12	13	PAHZZ	2520-00-706-1278	7061278	19207	BOOT,BELLOWS,RUBBER:SHIFTER SHAFT	EA	2
12	14	PAHZZ	5330-00-737-6534	7376534	19207	CUP RETAINING:TO ATTACH SHIFTER SHAFT BOOT D	EA	2
12	14	PAHZZ	5305-00-737-6534	7376534	19207	CUP RETAINING:TO ATTACH SHIFTER SHAFT BOOT C	EA	3
12	15	PAHZZ	5330-00-298-3947	7061272	19207	SEAL,SHIFTER:FORK ROD D	EA	2
12	15	PAHZZ	5330-00-298-3947	7061272	19207	SEAL,SHIFTER:FORK ROD C	EA	3
12	16	PAHZZ	4730-00-012-5947	125947	21450	PLUG,HOUSING	EA	1
12	17	PAHZZ	2520-00-706-1160	7061160	19207	CASE,HOUSING:POWER TAKE-OFF	EA	1
12	18	PAHZZ	3110-00-887-5524	MS19061-15	96906	BALL,BEARING:LOCK,SHIFTER SHAFT	EA	1
12	19	PAHZZ	2520-00-522-6128	5226128	19207	SPRING,HELICAL COMPRESSION:SHIFTER SHAFT LOCKING	EA	1
12	20	PFHZZ	2520-00-104-4603	5226140	19207	RETAINER,POPPET:SHIFTER SHAFT LOCKING	EA	1
12	21	PAHZZ	2520-00-706-1281	7061281	19207	WASHER,THRUST:REVERSE SHAFT GEAR	EA	2
12	22	PAHZZ	2520-00-706-1275	7061275	19207	GEAR,REVERSE SHAFT:POWER TAKE-OFF	EA	1
12	23	PAHZZ	5315-00-816-1794	MS24665-285	96906	PIN,COTTER:LOCKING PIN	EA	1
12	24	PAHZZ	5315-00-735-1406	7351406	19207	PIN,STRAIGHT HEADED:REVERSE GEAR SHAFT	EA	1
12	25	PAHZZ	5305-00-144-1495	10910354	19207	SCREW:COVER,ACCESS	EA	6
12	26	PAHZZ	5310-00-616-7998	MS35335-31	96906	WASHER,LOCK:COVER ACCESS	EA	6
12	27	PAHZZ	2520-00-354-0771	5323473	19207	COVER,ACCESS:HOUSING CASE	EA	1
12	28	PAHZZ	5330-00-234-3317	8327322	19207	GASKET:CASE AND COVER	EA	2
12	29	PAHZZ	2520-00-706-1267	7061267	19207	SHAFT,POWER TAKE-OFF:SHIFTER	EA	1
12	30	PAHZZ	5310-00-975-2075	MS35691-21	96906	NUT,PLAIN,HEXAGON:SHIFTER SHAFT ROD END CONNECTOR	EA	1
12	31	PAHZZ	5340-00-753-8679	7538679	19207	CONNECTOR,ROD END:SHIFTER SHAFT	EA	1
12	32	PAHZZ	2520-00-706-1269	7061269	19207	CAP,OUTPUT SHAFT:REAR BEARING D	EA	1
12	33	PAHZZ	3110-00-183-6723	MS17131-35	96906	BEARING,ROLLER,NEEDLE:GEAR SHAFT,ACCESSORY DRIVR	EA	1
12	34	PAHZZ	5310-00-732-0559	MS51968-8	96906	NUT,PLAIN,HEXAGON:POWER TAKE-OFF MOUNTING	EA	6
12	35	PAHZZ	5310-00-637-9541	MS122036	96906	WASHER,LOCK:POWER TAKE-OFF MOUNTING	EA	6
12	36	PAHZZ	5315-00-013-7258	MS24665-497	96906	PIN,COTTER:HOUSING	EA	1
12	37	PAHZZ	5307-00-753-8668	7538668	19507	STUD,PLAIN:POWER TAKE-OFF MOUNTING	EA	6
12	38	PAHZZ	2520-00-706-1266	7061266	19207	SHAFT,SHOULDERED:INOUT GEAR CLUSTER	EA	1

FIG. NO.	ITEM NO.	SMR CODE	NATIONAL STOCK NUMBER	PART NUMBER	FSCM	DESCRIPTION / USABLE ON CODE	U/M	QTY. INC. IN UNIT
						2004-POWER TAKE-OFF ASSEMBLY-CONTINUED		
12	39	PAHZZ	2520-00-706-1265	7061265	19207	SPACER POWER TAKE-OFF:OUTPUT SHAFT	EA	1
12	40	PAHZZ	3110-00-516-5490	7539700	19207	BEARING,BALL,ANNULAR:OUTPUT SHAFT REAR	EA	1
12	41	PAHZZ	5365-00-508-4691	7538660	19207	RING,RETAINING:OUTPUT SHAFT GEAR	EA	1
12	42	PAHZZ	3120-00-753-8661	7538661	19207	WASHER,THRUST:OUTPUT SHAFT HELICAL GEAR	EA	1
12	43	PAHZZ	3020-00-347-4549	7538672	19207	GEAR,HELICAL:OUTPUT SHAFT	EA	1
12	44	PAHZZ	3020-00-318-0908	7538671	19207	GEAR,SPUR:OUTPUT SHAFT	EA	1
12	45	PAHZZ	2520-00-753-8673	7538673	19207	SHAFT,OUTPUT	EA	1
12	46	PAHZZ	5315-00-616-5527	MS35756-18	96906	KEY:OUTPUT SHAFT (1) ACCESSORY DRIVE SHAFT(1)	EA	2
12	47	PAHZZ	3110-00-156-4704	714248	00000	BEARING,BALL,ANNULAR:FRONT,OUTPUT SHAFT	EA	1
12	48	PAHZZ	5330-00-532-3606	5323606	19207	GASKET:OUTPUT SHAFT FRONT BEARING RETAINER	EA	1
12	50	PAHZZ	2540-00-626-2467	6262467	19207	CAP:OUTPUT SHAFT FRONT BEARING	EA	1
12	51	PAHZZ	5306-00-238-5605	878355	19207	SCREW,CAP,HEXAGON HEAD:OUTPUT SHAFT FRONT BEARING RETAINER(4),ACCESSORY DRIVE HOUSING(5)	EA	9
12	52	PAHZZ	5310-00-514-6674	MS35335-20	96906	WASHER.LOCK:OUTPUT SHAFT FRONT BEARING RETAINER(4), ACCESSORY DRIVE HOUSING(5)	EA	9
12	53	PAHZZ	3120-00-753-8656	7538656	19207	WASHER:INPUT SHAFT GEAR	EA	2
12	54	PAHZZ	2520-00-347-4571	7538674	19207	GEAR CLUSTER,HELICAL:INPUT SHAFT	EA	1
12	55	PAHZZ	3110-00-227-3074	7538658	19207	ROLLER ASSEMBLY:INPUT GEAR SHAFT D	EA	2
12	56	PAHZZ		7521058	19207	SCREW:SHIFTER FORK C	EA	1
12	57	PAHZZ	2520-00-706-1274	7061274	19207	SHIFTER FORK:ACCESSORY DRIVE C	EA	1
12	58	PAHZZ	2520-00-706-1268	7061268	19207	SHAFT,STRAIGHT:ACCESSORY DRIVE SHIFTER C	EA	1
12	59	PAHZZ	5330-00-706-1270	7061270	19207	GASKET:ACCESSORY DRIVE REAR OUTPUT COVER	EA	1
12	60	XAHZZ		7061194	19207	HOUSING:POWER TAKE-OFF ACCESSORY DRIVE C	EA	1
12	61	PAHZZ	5365-00-769-6481	7696481	19207	PLUG,MACHINE THREADED:ACCESSORY DRIVE HOUSING C	EA	1
12	62	PAHZZ	3110-00-753-9028	MS19059-53	96906	SPRING HELICAL COMPRESSION:ACCESSORY DRIVE C SHIFTER SHAFT	EA	1
12	63	PAHZZ	3110-00-100-6155	MS19059-51	96906	BALL,BEARING:ACCESSORY DRIVE C SHIFTER SHAFT	EA	1
12	64	PAHZZ	2520-00-706-1279	7061279	19207	BOOT,DUST AND MOISTURE:ACCESORY DRIVE C SHIFTER SHAFT	EA	1
12	65	PAHZZ	5330-00-559-8733	7061271	19207	SEAL,PLAIN ENCASED:DRIVE SHAFT, C ACCESSORY DRIVE	EA	1
12	66	PAHZZ	5365-00-706-1280	7061280	19207	RING,RETAINING:ACCESSORY DRIVE C SHAFT SEAL AND BEARING	EA	2
12	67	PAHZZ	5340-00-706-1277	7061277	19207	RING:BALL BEARING RETAINING,ACCESSORY DRIVE C	EA	1
12	68	PAHZZ	3110-00-144-8518	700078	00000	BEARING,BALL,ANNULAR:ACCESSORY C DRIVE SHIFTER SHAFT	EA	1
12	69	PAHZZ	2520-00-706-1211	7061211	19207	SHAFT,SHOULDERED:ACCESSORY DRIVE GEAR C	EA	1
12	49	PAHZZ	2520-00-706-1210	7061210	19207	SLEEVE,SLIDING:CLUTCH C	EA	1
13	1	PAFZZ	5307-00-350-5532	8328010	19207	STUD,PLAIN:ASSEMBLY MOUNTING	EA	2
13	2	PAFZZ	5310-00-685-3228	MS35333-43	96906	WASHER,LOCK ASSEMBLY MOUNTING	EA	2
13	3	PAFZZ	5310-00-926-5916	MS51968-12	96906	NUT,PLAIN HEXAGON:ASSEMBLY MOUNTING	EA	2
13	4	PAHZZ	5305-00-071-1788	MS90728-87	96906	SCREW,CAP,HEXAGON HEAD:ASSEMBLY MOUNTING	EA	4

(1) ILLUSTRATION		(2)	(3)	(4)	(5)	(6)		(7)	(8) QTY.
(a) FIG. NO.	(b) ITEM NO.	SMR CODE	NATIONAL STOCK NUMBER	PART NUMBER	FSCM	DESCRIPTION USABLE ON CODE		U/M	INC. IN UNIT
						2004-POWER TAKE-OFF ASSEMBLY-CONTINUED			
13	5	PAHZZ	5310-00-209-0965	MS35338-47	96906	WASHER,LOCK:ASSEMBLY MOUNTING		EA	4
13	6	PAFFF	2520-00-229-5673	11609228	19207	POWER TAKE-OFF,TRANSFER:ROCKWELL STANDARD MODEL P-136-C, WITH PUMP ASSEMBLY		EA	1
14	1	PAFZZ	4710-00-137-2019	11621138	19207	LINE,OIL:TRANSFER CASE TO OIL PUMP		EA	1
14	2	PAFZZ	4730-00-278-4596	8328015	19207	ADAPTER,STRAIGHT,PIPE-TO-TUBE:CONNECTOR OIL LINE TO PUMP		EA	1
14	4	PAFZZ	5330-00-522-8428	8344200-1	19207	GASKET:CARRIER TO TRANSFER MOUNTING		EA	1
14	5	PAOZZ	4730-00-639-9921	444595	21450	PLUG,PIPE:SLIDING CLUTCH FORK SETSCREW ACCESS		EA	1
14	6	PAHZZ	5306-00-262-9619	7412211	19207	BOLT,MACHINE:SHIFTER SHAFT LOCK ACCESS		EA	1
14	7	PAHZZ	2520-00-492-4348	10896746	19207	WASHER,SEAL:SHIFTER LOCK SPRING		EA	1
14	8	PAHZZ	5340-00-321-5710	6156769	19207	SPRING,HELICAL COMPRESSION:SHIFTING LEVER LOCK		EA	1
14	9	PAHZZ	3110-00-100-6159	MS19059-59	96906	BALL,BEARING:SHIFTING LEVER LOCK		EA	1
14	10	PAHZZ	5330-00-741-3286	7413286	19207	GASKET:GOVERNOR DRIVE HOLE		EA	1
14	11	PAHZZ	5365-00-741-3289	7413289	19207	PLUG,MACHUNE THREAD:GOVERNOR DRIVE HOLE		EA	1
14	12	XAFZZ		7413288	19207	LEVER:SHIFTER SHAFT		EA	1
14	13	PAFZZ	5310-00-227-1921	8328008	19207	WASHER,FLAT:SHIFTER SHAFT		EA	1
14	14	PAFZZ	5315-00-842-3044	MS24665-283	96906	PIN,COTTER:SHIFT LEVER		EA	2
14	15	PAFZZ	5310-00-165-8488	10896750	19207	NUT,PLAIN,SLOTTED,HEXAGON:SHIFT LEVER		EA	1
14	16	PAFZZ	5310-00-809-5998	MS27183-18	96906	WASHER,FLAT:SHIFT LEVER		EA	1
14	17	PAHZZ	5330-00-808-7417	11621135	19207	GASKET:OIL PUMP MOUNTING		EA	1
14	18	PAFZZ	5310-00-407-9566	MS35338-45	96906	WASHER,LOCK:OIL PUMP MOUNTING		EA	4
14	19	PAFZZ	5306-00-165-8251	11609558	19207	SCREW,HEXAGON HEAD:OIL PUMP MOUNTING		EA	3
14	20	PAFZZ	5306-00-226-4826	MS90728-33	96906	SCREW,CAP,HEXAGON HEAD:OIL PUMP MOUNTING3		EA	1
14	21	PAFZZ	2520-00-808-7379	11621137	19207	PUMP ASSEMBLY, OIL:TRANSFER POWER TAKE-OFF		EA	1
14	22	PAFZZ	5310-00-809-4085	MS27183-16	96906	WASHER,FLAT:CLAMP,OIL LINE		EA	1
14	24	PAFZZ	4730-00-127-4461	MS39202-6	96906	ELBOW PIPE-TO-TUBE OIL LINE TRANSFER CASE TO PUMP		EA	1
14	25	XAFZZ		7413265	19207	CLAMP:LINE,OIL PUMP		EA	1
14	26	PAFZZ	5325-00-741-2204	7412204	19207	GROMMET,RUBBER:OIL PUMP LINE		EA	1
15	1	PAHZZ	5310-00-426-0809	10896749	19207	NUT,HEXAGON,SLOTED:RETAINER OUTPUT SHAFT		EA	1
15	2	PAHZZ	5315-00-298-1481	MS24665-357	96906	PIN,COTTER:RETAINER NUT, OUTPUT SHAFT		EA	1
15	3	PAHZZ	5330-00-231-0285	8757686	19207	RETAINER AND SEAL ASSEMBLY:SHAFT BEARING,OUTPUT		EA	1
15	3.1	PAHZZ	5310-00-919-8882	10938286	19207	SEAL,OIL:POWER TAKE-OFF SHAFT		EA	1
15	3.2	PAHZZ		7735601	19207	RETAINER ASSEMBLY:BEARING		EA	1
15	4	PAHZZ	5306-00-225-8497	MS90725-32	96906	SCREW,CAP,HEXAGON HEAD:RETAINER,BEARING AND SEAL		EA	4
15	5	PAHZZ	5310-00-407-9566	MS35338-45	19207	WASHER,LOCK:RETAINER,BEARING AND SEAL		EA	4
15	6	PAHZZ	5330-00-615-6756	6156756	19207	GASKET:BEARING CAP,OUTPUT SHAFT		EA	1
15	7	PAHZZ	3110-00-100-0799	137251	24617	CUP,TAPERD ROLLER BEARING:DRIVE SHAFT		EA	2
15	8	PAHZZ	3110-00-142-4364	148346	24617	CONE AND ROLLERS,TAPERED:DRIVE SHAFT		EA	2
15	9	PAHZZ	3020-00-741-3287	7413287	19207	GEAR,WORM:DRIVE GOVERNOR		EA	1
15	10	PAHZZ	5315-00-032-1872	5168861	19207	KEY,MACHINE:POWER TAKE-OFF SHAFT		EA	1
15	11	PAHZZ	5315-00-616-5514	MS35756-6	96906	KEY,WOODRUFF:POWER TAKE-OFF SHAFT		EA	1
15	12	PAHZZ	2520-00-355-7809	11621144	19207	SHAFT,POWER TAKE-OFF:DRIVE		EA	1

| (1) ILLUSTRATION | | (2) | (3) | (4) | (5) | (6) | (7) | (8) QTY. |
(a) FIG. NO.	(b) ITEM NO.	SMR CODE	NATIONAL STOCK NUMBER	PART NUMBER	FSCM	DESCRIPTION USABLE ON CODE	U/M	INC. IN UNIT
						2004-POWER TAKE-OFF ASSEMBLY-CONTINUED		
15	13	PAFZZ	9505-00-248-9850	MS20995F47	96906	LOCKWIRE:SETSCREW,SHIFTER FORK(1),SETSCREW SHIFTER FORK (1)	FT	V
15	14	PAHZZ	5305-00-270-7328	8328007	19207	SETSCREW:LOCK,SHIFTER FORK	EA	1
15	15	PAHZZ	2520-00-734-9606	7349606	19207	SHIFTER FORK:CLUTCH	EA	1
15	16	PAHZZ	5305-00-490-1850	10896745	19207	SETSCREW:DRIVE CLUTCH	EA	1
15	17	PAHZZ	3040-00-077-1878	11621139	19207	COUPLING,SHAFT,RIGID:TRANSFER POWER TAKE-OFF CLUTCH	EA	1
15	18	PAHZZ	2520-00-808-7385	11621140	19207	CLUTCH,SLIDING SLEEVE:TRANSFER POWER TAKE-OFF	EA	1
15	19	PAHZZ	3120-00-516-7516	5284913	19207	BEARING SLEEVE:GOVERNOR SHAFT	EA	1
15	20	PAHZZ	5330-00-585-1064	MS29561-12	96906	PACKING,PREFORMED: "O" RING,SHIFTER SHAFT	EA	1
12	21	PAHZZ	2520-00-030-7262	7349680	19207	SHAFT,SHIFTER:SLIDING CLUTCH	EA	1
15	22	PAHZZ	5365-00-769-9003	7699003	19207	SHIM SET:POWER TAKE-OFF END PLAY	EA	1
				6143903	19207	SHIM:0=.003	EA	4
				5186593	19207	SHIM:.005	EA	2
				5186592	19207	SHIM:.010	EA	1
15	23	XAHZZ		11621143	19207	CARRIER ASSEMBLY	EA	1

SPECIAL TOOLS LIST

(1) ILLUSTRATION		(2)	(3)	(4)	(5)	(6)	(7)	(8) QTY.
(a) FIG. NO.	(b) ITEM NO.	SMR CODE	NATIONAL STOCK NUMBER	PART NUMBER	FSCM	DESCRIPTION USABLE ON CODE	U/M	INC. IN UNIT
						GROUP 26-TOOLS AND TEST EQUIPMENT		
						2604-SPECIAL TOOLS		
16	1	PEHHH	5340-00-610-0919	7010362	19207	BRACKET,ANGLE:LEFT SIDE TRANSFER CASE TO STAND	EA	1
16	2	PEHHH	5340-00-610-0920	7010363	19207	BRACKET,ANGLE:RIGHT SIDE TRANSFER CASE TO STAND	EA	1
16	3	PEHHH	4910-00-694-4777	8708279	19207	FIXTURE,TRANSFER CASE:REMOVING AND/OR REPLACING TRANSFER CASE	EA	1
16	4	PEHHH	5120-00-338-6721	8708724	19207	PULLER KIT,MECHANICAL:TRANSMISSION AND TRANSFER	EA	1
16	5	PEHHH	5120-00-708-3241	7083241	19207	HANDLE,REMOVER AND REPLACER:TRANSFER (USED WITH REMOVER-REPLACER 7083241)	EA	1
16	6	PEHHH	5120-00-708-3247	7083247	19207	REMOVER AND REPLACER:TRANSFER IDLER SHAFT FRONT BEARING CUP	EA	1
16	7	PEHHH	5120-00-708-3254	7083254	19207	ADAPTER,MECHANICAL PULLER:TRANSMISSION IDLER GEAR SHAFT "NON-ILLUSTRATED"	EA	1
		PEFHH	4910-00-795-0356	7950356	19207	TOOL KIT:SPECAIL POWER TRAIN REBUILD	EA	1

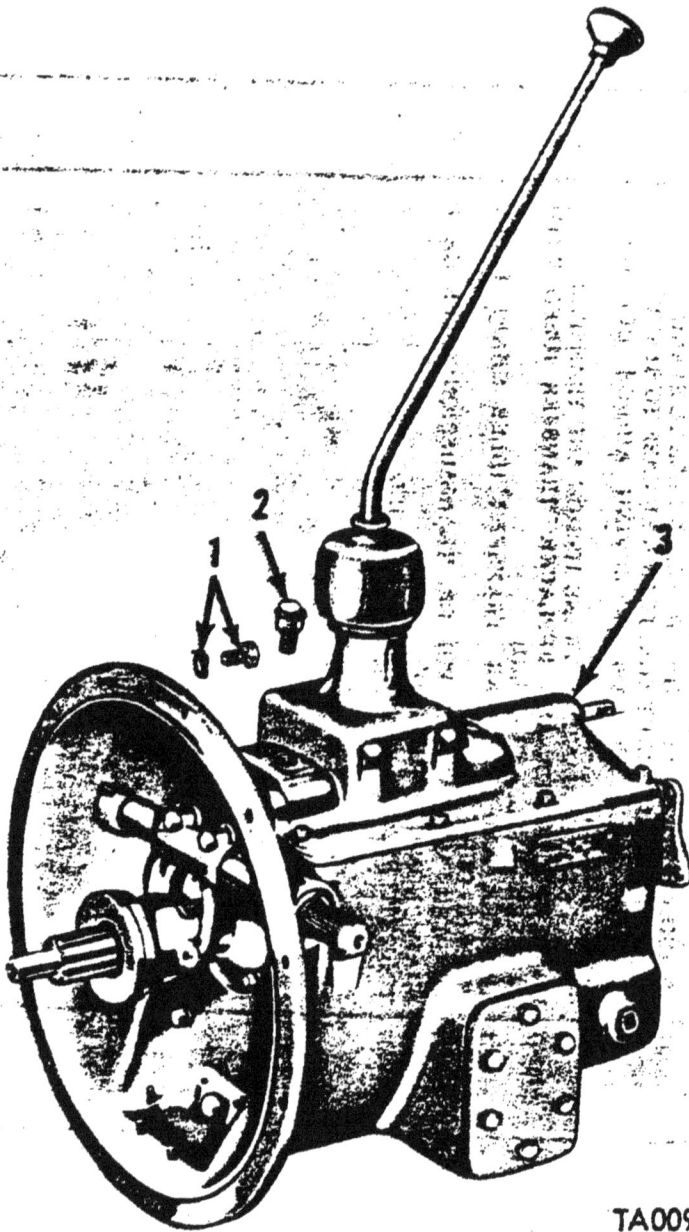

TA009623

Figure 1. Transmission assembly

Figure 2. Transmission case, clutch housing and attaching parts.

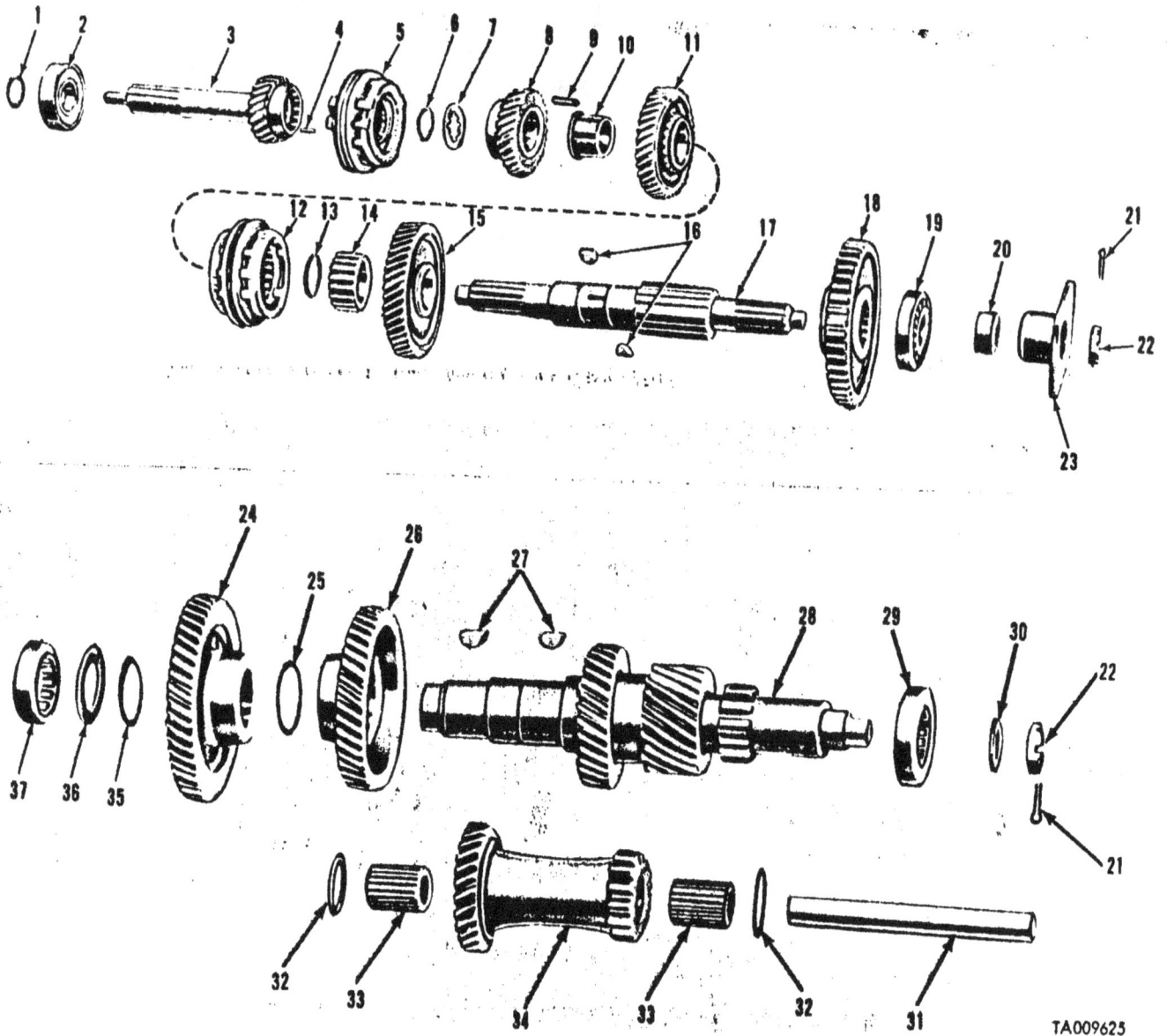

Figure 3. Transmission input and output shafts, countershafts and idler shafts with gears.

TA009625

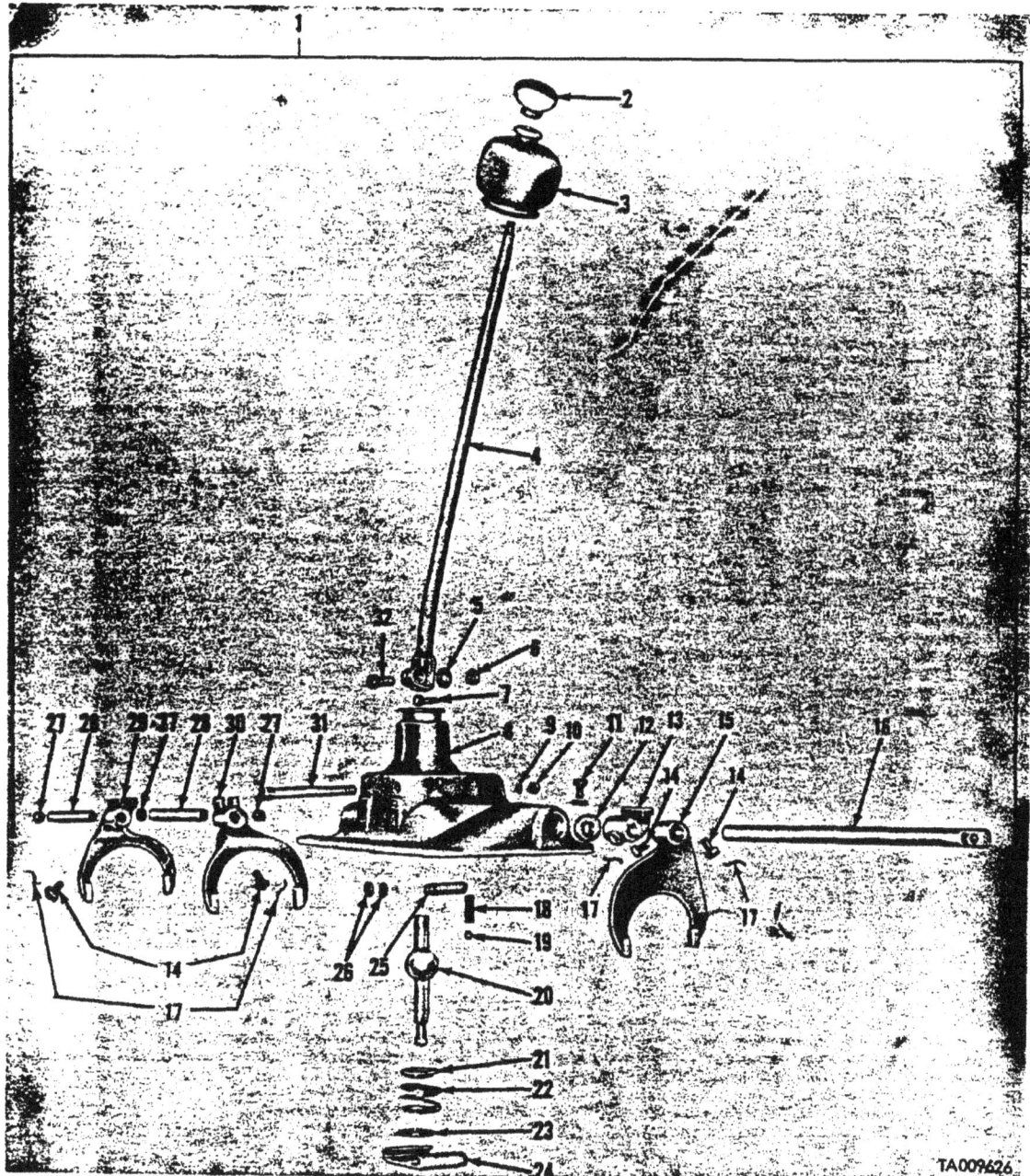

Figure 4. Transmission cover assembly.

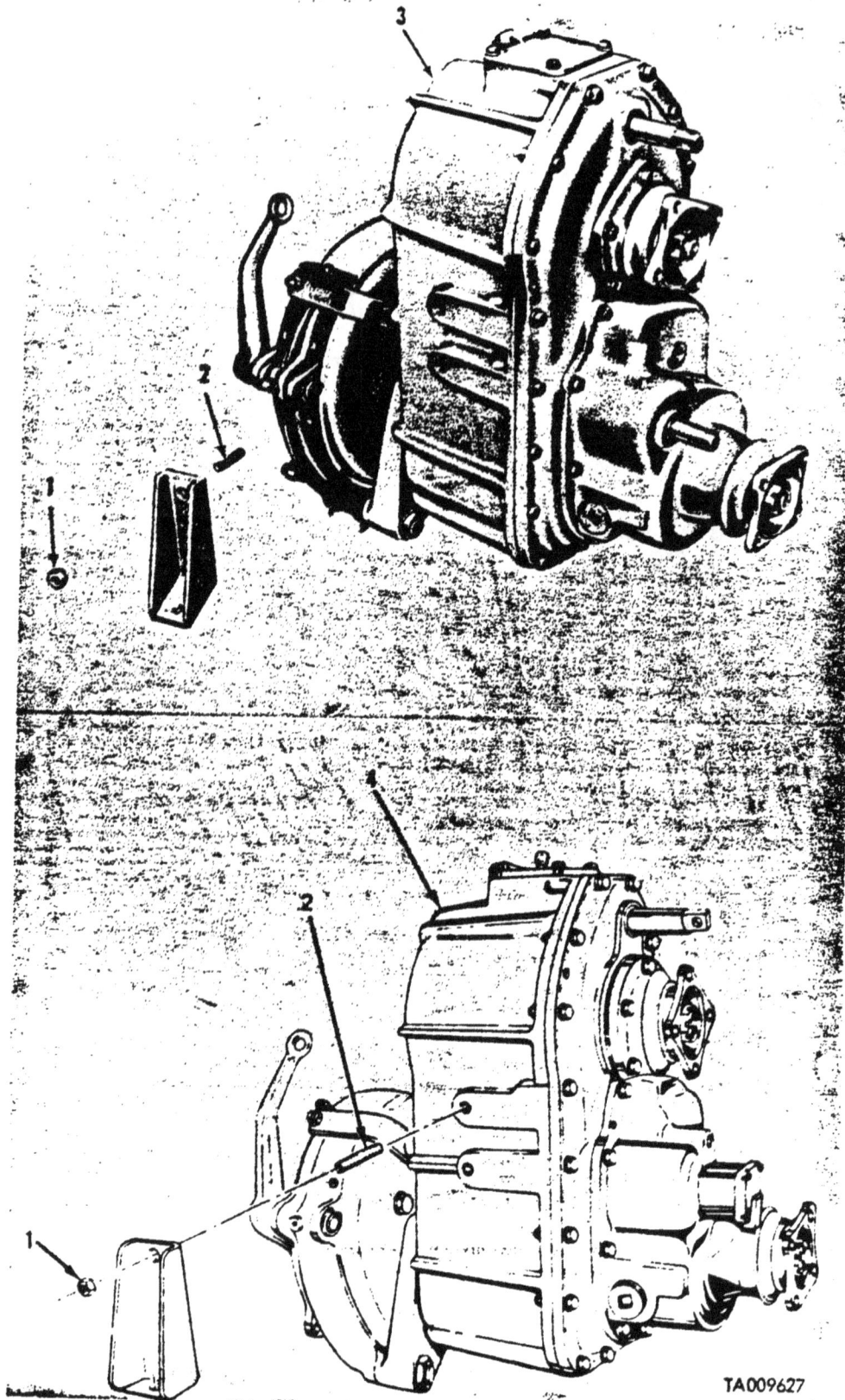

Figure 5. Power transfer assembly and attaching parts.

TA009627

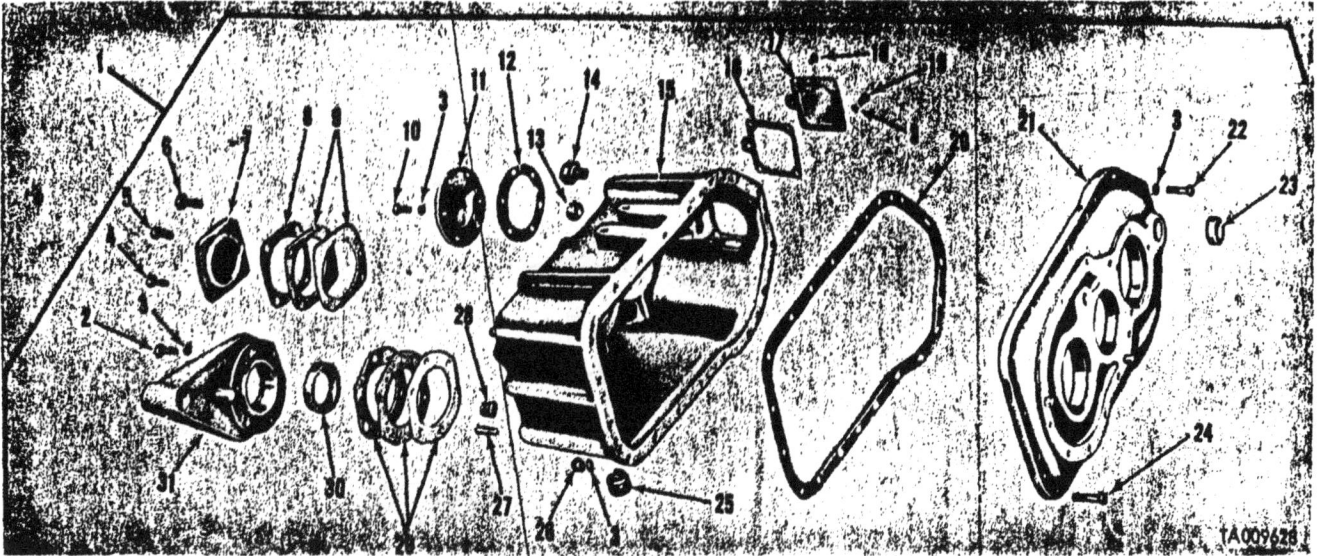

Figure 6. Housing, cover and related parts.

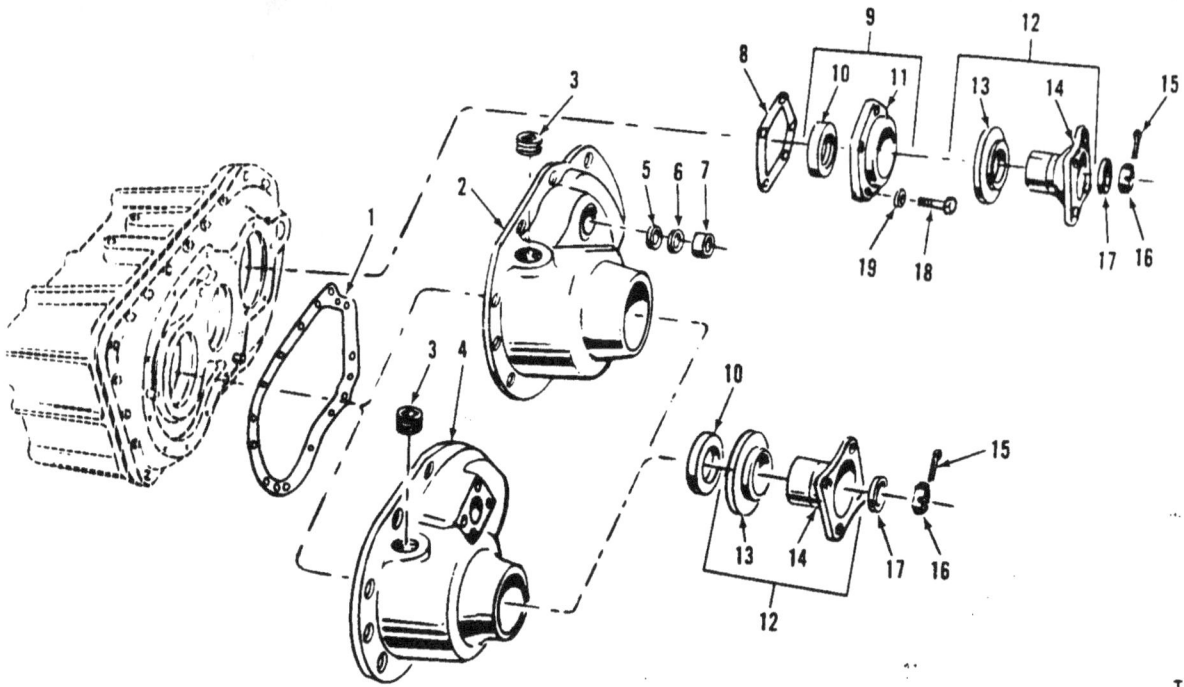

Figure 7. *Transfer case and covers.*

TA021438

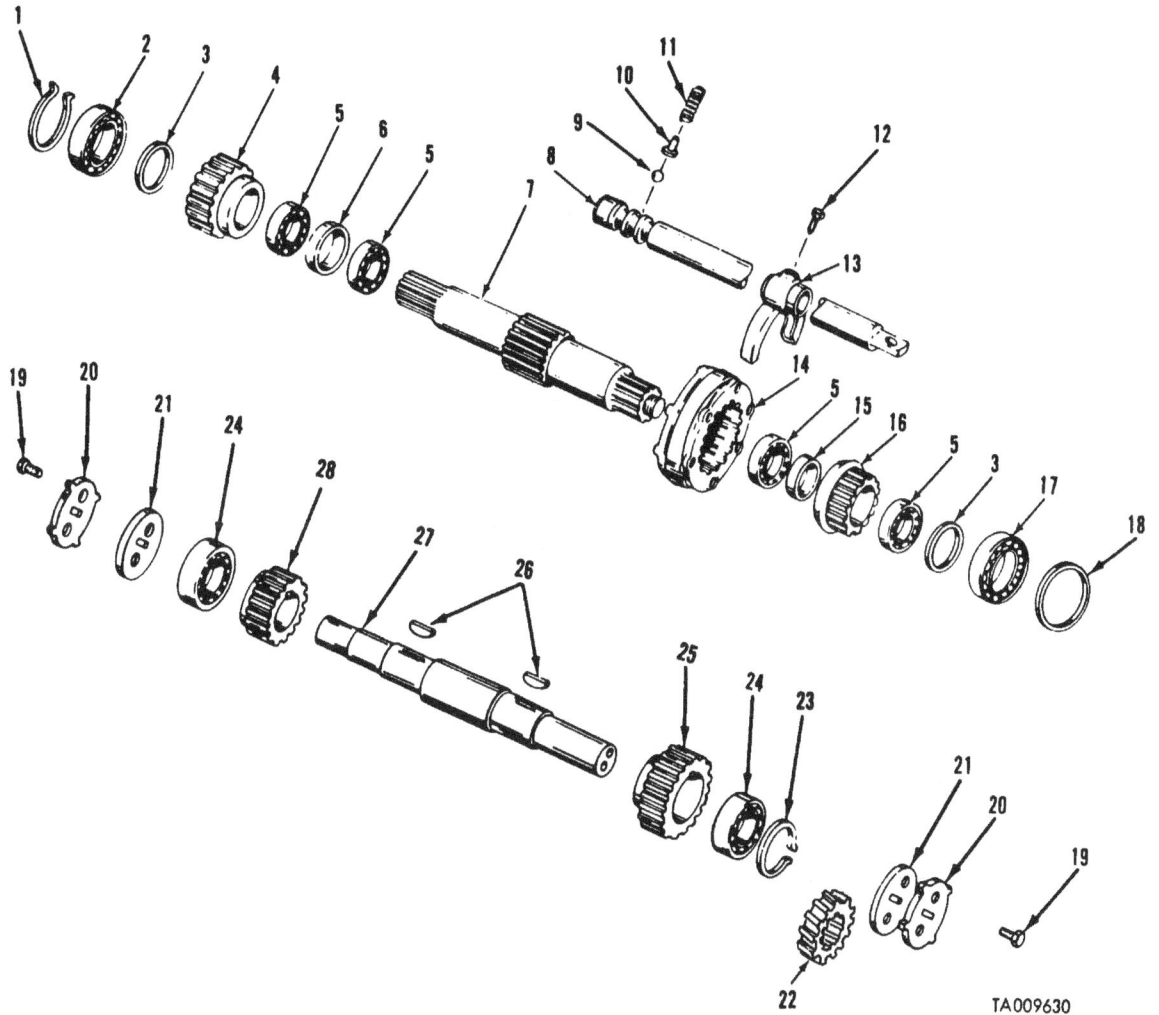

TA009630

Figure 8. Input shaft, countershafts and related parts.

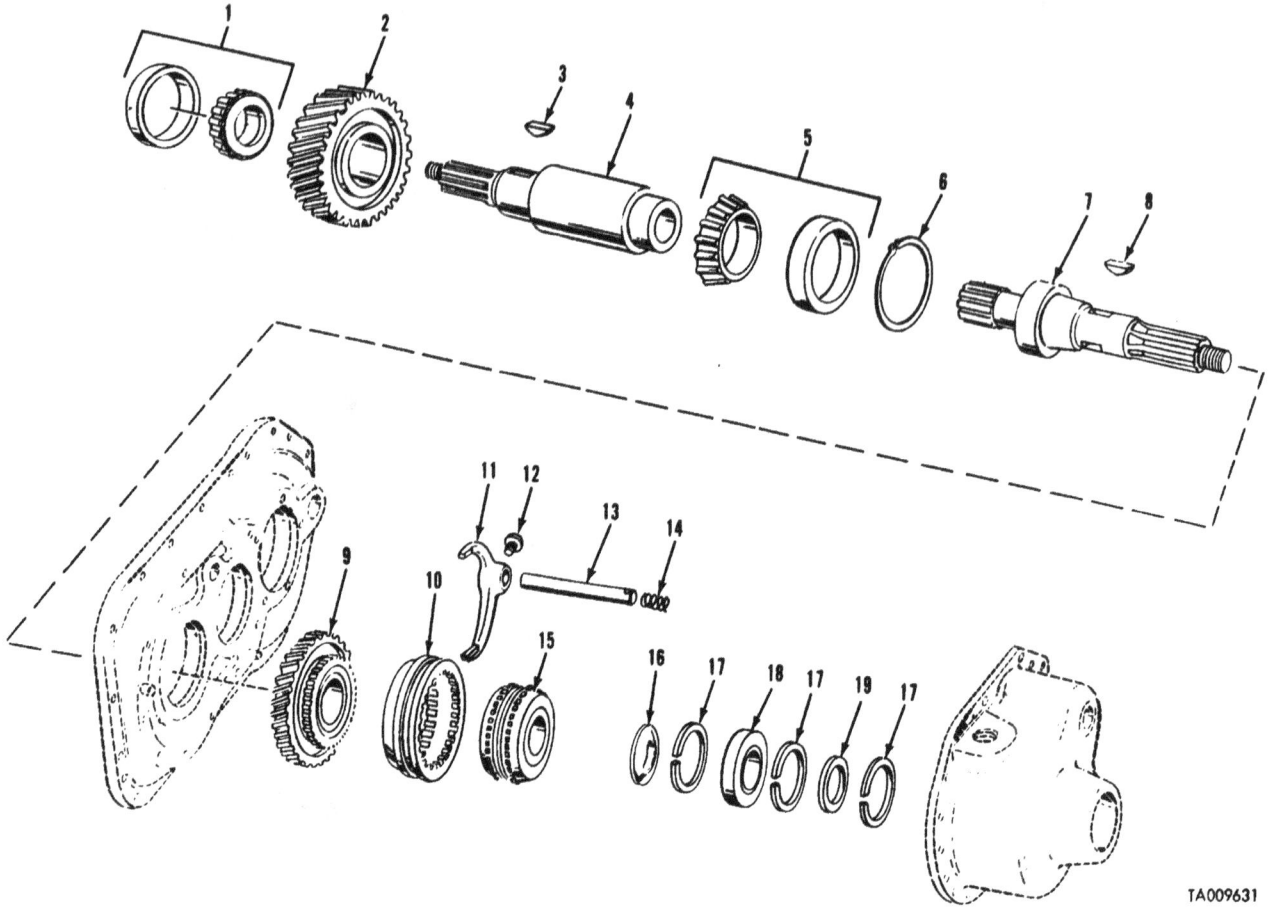

TA009631

Figure 9. Transfer gears and related parts.

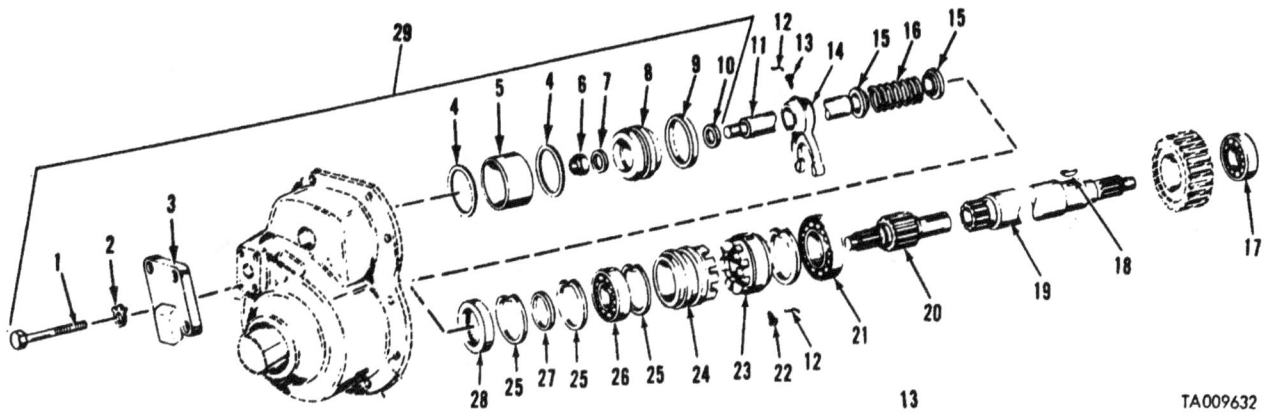

Figure 10. Transfer output shaft gears and air cylinders.

TA009632

31

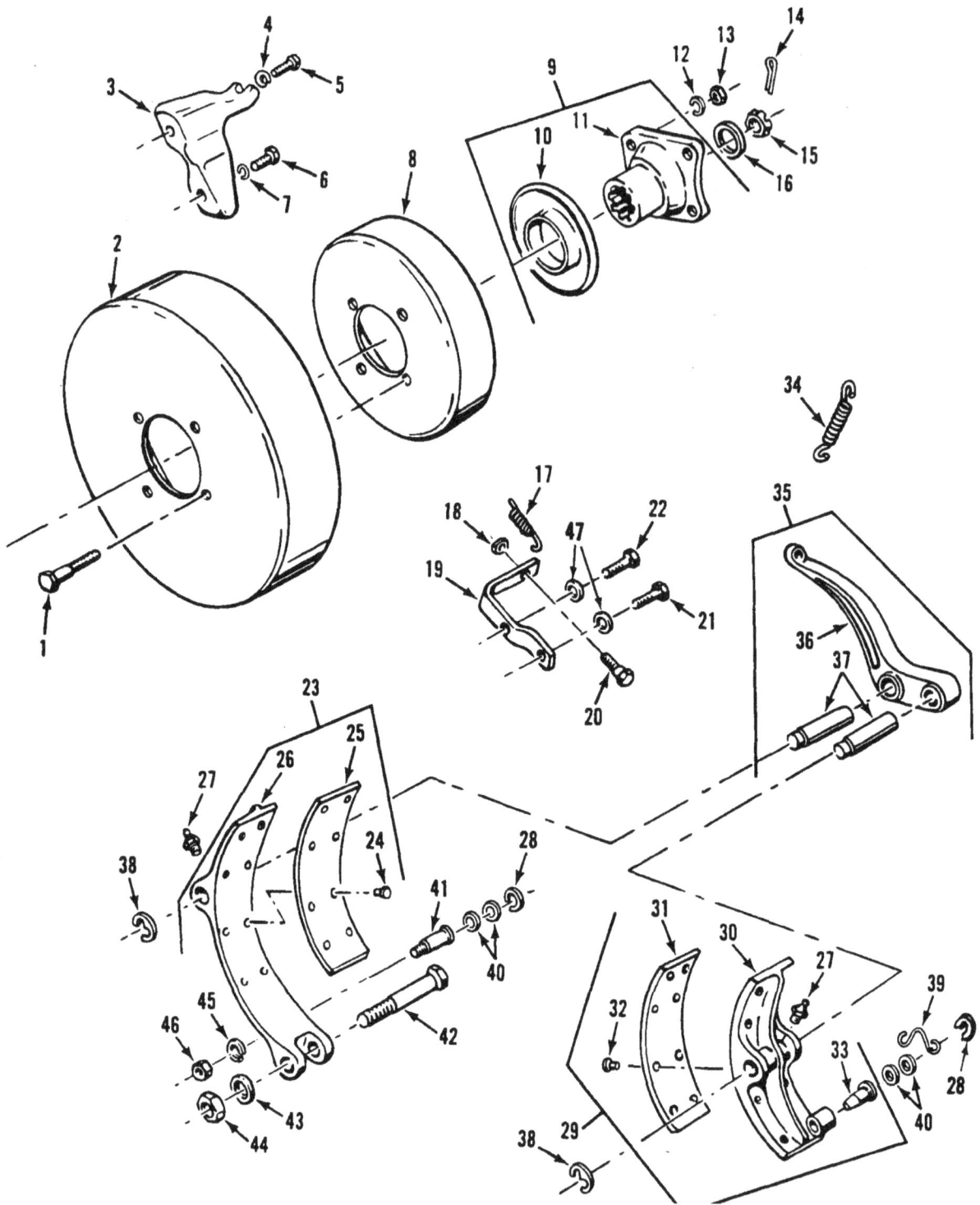

Figure 11. Parking brake drum, shoes and related parts.

TA021439

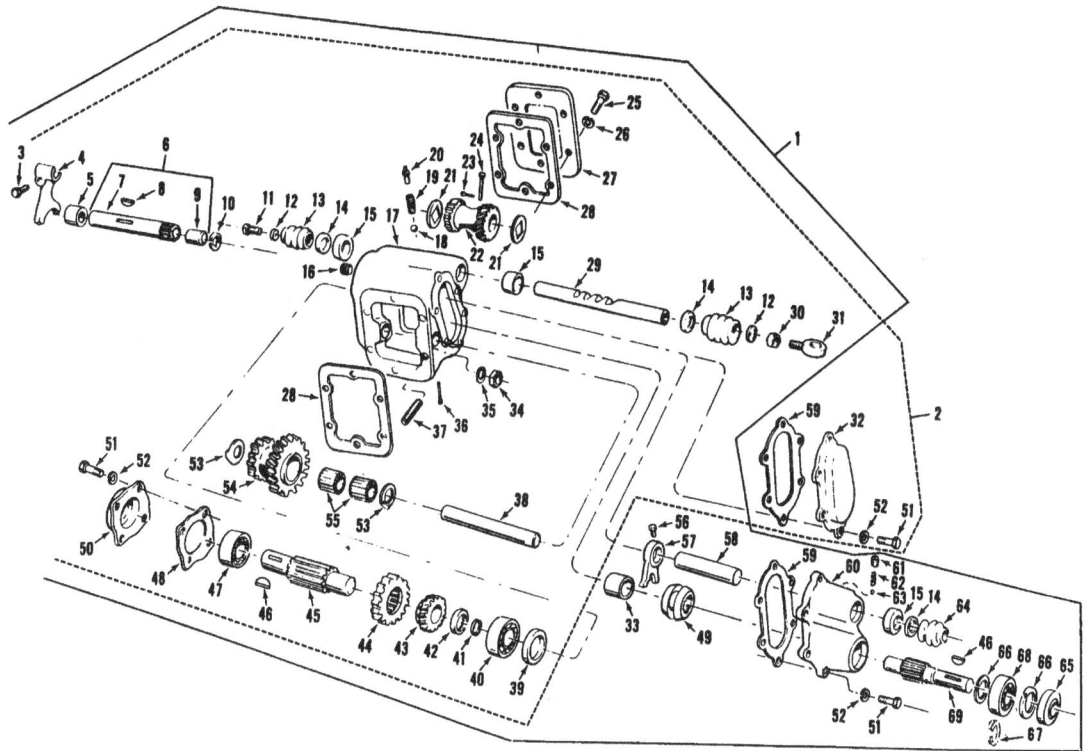

Figure 12. Transmission mounted power take-off assembly.

3 3

TA025462

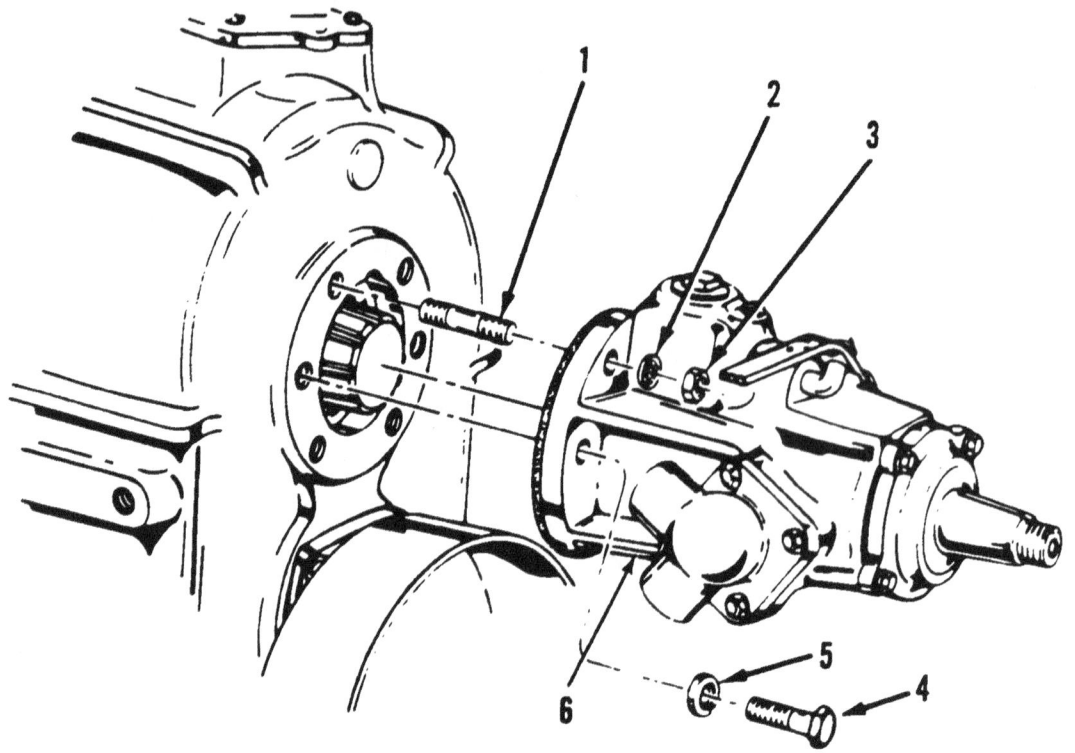

Figure 13. Transfer power take-off assembly.

TA009634

34

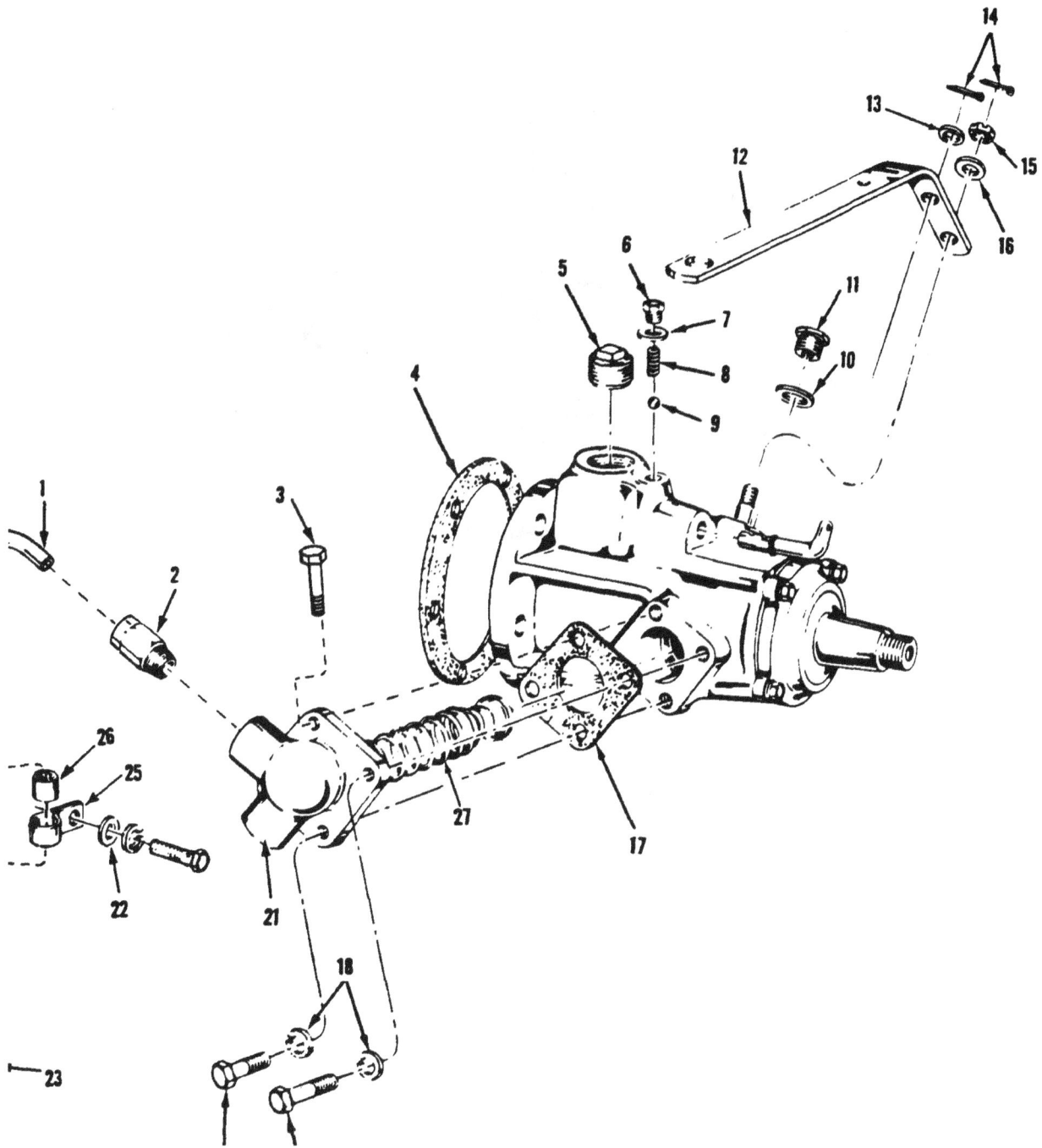

Figure 14. Transfer power *take-off and related parts.*

TA009635

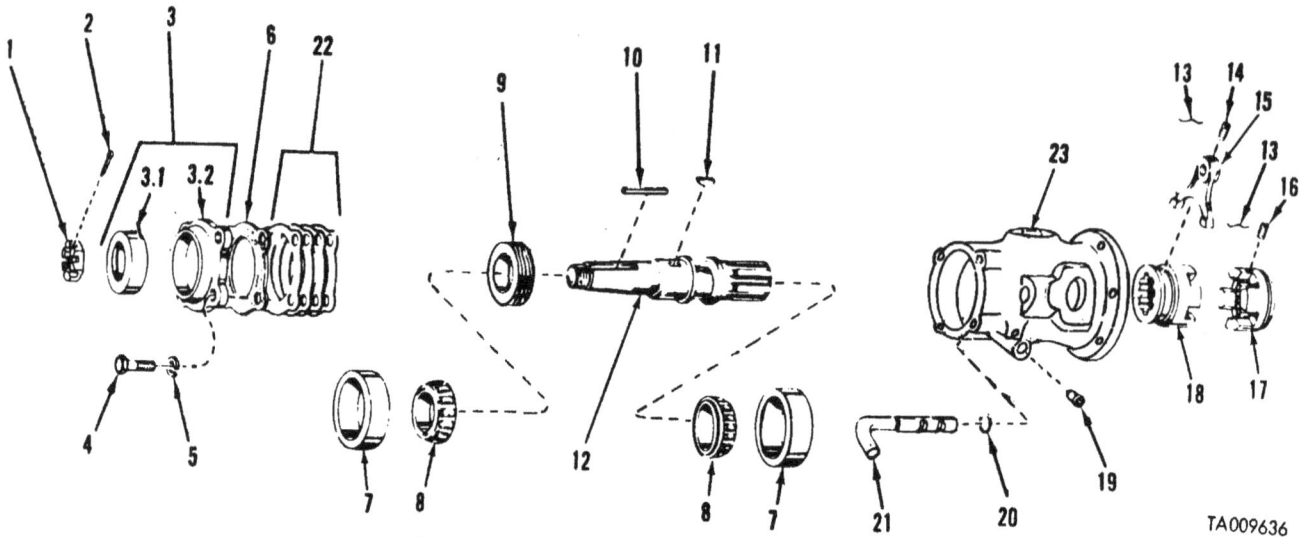

Figure 15. Transfer power take-off assembly and related parts.

TA009636

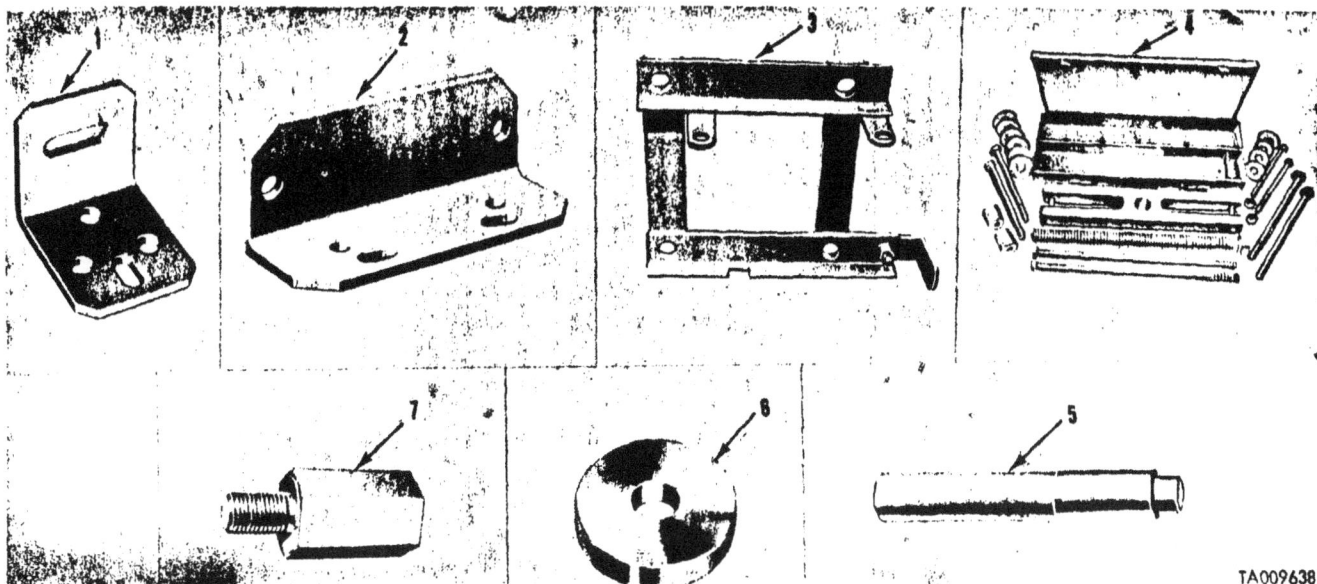

Figure 16. Special tools.

TA009638

NATIONAL STOCK NUMBER AND PART NUMBER INDEX
National Stock Number Cross Referenced to Figure and Item Number

National Stock Number	Figure No.	Item No.	National Stock Number	Figure No.	Item No.
2520-00-001-7855	5	1	2520-00-752-0965	2	35
2520-00-024-6540	6	31	2520-00-752-0966	2	2
2520-00-030-7262			2520-00-752-0967	2	6
2520-00-089-8287	5	4	2550-00-752-0979	2	29
2520-00-090-5999	2	16	5330-00-752-0987	2	NI
2520-00-104-4603	12	20	2520-00-752-0994	3	3
2520-00-134-5124	7	14	2520-00-752-1003	3	23
2520-00-134-5124	11	11	2520-00-752-1019	3	17
2520-00-229-5673	13	6	2520-00-752-1022	3	5
2520-00-301-7734	9	5	2520-00-752-1023	3	12
2520-00-347-4520	1	3	2520-00-752-1032	2	28
2520-00-347-4571	12	54	2520-00-752-1037	3	28
2520-00-347-4591	4	1	2520-00-752-1039	2	25
2520-00-354-0771	2	27	2520-00-752-1047	3	31
2520-00-355-7809	15	12	2520-00-752-1057	4	29
2520-00-492-4348	14	7	2520-00-752-1059	4	30
2520-00-508-4664	2	13	2520-00-752-1060	4	15
2520-00-513-5811	9	17	2520-00-752-1093	4	20
2520-00-514-3782	8	10	2520-00-752-1094	4	13
2520-00-522-6128	12	19	2520-00-752-1096	4	21
2520-00-692-5753	6	17	2520-00-752-1097	4	16
2520-00-627-5978	4	8	2520-00-752-1099	4	28
2520-00-692-6065	2	19	2520-00-752-1276	7	12
2520-00-692-6070	2	23		11	9
2520-00-692-6072	4	3	2520-00-752-1340	8	28
2520-00-692-6080	7	13	2520-00-752-1341	9	15
2520-00-692-6091	7	9	2520-00-752-1373	8	21
2520-00-706-1136	12	1	2520-00-752-1385	8	8
2520-00-706-1137	12	2	2520-00-752-1439	6	7
2520-00-706-1160	12	17	2520-00-752-1441	6	8
2520-00-706-1210	12	70	2520-00-752-1446	6	9
2520-00-706-1211	12	69	2520-00-752-1581	8	14
2520-00-706-1265	12	39	2520-00-752-1585	6	11
2520-00-706-1266	12	38	2520-00-753-8673	12	45
2520-00-706-1267	12	29	2520-00-781-6264	6	21
2520-00-706-1268	12	58	2520-00-781-6265	7	11
2520-00-706-1269	12	32	2520-00-786-0209	8	13
2520-00-706-1273	12	4	2520-00-808-7379	14	21
2520-00-706-1274	12	57	2520-00-808-7385	15	18
2520-00-706-1275	12	22	2520-00-808-7424	10	24
2520-00-706-1278	12	13	2520-00-808-7428	10	15
2520-00-706-1279	12	64	2520-00-808-7445	10	3
2520-00-706-1281	12	21	2520-00-832-7794	10	19
2520-00-734-9606	15	15	2520-00-832-8071	10	14
2520-00-737-1995	9	13	2520-00-832-8074	10	11
2520-00-737-2010	9	12	2520-00-832-8075	10	20
2520-00-737-2011	9	9	2520-00-832-8235	10	9
2520-00-737-2012	9	11	2520-00-832-8236	10	8
2520-00-752-0952	2	9	2520-00-832-8237	10	23
2520-00-752-0956	2	30	2520-00-884-4832	3	26
2520-00-752-0965	2	34	2520-00-884-4833	1	3
2520-00-752-0966	3	2	2520-00-884-4834	3	8
2520-00-737-2010	9	12	2520-00-884-4835	3	24
2520-00-737-2011	9	9	2520-00-885-3423	3	3
2520-00-737-2012	9	11	2520-00-930-3138	4	4
2520-00-737-2020	9	16	2520-00-933-3112	4	24
2520-00-752-0952	2	9	2530-00-289-7207	11	25
2520-00-752-0956	2	30	2530-00-622-3949	11	3
2520-00-752-0958	2	15	2530-00-693-0679	11	29
2520-00-752-0959	2	7	2530-00-693-0680	11	23

National Stock Number	Figure No.	Item No.	National Stock Number	Figure No.	Item No.
2530-00-736-8682	11	31	3120-00-753-8656	12	53
2530-00-736-8683	11	35	3120-00-753-8661	12	42
2530-00-752-1488	11	19	3120-00-753-9011	8	15
2530-00-752-1488	11	19	3120-00-832-7334	10	5
2530-00-752-1490	11	2	4710-00-137-2019	14	1
2530-00-752-9383	11	8	4730-00-012-5947	12	16
2540-00-626-2467	12	50	4730-00-044-4655	6	28
3020-00-303-5106	8	22	4730-00-050-4208	2	36
3020-00-312-8348	8	26		11	27
3020-00-312-8350	9	3	4730-00-127-4461	14	24
3020-00-318-0908	12	44	4730-00-172-0034	2	10
3020-00-347-4549	12	43	4730-00-278-3380	2	24
3020-00-692-6084	8	29	4730-00-278-4596	14	2
3020-00-741-3287	15	9	4730-00-639-9921	14	5
3020-00-752-1005	3	14	4730-00-737-5248	7	3
3020-00-752-1006	3	18	4730-00-737-5249	6	25
3020-00-752-1015	3	8	4730-00-884-4838	3	8
3020-00-752-1016	3	11	4820-00-726-4719	1	2
3020-00-752-1017	3	15		6	18
3020-00-752-1029	3	26	4820-00-808-7442	6	14
3020-00-752-1030	3	34	4910-00-694-4777	16	3
3020-00-752-1031	3	24	4910-00-795-0356	NOT ILLUSTRATED	GROUP 2604
3020-00-786-0211	8	4	5120-00-338-6721	16	4
3020-00-786-0213	8	16	5120-00-708-3241	16	5
3040-00-077-1878	15	17	5120-00-708-3247	16	6
3040-00-706-1159	12	6	5120-00-708-3254	16	7
3040-00-808-7441	8	7		4	11
3040-00-808-7445	8	28	5305-00-071-1786	8	19
3110-00-013-7743	8	25	5305-00-071-1785	6	10
3110-00-100-0799	15	7	5305-00-071-1788	6	2
3110-00-100-6155	2	63		7	18
3110-00-100-6156	8	9		11	6
3110-00-100-6159	14	9		13	4
3110-00-101-6480	10	21	5305-00-071-1789	6	22
3110-00-117-0693	3	37		11	22
3110-00-120-4276	3	33	5305-00-071-2055	6	6
3110-00-142-4364	15	8	5305-00-071-2235	10	1
3110-00-142-6040	4	9	5305-00-139-4620	6	24
3110-00-144-8518	12	68	5305-00-144-1495	12	25
3110-00-144-8668	10	26	5305-00-165-8248	10	22
3110-00-155-6686	3	29	5305-00-165-8250	6	5
3110-00-155-6706	3	19	5305-00-206-0932	11	41
3110-00-156-4704	12	47	5305-00-270-7328	15	14
3110-00-165-6329	10	17	5305-00-317-3102	1	1
	9	1	5305-00-490-1850	15	16
3110-00-183-6723	12	33	5305-00-543-4372	6	19
3110-00-198-1502	2	1	5305-00-695-7174	8	12
3110-00-227-3074	12	55		11	20
3110-00-227-4123	3	4	5305-00-710-4195	4	32
3110-00-293-9305	8	17	5305-00-716-8184	2	33
3110-00-516-5490	12	40	5305-00-737-6534	12	14
3110-00-516-5491	9	20	5305-00-752-1058	4	14
	10	26	5305-00-752-1391	9	14
3110-00-554-3078	8	5	5306-00-021-4077	10	13
3110-00-753-9028	12	62		4	11
3110-00-554-5960	3	2		2	22
3110-00-887-5524	12	18	5306-00-044-5319	2	12
3110-00-902-1686	12	5	5306-00-009-3837	11	5
3120-00-516-7516	15	19	5306-00-165-8251	14	19
3120-00-737-3115	12	9	5306-00-174-4202	2	3
3120-00-752-1020	3	10	5306-00-225-8494	12	11
3120-00-752-1049	3	32	5306-00-225-8497	15	4
3120-00-752-1487	8	18	5306-00-225-8498	14	3
	9	21	5306-00-226-4826	14	20
	10	27	5306-00-238-5605	12	51
3120-00-752-1583	8	3	5306-00-262-9619	14	6

National Stock Number	Figure No.	Item No.	National Stock Number	Figure No.	Item No.
5306-00-312-0845	11	42	5315-00-058-8581	12	8
5306-00-404-0075	12	3	5315-00-058-8583	3	27
5306-00-752-1014	11	1	5315-00-281-7839	9	10
5306-00-752-1055	4	31	5315-00-298-1481	3	21
5307-00-350-5532	13	1		15	2
5307-00-753-8668	12	37	5315-00-312-0831	11	37
5307-00-695-7213	5	2	5315-00-316-0992	11	33
5310-00-080-6004	12	12	5315-00-616-5514	15	11
5310-00-145-2085	10	2	5315-00-616-5527	12	46
5310-00-165-8488	14	15	5315-00-692-6101	4	7
5310-00-167-0820	4	26	5315-00-699-8458	3	9
5310-00-177-0973	10	7	5315-00-699-8460	6	27
5310-00-209-0965	2	4	5315-00-735-1406	12	24
	4	5	5315-00-816-1794	12	23
	6	3	5315-00-842-3044	14	14
	7	19	5315-00-850-7038	8	24
	11	7		9	4
	11	47		10	18
	13	5	5320-00-443-5065	11	24
5310-00-227-1921	14	13		11	32
5310-00-225-6993	10	6	5325-00-741-2204	14	26
5310-00-241-6664	5	1	5330-00-143-8666	6	30
5310-00-285-2169	3	7		7	10
5310-00-279-3314	11	40		10	28
5310-00-333-7519	11	28	5330-00-171-9052	12	49
	14	18	5330-00-231-0285	15	3
	15	5	5330-00-234-3317	2	25
5310-00-426-0809	15	1		12	28
5310-00-514-6674	11	4	5330-00-285-5121	7	6
	12	52	5330-00-298-3947	12	15
5310-00-527-4222	10	10	5330-00-311-7774	6	16
5310-00-529-4103	3	22	5330-00-350-9958	2	21
5310-00-584-5272	2	32	5330-00-353-2465	7	7
	11	12	5330-00-454-0338	4	12
5310-00-616-7998	12	26	5330-00-522-8421	14	4
	12	35	5330-00-532-3606	12	48
5310-00-661-9568	2	5	5330-00-559-8733	12	65
5310-00-685-3228	13	2	5330-00-579-8156	7	5
5310-00-732-0558	11	46	5330-00-585-1064	15	20
	12	34		14	4
5310-00-732-0560	11	13	5330-00-594-8953	2	14
5310-00-752-1036	3	30	5330-00-615-6756	15	6
5310-00-752-1048	3	36	5330-00-706-1270	12	59
5310-00-752-1234	7	16	5330-00-737-6534	12	14
	11	15	5330-00-741-3286	14	10
5310-00-752-1235	7	17	5330-00-752-0957	2	30
	11	16	5330-00-752-0958	2	15
5310-00-753-9169	11	44	5330-00-752-0959	2	7
5310-00-809-4085	14	22	5330-00-752-1061	2	17
5310-00-809-5998	14	16	5330-00-752-1437	7	
5310-00-832-7904	10	19	5330-00-752-7750	4	9
	8	20	5330-00-808-7417	14	17
5310-00-880-7745	6	26	5340-00-050-1593	4	27
	4	6	5340-00-321-6375	11	39
	11	18	5340-00-392-4017	2	34
5310-00-919-8882	15	3.1	5340-00-610-0919	16	1
5310-00-984-3807	4	10	5340-00-610-0920	16	2
5310-00-926-5916	13	3	5340-00-706-1277	12	67
5310-00-975-2075	11	17	5340-00-752-0976	2	8
	12	30	5340-00-752-1372	6	13
5310-00-982-6809	5	1	5340-00-753-8679	12	31
5315-00-012-4553	3	16	5355-00-962-3018	4	2
5315-00-013-7214	7	15	5360-00-342-7071	11	34
	11	14	5360-00-347-4563	4	18
5315-00-013-7258	12	36	5360-00-692-6074	4	25
5315-00-032-1872	15	10	5360-00-692-6075	4	22

National Stock Number Cross Referenced to Figure and Item Number-Contin

National Stock Number	Figure No.	Item No.	National Stock Number	Figure No.	Item No.
5360-00-692-6089	8	11	5365-00-752-1021	3	20
5360-00-693-0615	11	17	5365-00-752-1045	3	35
5360-00-737-2020	9	16	5365-00-752-1046	3	25
5360-00-832-8072	10	16	5365-00-752-1277	6	29
5365-00-508-4675	3	1	5365-00-752-1374	8	23
5365-00-508-4691	12	41	5365-00-752-1575	8	6
5365-00-696-0267	9	19	5365-00-752-1584	8	1
5365-00-664-2779	10	25	5365-00-769-6481	12	61
	11	38	5365-00-769-9003	15	22
5365-00-699-8456	3	6	5365-00-832-7774	10	4
5365-00-699-8457	3	13	5365-00-808-7423	GROUP 0801	
5365-00-699-8459	4	23	9505-00-248-9850	4	17
5365-00-706-1276	12	10		10	12
5365-00-706-1280	12	66		15	13
5365-00-741-3289	14	11			

National Stock Number Cross Referenced to Figure and Item Number-Contin

National Stock Number	Figure No.	Item No.	National Stock Number	Figure No.	Item No.

42

Part Number	FSCM	Fig. No.	Item No.	Part Number	FSCM	Fig. No.	Item No.
MS15003-1	96906	11	27	MS90725-33	96906	14	3
		2	36	MS90726-90	96906	4	32
MS15003-7	96906	2	10	MS90726-112	96906	2	33
MS16633-1098	96906	11	38	MS90728-33	96906	14	20
MS17131-35	96906	12	33	MS90728-58	96906	6	19
MS19059-51	96906	12	63	MS90728-84	96906	6	10
MS19059-53	96906	12	62	MS90728-85	96906	8	19
MS19059-55	96906	8	9	MS90728-87	96906	6	2
MS19059-59	96906	14	9			11	6
MS19061-11	96906	4	19			11	22
MS19061-15	96906	12	18			13	4
MS20995-F47	96906	4	17			7	18
		10	12	MS90728-88	96906	6	22
		15	13	MS90728-89	96906	6	6
MS24665-283	96906	14	14			7	18
MS24665-285	96906	12	23	ST202	31007	9	20
MS24665-357	96906	3	21			10	26
		15	2	1KS5021	44731	10	21
MS24665-359	96906	7	15	10872096	19207	1	3
		11	14	10896745	19207	15	16
MS24665-497	96906	12	36	10896746	19207	14	7
MS27183-14	96906	12	12	10896748	19207	11	24
MS27183-16	96906	14	22	10896748	19207	11	32
MS27183-18	96906	14	16	10896749	19207	15	1
MS28775-212	96906	7	5	10896750	19207	14	5
MS29561-12	96906	15	20	10910354	19207	12	25
MS35333-43	96906	13	2	10913209	19207	10	4
MS35335-31	96906	12	26	10924755	19207	8	NI
MS35335-34	96906	11	4	10937790	19207	4	4
MS35335-20	96906	12	52	10937974	19207	4	24
	96906	11	10	10938286	19207	15	3.1
MS35338-47	96906	2	4	10914635	19207	3	3
		4	5	10914636	19207	3	24
		6	3	10914637	19207	3	8
		7	19	10914638	19207	3	26
	96906	11	7	10937391	19207	10	22
		11	47	10937638	19207	10	19
		13	5	10938287	19207	6	23
MS35338-48	96906	2	32	10945186	19207	10	16
MS35648-5	96906	4	27	10948042	19207	10	3
MS35691-21	96906	12	20	10948043	19207	10	8
MS35756-6	96906	15	11	10948080	19207	10	17
MS35756-17	96906	3	16	10948080	19207	10	17
MS35756-18	96906	12	46			9	1
MS35756-105	96906	8	27	10948082	19207	8	25
		9	4	10948099	19207	10	11
		10	18	10948100	19207	10	14
MS35756-109	96906	12	8	10948101	19207	10	20
MS35756-110	96906	3	27	10948103	19207	8	24
MS35769-4	96906	4	9			8	NI
MS39202-6	96906	14	24	10948105	19207	10	24
MS49005-10	96906	2	24	10948106	19207	10	23
MS51915-21-1	96906	4	12	10948108	19207	10	15
MS51922-13	96906	4	10	10948109	19207	7	2
MS51943-44	96906	5	1	11609221	19207	10	7
MS51967-8	96906	11	46	11609224	19207	5	4
MS51968-8	96906	12	34	11609226	19207	5	3
MS51968-11	96906	4	6	11609228	19207	13	6
	96906	11	18	11609328-1	19207	6	5
MS51968-12	96906	13	3			11	21
		11	18	11609328-2	19207	6	24
MS51968-14	96906	11	13			11	20
MS90725-16	96906	10	1	11609558	19207	14	19
MS90725-31	96906	12	11	11621135	19207	14	17
MS90725-32	96906	15	4	11621137	19207	14	21

Part Number	FSCM	Fig. No.	Item No.	Part Number	FSCM	Fig. No.	Item No.
11621138	19207	14	1	7061268	19207	12	58
11621139	19207	15	17	7061269	19207	12	32
11621140	19207	15	18	7061270	19207	12	59
11621143	19207	15	23	7061271	19207	12	65
11621144	19207	15	12	7061272	19207	12	15
11621145	19207	8	7	7061273	19207	12	4
11621147	19207	6	15	7061274	19207	12	57
11621191	19207	8	28	7061275	19207	12	22
11621193	19207	6	14	7061276	19207	12	10
125947	21450	12	16	7061278	19207	12	13
137251	19207	15	7	7061280	19207	12	66
148346	24617	18	8	7061281	19207	12	21
2135-1	08162	2	1	7064466	19207	11	39
444595	21450	14	5	7064467	19207	11	40
444655	19207	6	28	7064468	19207	11	28
48488	62983	4	26	7064469	19207	11	41
500037	21450	12	49	7064470	19207	11	33
5168861	19207	15	10	7083241	19207	16	5
5143782	19207	8	10	7083247	19207	16	6
5164525	19207	7	6	7083254	19207	16	7
5196397	19207	1	2	713925	21450	8	2
		6	18	7349606	19207	15	15
5186592	19207			7349680	19207	15	21
5186593	19207			7351406	19207	12	24
5214784	19207	12	62	7363016	19207	8	16
5214807	19207	2	3	7363017	19207	8	4
521606	12204	3	33	7368681	19207	11	25
5226033	19207	2	5	7368682	19207	11	31
5226140	19207	12	20	7368683	19207	11	35
5384913	19207	15	19	7371995	19207	9	13
5294103	19207	3	22	7372010	19207	9	12
5323606	19207	12	48	7372011	19207	9	9
5323473	19207	12	27	7372012	19207	9	11
5226128	19207	12	19	7372020	19207	9	16
583509	21450	2	21	7373115	19207	12	9
6143903	19207			7373243	19207	11	34
594119	21450	1	1	7375249	19207	6	25
6156769	19207	14	8	7376534	19207	12	14
6156756	19207	15	6	7397808	19207	11	3
6184279	19207	4	2	7412204	19207	14	26
6262467	19207	12	50	7412211	19207	14	6
7061277	19207	12	67	7413265	19207	14	25
700078	00000	12	68	7413286	19207	14	10
700603	21450	8	5	7413287	19207	15	9
700771	21450	3	9	7413289	19207	14	11
700795	00000	8	17	7520952	19207	2	9
700955	21450	3	2	7520956	19207	2	31
707617	0000	3	37	7520957	19207	2	30
708231	00000	3	33	7520958	19207	2	15
7010362	19207	16	1	7520959	19207	2	7
7010363	19207	16	2	7520965	19207	2	35
712663	00000	9	6	7520966	19207	2	2
		10	21	7520967	19207	2	6
713421	00000	12	5	7520976	19207	2	8
714248	00000	12	47	7520979	19207	2	29
7061136	19207	12	1	7520980	19207	2	34
7061137	19207	12	2	7520987	19207	2	NI
7061159	19207	12	6	7520988	19207	2	16
7061160	19207	12	17	7520989	19207	2	13
7061194	19207	12	60	7520990	19207	2	14
7061210	19207	12	70	7520993	19207	3	1
7061211	19207	12	69	7520994	19207	3	3
7061265	19207	12	39	7520995	19207	1	3
7061266	19207	12	38	7521003	19207	3	23
7061267	19207	12	29	7521004	19207	2	18

Part Number	FSCM	Fig. No.	Item No.	Part Number	FSCM	Fig. No.	Item No.
7521005	19207	3	14	7521284	19207	11	17
7521006	19207	3	18	7521285	19207	11	28
7521007	19207	2	19	7521340	19207	8	28
7521008	19207	3	6	7521241	19207	9	15
7521009	19207	3	13	7521263-1	19207	6	1
7521010	19207	3	4	7521264	19207	6	20
7521014	19207	11	1	7521272	19207	6	13
7521015	19207	3	8	7521373	19207	8	23
7521016	19207	3	11	7521375	19207	8	29
7521017	19207	3	15	7521376	19207	8	26
7521018	19207	3	9	7521383	19207	8	8
7521019	19207	3	17	7521387	19207	8	11
7521020	19207	3	10	7521390	19207	8	12
7521021	19207	3	20	7521391	19207	9	14
7521022	19207	3	5			10	13
7521023	19207	3	12	7521433	19207	9	19
7521024	19207	3	7			10	25
7521029	19207	3	26	7521434	19207	9	8
7521032	19207	2	28	7521437	19207	10	
7521036	19207	3	30	7521438	19207	8	22
7521038	19207	2	23	7521439	19207	6	7
7521030	19207	3	34	7521441	19207	6	8
7521031	19207	3	24	7521446	19207	6	9
7521032	19207	2	27	7521448	19207	11	37
7521037	19207	3	28	7521487	19207	8	18
7521039	19207	2	25			9	21
7521045	19207	3	35			10	27
7521046	19207	3	25	7521488	19207	11	19
7521047	19207	3	31	7521490	19207	11	2
7521048	19207	2	36	7521573	19207	7	9
7521049	19207	3	32			7	8
7521055	19207	4	31	7521575	19207	8	6
7521056	19207	4	1	7521581	19207	8	14
7521057	19207	4	29	7521583	19207	8	3
7521058	19207	12	56	7521584	19207	8	1
		4	14	7521585	19207	6	11
7521059	19207	4	30	7529383	19207	11	8
7521060	19207	4	15	7538656	19207	12	53
7521061	19207	2	17	7538658	19207	12	55
7521062	19207	4	3	7538660	19207	12	41
7521093	19207	4	20	7538661	19207	12	42
7521094	19207	4	13	7538668	19207	12	37
7521095	19207	4	23	7538671	19207	12	44
7521096	19207	4	21	7538672	19207	12	43
7521097	19207	4	16	7538673	19207	12	45
7521099	19207	4	28	7538674	19207	12	54
7521100	19207	4	18	7538679	19207	12	31
7521101	19207	4	25	7539011	19207	8	15
7521102	19207	11	15	7539023	19207	4	8
	19207	7	1	7539146	19207	4	7
7521234	19207	7	16	7539169	19207	11	44
7521235	19207	7	17	7539700	19207	12	40
7521237	19207	11	16	7696481	19207	12	61
		6	27	7699003	19207	15	22
7521240	19207	9	3	7700135	19207	7	11
7521241	19207	6	30	7702649	19207	7	7
		7	10	7735601	19207	15	3.2
		10	28	7735641	19207	6	17
7521242	19207	7	13	7748541	19207	2	20
		11	10	7950356	19207	2604	
7521275	19207	6	31	7971286	19207	6	16
7521276	19207	7	12	8327009	19207	10	12
		11	9			12	28
7521277	19207	6	29	8327904	19207	8	20
7521283	19207	11	42	8328007	19207	15	14

Part Number Cross Referenced to Figure and Item Number-Continued

Part Number	FSCM	Fig. No.	Item No.	Part Number	FSCM	Fig. No.	Item No.
8328008	19207	14	13	8743044	19207	12	7
8328010	19207	13	11	8743049	19207	2	22
8328015	19207	14	2			4	11
8329891	19207	9	10	8757599	19207	11	5
8329892	19207	9	17	8757674	19207	7	14
		9	18			11	11
8331791	19207	5	2	8757686	19207	15	3
8344155	19207	12	3	8757704	19207	7	3
8344200-1	19207	6	12	8757705	19207	11	43
		14	4	8757706	19207	11	26
8376462	19207	9	5	8757707	19207	11	30
8380558	19207	11	29	8757709	19207	11	36
8380559	19207	11	23	8757711	19207	11	8
8675959	19207	6	21	8757713	19207	7	4
8380559	19207	11	23	8758292-2	19207	10	5
8675959	19207	6	21	8758294	19207	10	9
8708724	19207	6	4	8758299	19207	10	10
8708279	19207	16	3	878355	19207	12	51
8712073	19207	2	12				

By Order of the Secretary of the Army:

FRED C. WEYAND
General, United States Army
Chief of Staff

Official:

PAUL T. SMITH
Major GeneraL United States Army
The Adjutant General

Distribution:

To be distributed in accordance with DA Form 12-38 (qty rqr block No. 45) Direct and General Support maintenance requirement for 2 1/2 Ton, 6x6, M34; M35A1; A2 and A2C; M36, A2 and A3, M36C.

☆ U.S. GOVERNMENT PRINTING OFFICE : 1993 0 - 342-421 (63300)

015415

www.ingramcontent.com/pod-product-compliance
Lightning Source LLC
Chambersburg PA
CBHW080415030426
42335CB00020B/2452